Telecommunication Networks and Computer Systems

Series Editors

Mario Gerla
Aurel Lazar
Paul Kühn
Hideaki Takagi

Springer
London
Berlin
Heidelberg
New York
Barcelona
Budapest
Hong Kong
Milan
Paris
Santa Clara
Singapore
Tokyo

Marco Conti, Enrico Gregori and
Luciano Lenzini

Metropolitan Area Networks

With 161 Figures

Springer

Marco Conti
Enrico Gregori
Italian National Research Council
CNUCE Institute
Via S. Maria 36, 56126 Pisa, Italy

Luciano Lenzini
Department of Information Engineering
University of Pisa
via Diotisalvi, 2
I-56126 Pisa, Italy

Series Editors

Mario Gerla
Department of Computer Science
University of California
Los Angeles
CA 90024, USA

Paul Kühn
Institute of Communications
Switching and Data Technics
University of Stuttgart
D-70174 Stuttgart, Germany

Aurel Lazar
Department of Electrical Engineering and
Center for Telecommunications Research
Columbia University
New York, NY 10027, USA

Hideaki Takagi
Institute of Policy and Planning Sciences
University of Tsukuba
1-1-1 Tennoudai, Tsukuba-shi
Ibaraki 305, Japan

ISBN 978-1-4471-1232-7   ISBN 978-1-4471-0909-9 (eBook)
DOI 10.1007/978-1-4471-0909-9

Springer-Verlag Berlin Heidelberg New York
British Library Cataloguing in Publication Data
A catalogue record for this book is available from the British Library

Typesetting: Camera ready by authors
Printed and bound at the Athenæum Press Ltd., Gateshead, Tyne and Wear
69/3830-543210 Printed on acid-free paper

Preface

With the continuing success of Local Area Networks (*LANs*), there is an increasing demand to extend their capabilities towards higher data rates and wider areas. This, together with the progress in fiber-optic technology, has given rise to the so-called Metropolitan Area Networks (*MANs*). MANs can span much greater distances than current LANs, and offer data rates on the order of hundreds of Megabits/sec (*Mbps*).

The success of MANs is mainly due to the opportunity they provide to develop new networking products capable of providing high-speed communications between applications at competitive prices, which nonetheless give an adequate return on the manufacturers' investments. A major factor in achieving this goal is the availability of appropriate networking standards. Fiber Distributed Data Interface (*FDDI*) and Distributed Queue Dual Bus (*DQDB*) are the two standard technologies for MANs for which industrial products are already available. For this reason, this book focuses mainly on these two standards.

Nowadays there are several books dealing with MANs, and these look mainly at FDDI (e.g., [2], [92], [118], [141]). These books focus primarily on the architectures and protocols, whereas they pay little attention to performance analysis. Due to the capability of MANs to integrate services, a quantitative analysis of the Quality of Service (*QoS*) provided by these technologies is a relevant issue, and is thus covered in depth in this book.

LAN were designed to support Electronic Data Processing (*EDP*) applications. These applications require a reliable transportation service, but they do not put stringent constraints on other performance measures such as throughput and access delay. High-level protocols (e.g., the transport protocol) are commonly used to enhance data-transfer reliability to meet the EDP application requirements.

With the development of new broadband technologies for MANs, new applications are becoming viable and the situation is changing drastically. Much of the traffic will be time-constrained (e.g., voice, video and alarm messages) and, therefore, MANs must be capable of providing service with guaranteed performance. Performance requirements of real-time applications can not be achieved using high-level protocols if the underlying network does not offer some guarantees. For example, with an Ethernet LAN it will never be possible to guarantee an upper bound on the information transfer delay, even if upper-layer protocols are added. Hence, to evaluate a MAN's suitability to support time-constrained applications, performance figures such as rate of packet loss, packet transfer delay and throughput must be quantified.

Research in this field is facing a wide range of performance-related problems, such as tuning the network parameters, dimensioning the key network components (e.g., buffer size), and determining the relationship between bandwidth allocation schemes, throughput, delay distribution and packet-loss rate.

The importance of FDDI and DQDB has caused their performance to be analyzed extensively. Most of the existing results have been obtained via simulation, as it is extremely difficult to solve detailed models of either MAC protocol analytically. In fact, FDDI has a more complex behavior than a polling system with an exhaustive-limited service discipline, while DQDB behaves like a round-robin scheduling algorithm for very small networks, but deviates significantly from that pattern of behavior when the length of the network increases. Due to the complexity of these protocols, exact models of the FDDI and DQDB MAC protocols have only been solved through simulative analysis, while models with analytical solutions have been developed to approximate each protocol's behavior under specific network configurations and workload conditions.

One aim of this book is to present a structured view of the published performance modeling activity concerning FDDI and DQDB. For each MAC protocol, a taxonomy of analytical models is proposed. Some relevant models from each class will be discussed by presenting the main simplifying assumptions, the techniques used for solving these models, and the performance indices analyzed. In the presentation of each model, the original notation has been modified to provide a uniform use of symbols throughout the book.

The evaluation of the FDDI and DQDB models requires advanced stochastic methods which are generally found in specialized, theoretical texts. Furthermore, some methodologies are currently under development, and thus are available only in journals and technical reports. This book collects these advanced results into a single, structured volume and guides the reader, by degrees, to an understanding of the more complex material. The stochastic background required to benefit from this book is a knowledge of basic stochastic processes (continuous-time and discrete-time Markov processes) and elementary queueing theory concepts (e.g., [54] and [99]). In the proofs and explanations, we favoured clarity and simplicity over formal rigor. Proofs requiring a very advanced mathematical background have been omitted.

The coverage of network protocols, models and solution methodologies makes this book useful to readers with different needs and interests.

Chapter 1 introduces the Metropolitan Area Networks, discussing the evolution from LANs to MANs, and explaining the need to design a new class of MAC protocols for MANs. This chapter is always recommended as the starting point in reading this book; it only assumes a knowledge of the Ethernet and Token Ring MAC protocols. On the other hand, the other chapters can be read in an order determined by the reader's interests. Specifically, the content of the other chapters can be subdivided into three parts:

- PART I: advanced stochastic concepts (Chapters 2, 3 and 4),
- PART II: architectures and protocols (Chapters 5, 7 and 9), and
- PART III: FDDI and DQDB performance evaluation (Chapters 6, and 8).

The entire book can be used for an advanced course on architecture, modeling and performance evaluation of MANs. Part III requires an understanding of the concepts developed in both Part I and Part II.

No prerequisite (other than Chapter 1) exists for reading Part II; Part II can therefore be used for an introductory course on MANs. Furthermore, results presented in Part I can be applied in a much more general setting than

MAN performance evaluation; for instance, this part can be used for a grad-
uate course on vacation systems.

ACKNOWLEDGMENTS. The authors would like to acknowledge Enrico Zuc-
chelli (CNUCE) for preparing and producing the accompanying figures. His
help has been invaluable.

Contents

1 Introduction

Several criteria can be used to classify packet-switching networks, e.g., geographical coverage, transmission speed, transmission technology and type of traffic supported. One of the most commonly used classifications is based on geographical coverage. At one end there are Wide Area Networks (*WANs*), which can span distances of thousands of kilometers, whereas Local Area Networks (*LANs*) only cover short distances, usually not more than one or two kilometers. The difference in geographical coverage has a significant impact on the way these networks are designed. WANs make use of the "store-and-forward" packet-switching technique. The network is made up of a set of packet-switching nodes interconnected via transmission links. At each node, packets[1] are processed and, on the basis of their destination address, are routed towards their final destination. LANs are typically based on a high-speed link which is shared by all the stations connected to the network. Information broadcasting is thus easily achieved, and routing is not necessary. In an office building or a university campus, LANs represent an efficient and cost-effective way to access servers, share expensive devices, exchange electronic mail, etc.

With the continuing success of LANs, demands evolved in the direction of extending their capabilities toward higher data rates and wider areas. This, together with progress in fiber-optic technology, has produced the so-called Metropolitan Area Networks, or *MANs* [101]. MANs thus represent the evolution of LANs toward higher data rates, e.g., 100-155 Mbps, and coverage up to 100 km.

Among the prominent features of MANs is service integration. The net-

---

1. Hereafter, unless otherwise stated, the terms "packet" and "message" are used interchangeably.

work provides a low-cost packet transport service attracting customers with different types of traffic (e.g., data, voice and video). By properly coordinating access to the network by different sources, a MAN technology must ensure sufficiently high utilization of the medium to justify its cost, while still guaranteeing the Quality of Service (*QoS*) required by each type of traffic.

Since a LAN or a MAN relies on a common transmission medium, Medium Access Control (*MAC*) protocols have been designed to manage the sharing of the transmission medium. From an architectural standpoint, the MAC protocol provides functions which are located in the MAC sublayer of Layer 2 of the OSI Reference Model (OSI/RM), i.e., the data link layer [91].

The aim of a MAC protocol is to control interference and competition among users while optimizing overall system performance, that is, to share resources efficiently among several users. This chapter introduces the "metrics" commonly used to evaluate MAC protocols, i.e., capacity, fairness and user-oriented performance figures (e.g., delay, throughput, packet loss, etc.).

1.1 CAPACITY

The transmissions of the network stations[1] must be coordinated, to some degree, if the stations are to share a common transmission medium. This coordination is always achieved by means of control information which can either be carried explicitly by control messages travelling along the medium (e.g., reservations, tokens, etc.), or can be provided implicitly by the medium itself by the channel being either active or idle. Control messages or message retransmission due to collision (the latter occurs in MAC protocols which make use of implicit control information) subtract channel bandwidth from that available for successful message transmission. Therefore, the fraction of channel bandwidth used by successfully transmitted messages gives a good indication of the overhead required by a MAC protocol to perform its coordination task among stations. This fraction is known as the utilization of the channel, and the maximum value it can attain (ρ_{max}) over all possible offered loads (see Section 1.4) is known as the *capacity* of the MAC proto-

---

1. Throughout this book the terms "station" and "node" are used interchangeably.

col ([1], [104]). From this definition it follows that the capacity of a MAC protocol will be, at most, equal to one.

Protocol capacity varies across the various MAC protocols, but it is also influenced by several other parameters, such as the number of active stations and the way active stations contribute to the offered load. Throughout this book, the MAC protocol capacity with only one active node will be denoted as ρ_{single}. In a MAC protocol which is ideal from the utilization standpoint, both ρ_{max} and ρ_{single} must be equal to 1.

The capacities of MAC protocols described in this book frequently depend upon the value of the ratio (commonly denoted by a)

$$a = \frac{\tau}{m} \; , \tag{1.1}$$

where τ is the end-to-end channel propagation delay, and m is the average message transmission time. Typically, a values range from 0.01 to 0.05 for LANs and from 0.5 to 50 for MANs. To illustrate the dependency of protocol capacity on a it may be useful, for tutorial reasons, to consider a bus with only two nodes A and B, one located at each end of the bus; the node take turn transmitting. This means that a node transmits its message right after

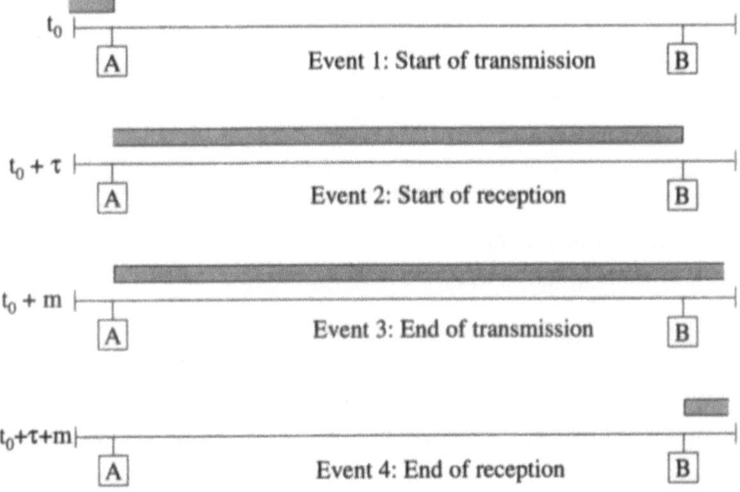

Figure 1.1: Effect of a on protocol capacity ($a < 1$)

receiving a message transmitted by the other node. For the sake of simplicity, it is also assumed that

(i)   message overhead (due to control information carried in the header) is negligible compared to useful information;

(ii)   messages are of a constant length, and m indicates the message transmission time.

Although assumptions (i) and (ii) are formulated in the context of this example, they are assumed to hold for all the MAC protocols described in this chapter.

Let t_0 denote the time instant at which the left-hand station starts trans-

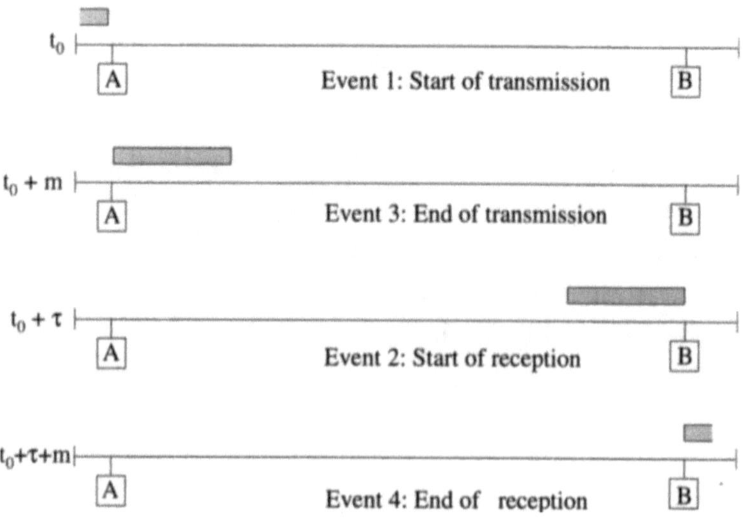

Figure 1.2: Effect of a on protocol capacity ($a > 1$)

mitting a message. If $a < 1$, the sequence of events which occurs from t_0 onwards (see Figure 1.1) is:

1. the left-hand station, A, begins transmission at time t_0;

2. the right-hand station, B, begins reception at time $t_0 + \tau$;

3. A completes its transmission at time $t_0 + m$;

4. B completes reception at time $t_0 + m + \tau$, and can start transmitting its own message.

As shown in Figure 1.2, if $a > 1$, event 3 occurs before event 2. From Figures 1.1 and 1.2 it appears that, for any value of a, the time it takes to perform a turn (i.e., the time between the moment when A starts transmitting and the earliest moment at which B starts transmitting) between A and B is equal to $m + \tau$, although the bus is only kept busy for useful transmissions for a time equal to m. Hence, the protocol capacity for this ideal protocol is $\rho_{max} = m/(\tau + m)$. By dividing the numerator and denominator by m and taking into consideration the definition of a (1.1),

$$\rho_{max} = \frac{1}{1 + a} \quad . \tag{1.2}$$

From (1.2) it follows that the larger the value of a (i.e., the larger the value of τ compared to m), the smaller the protocol capacity. This tendency is easy to see when one consider that, with m kept constant, the fraction of time the bus remains idle increases as τ increases.

1.2 FROM LANS TO MANS

MAC protocols designed for the two most popular IEEE standards for LANs, Ethernet [87] and Token Ring [89], are not suitable for high transmission speeds and/or long distances. This is because the Ethernet and Token Ring MAC protocol capacities become very small, i.e., $\rho_{max} \to 0$, for lengths (of up to 100 Km) and channel speeds (greater than 100 Mbps) typical of a MAN environment. This is proven below

ETHERNET. The Ethernet protocol capacity is computed by evaluating, in the worst case scenario, the average time the channel is occupied to successfully transmit a message, i.e., virtual transmission time (t_V) . This section follows the line of reasoning used in [137]. A more accurate analysis can be found in [105].

In the general case, by denoting with i the station performing the tagged successful transmission, the following relation holds:

$$t_V = N_c \cdot T_{coll} + m + \tau_i \quad , \tag{1.3}$$

where
- τ_i is the time the node$\{i\}$ successfully transmitted message prevents the other nodes to start a transmission due to the carrier sensing mechanism; the worst-case value for this figure is τ;

- N_c is the average number of collisions before node$\{i\}$ begins a successful transmission; and

- T_{coll} is the time the channel is kept busy due to a collision.

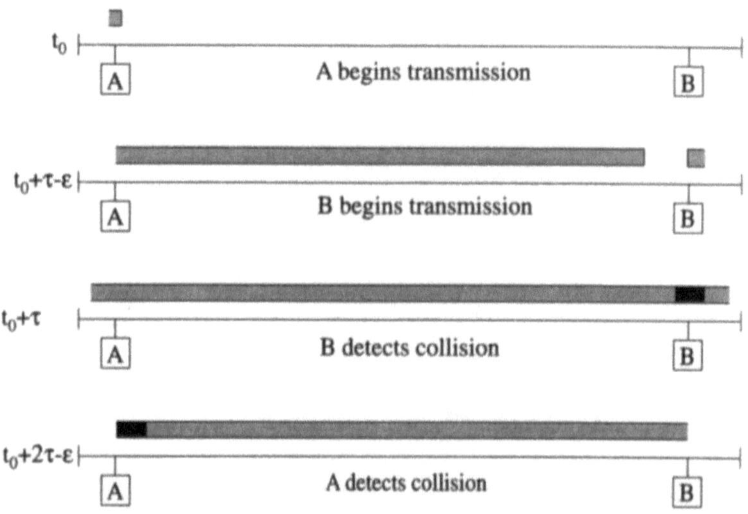

Figure 1.3:   Sequence of events in collision detection $(a<1)$

The upper bound of T_{coll} is obtained by considering a scenario in which the two colliding nodes, say node$\{A\}$ and node$\{B\}$, are positioned at the opposite ends of the bus. For this scenario, as can be seen from Figure 1.3, it takes at most twice the bus propagation delay before a collision is detected. The events which lead to a collision detection time equal to 2τ are

- the left-hand node, say node$\{A\}$, starts transmitting a message at time t_0;

- the right-hand node, say node$\{B\}$, decides to transmit a message just before the node$\{A\}$ message arrives at the other end of the bus, i.e., at time $t_0 + \tau - \varepsilon$;

- at time $t_0 + \tau$ there is a collision which is detected by node$\{B\}$; this collision is detected by node$\{A\}$ $\tau - \varepsilon$ seconds later. After the collision is detected a station stops immediately its transmission but the bits transmitted immediately before the collision detection remain in the bus until they are absorbed at the end of the bus. This time is always less or equal to τ. Hence $T_{coll} \leq 3 \cdot \tau$.

From the observation above it follows that

$$t_V = m + \tau + 3\tau N_c = m[1 + a(1 + 3N_c)] \quad . \tag{1.4}$$

The only unknown quantity in equation (1.4) is N_c, the value of which depends upon the retransmission strategy. Simulation studies carried out by Lam [105] indicate that it is not necessary to consider the retransmission strategy, and that accurate results can be obtained by assuming that the number of collisions before a successful transmission is geometrically distributed with parameter v. Hence,

$$N_c = \sum_{j=1}^{\infty} j v (1-v)^j = (1-v)/v \quad . \tag{1.5}$$

The probability v can be calculated by reasoning as follows. If K $(K \gg 1)$ nodes are ready to transmit a message, and p is the probability that any given node wants to transmit in a 2τ interval, the probability that one and only one node is successful in transmitting is given by

$$v = Kp (1-p)^{K-1} \quad , \tag{1.6}$$

which has a maximum when $p = 1/K$. Using this value of p, which maximizes the right-hand term in (1.6) and hence provides the greatest chance of success, and with $K \gg 1$ as assumed, (1.6) becomes

$$v_{max} = \left(1 - \frac{1}{K}\right)^{K-1} \to \frac{1}{e}, \quad K \to \infty \quad . \tag{1.7}$$

Therefore, the value of v to be used in (1.5) is e^{-1}, and (1.4) becomes

$$t_V = m [1 + a (1 + 3 \cdot (e-1))] \quad . \tag{1.8}$$

Thus, to transmit a message of duration m, it takes, on average, a time interval whose length is given by equation (1.8). Hence

$$\rho_{max} = \frac{m}{m [1 + a (1 + 3 \cdot (e-1))]} = \frac{1}{[1 + a (1 + 3 \cdot (e-1))]} \quad . \tag{1.9}$$

Using the numerical value of e, equation (1.9) can also be written as

$$\rho_{max} = \frac{1}{1 + 6.15a} \quad . \tag{1.10}$$

Other researchers, using slightly different approximations in the ρ_{max} computation, have obtained formulas with the same structure but with different values for the factor which gets multiplied by a. For example, this factor becomes 3.44 in [140], and 7.34 in [82].

TOKEN RING. The protocol capacity for the Token Ring MAC protocol is derived by computing the total transmission time between two consecutive token arrivals at the same node, which, in the following discussion, is assumed to be node$\{1\}$.

Let K be the number of active nodes connected to the LAN and let $\tau_{i,j}$ be the propagation delay between node$\{i\}$ and node$\{j\}$, with the convention that $\tau_{i,i} = \tau$ and that the token transmission time is negligible.

Two scenarios need to be considered, depending upon whether $a < 1$ or $a > 1$. For $a > 1$ (i.e., $\tau > m$), the following sequence of events takes place (see left side of Figure 1.4):

1. node$\{1\}$ begins message transmission at time t_0 and completes it at time $t_0 + m$;

2. node$\{1\}$ starts removing its own message from the ring at time $t_0 + \tau$ and it immediately releases the token, which will be captured by the next downstream node (node$\{2\}$);

3. node$\{2\}$ begins message transmission at time $t_0 + \tau + \tau_{1,2}$ and then waits to receive its own transmission before releasing the token.

4. node$\{i\}$, $i = 3, ..., K-1$, begins message transmission at time $t_0 + (i-1)\tau + \tau_{1,i}$ and then waits to receive its own transmission before releasing the token;

5. node$\{K\}$ begins transmission at time $t_0 + (K-1)\tau + \tau_{1,K}$, and it receives the leading edge of its own frame (and so releases the token) at time $t_0 + K\tau + \tau_{1,K}$; and

6. node$\{1\}$ begins transmission again at time $t_0 + (K+1)\tau$.

At this point in time, events 1 to 6 will repeat over and over again. From this sequence of events, it follows that over a time interval of duration $(K+1)\tau$, each of the K active nodes transmits one message of duration m. Hence, the fraction of useful time for message transmission, i.e., the Token Ring capacity, is given by the ratio

$$\rho_{max} = \frac{Km}{(K+1)\tau} = \frac{1}{\left(1+\dfrac{1}{K}\right)a} \, . \tag{1.11}$$

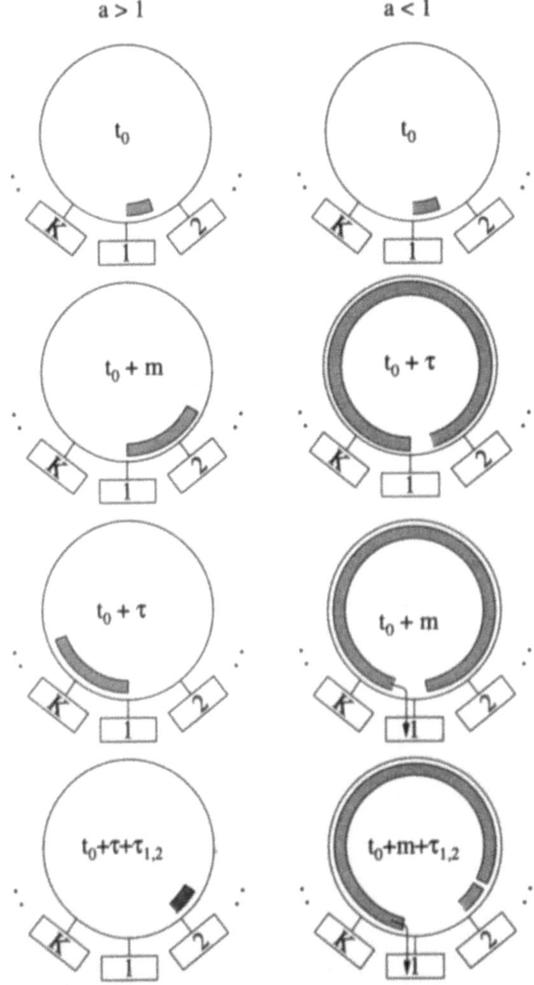

Figure 1.4: Effect of a on the capacity of the Token Ring MAC protocol

For $a < 1$ (i.e., $\tau < m$), the derivation of the protocol capacity follows the same line of reasoning already shown for the case $a > 1$. Specifically, the following sequence of events takes place (see right side of Figure 1.4):

1. node$\{1\}$ begins message transmission at time t_0 and message removal begins at time $t_0 + \tau$, while message transmission is still underway;

2. node$\{1\}$ completes message transmission at time $t_0 + m$, and releases the token immediately;

3. node$\{2\}$ begins message transmission at time $t_0 + m + \tau_{1,2}$;

4. node{2} message completes transmission at time $t_0 + 2m + \tau_{1,2}$, and it releases the token immediately thereafter;

5. node{i}, $i = 3, ..., K-1$, begins and completes message transmission at times $t_0 + (i-1)m + \tau_{1,i}$ and $t_0 + im + \tau_{1,i}$, respectively. After message transmission, node{i} releases the token;

6. node{K} completes message transmission at time $t_0 + Km + \tau_{1,N}$; and

7. node{1} begins transmission again at time $t_0 + Km + \tau$.

At this point in time, events 1 to 7 will repeat over and over again. From this sequence of events (for $a < 1$), it follows that over a time interval of duration $Km + \tau$, each of the K active nodes transmits one message of duration m. Hence,

$$\rho_{max} = \frac{Km}{Km + \tau} = \frac{1}{1 + \dfrac{a}{K}} \; . \tag{1.12}$$

In summary, for the Token Ring MAC protocol,

$$\rho_{max} = \begin{cases} \dfrac{1}{1 + \dfrac{a}{K}} & a < 1 \\[4ex] \dfrac{1}{\left(1 + \dfrac{1}{K}\right)a} & a > 1 \end{cases} \; . \tag{1.13}$$

Formulas (1.10) and (1.13) are now used to show the inadequacy of the

Table 1.1 Values of a for various network configurations and transmission speeds

Channel Bandwidth	Coverage		
	1 Km	10 Km	100 Km
10 Mbps	a=0.05	a=0.5	a=5
100 Mbps	a=0.5	a=5	a=50

Ethernet and Token Ring MAC protocols for MANs, i.e., for networks employing a fiber-optic bus or ring with a length of up to 100 Km and a channel bandwidth of 100 Mbps and above. Assuming messages of 1000 bits and a propagation speed of the light in the fiber of 200,000 Km/sec, a values for various medium configurations and transmission speeds are shown in Table 1.1.

Table 1.2 Capacity values

a value	Ethernet Capacity	Token Ring Capacity
0.05	0.76	≈ 1
0.50	0.24	≈ 1
5.0	0.03	0.20
50	0.003	0.02

A typical a value for a LAN (e.g., coverage of 1 Km and transmission speed of 10 Mbps) is 0.05, while a typical a value for a MAN (e.g., coverage of 100 Km and transmission speed of 100 Mbps) is 50.

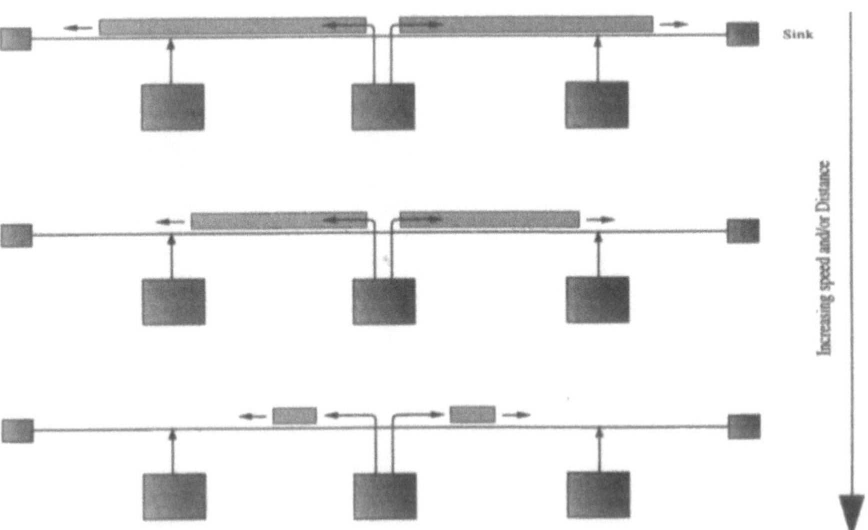

Figure 1.5: Message shrinkage with increasing speed and/or bus length in Ethernet

Table 1.2 shows the protocol capacities for various a values, assuming a network with 50 active nodes.[1] The Ethernet and Token Ring protocol capacities vary from acceptable values (0.76 for Ethernet and 1 for Token Ring) to

____

1. Ethernet capacity with infinite active nodes is a good approximation for the case with 50 active nodes.

unacceptable values (0.003 for Ethernet and 0.02 for Token Ring). This result can be visualized with the help of Figure 1.5 and Figure 1.6. If one increases the channel bandwidth of the medium and/or increases the medium length, while keeping the number of bits in a message constant (i.e., if one increases the a value), a sort of "message shrinkage" results under both the Ethernet and Token Ring MAC protocols.

In Ethernet, message shrinkage causes an increase in the time the cable is erroneously sensed as being idle, and this causes an increase in the probability of message collision. In Token Ring, message shrinkage causes an increase in the time during which no node is authorized to transmit.For this reason, new MAC protocols have been designed to operate efficiently in MAN environments. Two (out of many) MAC protocols will be described at length in this book: FDDI and DQDB. The former is an ANSI standard [4], while the latter is an IEEE standard [90].

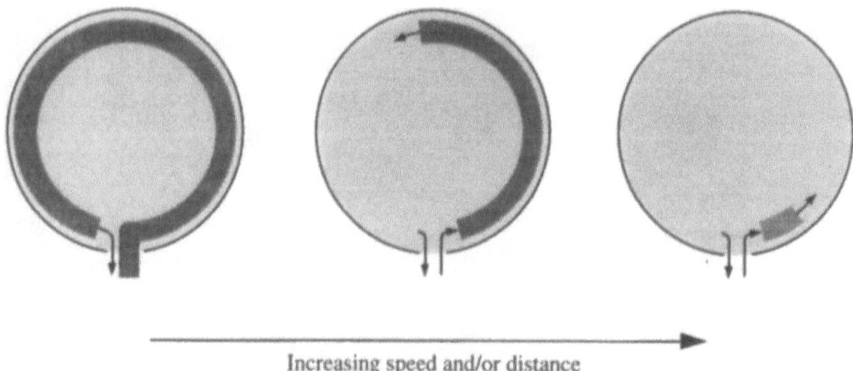

Figure 1.6: Message shrinkage with increasing speed and/or ring length in Token Ring

1.3 FAIRNESS

Capacity is a measure of the aggregate bandwidth that the network nodes can rely upon. However, it does not indicate anything about the way this bandwidth is subdivided among the nodes. Therefore an additional performance measure, named *fairness*, is introduced. Fairness means that the network does not differentiate between stations in granting them access rights to the transmission bandwidth [72].

Generally speaking, one would like a network to behave like a "black box", and the quality of service achieved by a node to be unaffected by its physical location (*ideal fair behavior*). To clarify this concept, it is useful to

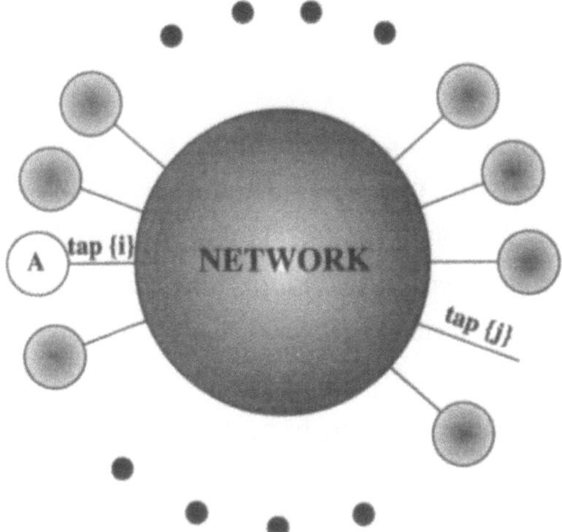

Figure 1.7: A network and its nodes

refer to Figure 1.7, which shows a network and its nodes. It is assumed that node{A} changes its physical location, while the shaded nodes do not. According to the above definition of ideal fair behavior, if the offered load of every node remains constant, node{A} must experience the same quality of service when it is at tap$\{i\}$ as when it is at tap$\{j\}$ for every $\{i\}$ and $\{j\}$. Throughout this book, a *fair network* is a network which exhibits ideal fair behavior.

Fairness measures highlight differences, if any, in the QoS achieved by network nodes. Since QoS includes measures such as delay, throughput and packet loss, a fairness metric can be defined for each QoS index. A network can be fair with respect to one or more of the QoS figures. An *ideal fair network* is fair with respect to all the QoS indices.

1.4 USER-ORIENTED PERFORMANCE FIGURES

Capacity and fairness are commonly used to evaluate the MAC protocol algorithms. However, additional performance measures are needed to determine the quality of service the user can rely upon.

As mentioned earlier, a prominent feature of MANs is service integration. The network provides a low-cost packet transport service, attracting customers with many different types of traffic. Traffic sources differ in their requirements for bandwidth, guarantees for the end-to-end delay, acceptable packet-loss rates, etc.

Table 1.3 shows a sample of applications for a future integrated-service networking environment. For each application, both the main traffic characteristics (*source bit rate* and *burstiness*) and the Quality of Service (*QoS*) requirements (*response time* and *reliability*) are shown. These characteristics and requirements are described below.

SOURCE BIT RATE. The source bit rate indicates the average number of bits per time unit produced by an application and sent to the network for transmission; hereafter the source bit rate is denoted by λ. The amount of the bit rate correctly delivered to the receiver is named *throughput, γ*. The percentage of the bit rate which is not delivered to the receiver is called the *packet loss probability, P_L*. The bit rate normalized with respect to the channel capacity is named *Offered Load, OL*.

BURSTINESS. Burstiness indicates the variability of the generated traffic, and it is generally measured by making reference to an interval of fixed length, Δ. Burstiness is the ratio between the maximum value of the bit rate in interval of length Δ and the bit rate. "Low" (in Table 1.3) indicates burstiness close to 1 (Constant Bit Rate or CBR applications), while "High" stands for burstiness exceeding 10. Applications with burstiness greater than one are named Variable Bit Rate or VBR applications.

RESPONSE TIME. The response time is defined as the time between the generation of a message at the sending station and its reception at the destination station. Response time in packet-switching networks is highly variable, and ranges from milliseconds up to several seconds. Applications tagged "High" can accept a response time of seconds; those tagged "Low", such as voice, can operate correctly only if the response time is on the order of milliseconds (e.g., 20-40 msec). In addition, an application is defined as *real-time* if packets experiencing a response time higher than a predefined threshold may, for all practical purposes, be considered lost. The best known examples of real-time applications are interactive voice (having a threshold of about 30 msec), entertainment video (threshold of about 1 sec), and alarm messaging (threshold of about 50 msec). Due to their sensitiveness to response time, real-time applications are also referred to as *timed-constrained* applications.

Among real-time applications, there is a further subdivision into *synchronous* and *asynchronous* applications. Synchronous applications have the additional requirement that the sending and receiving stations be synchronized. In other words, the bit pattern at the sending and receiving sides must be the same with only a constant temporal shift. Since the response time in packet-switching networks is variable (*delay jitter*), a *jitter absorption* mechanism is required to recover from this variability. Typically, the jitter absorption mechanism artificially increases the response time, and this requires an adequate buffering capacity in the receiving equipment [97]. For example, assuming a periodic application with a constant bit rate and period of δ time units (e.g., telephone), synchronization is achieved at the receiving station by delaying the first packet until its response time is equal to the 99-th percentile of the response time distribution. From that point on, a packet is removed from the receiving buffer every δ time units. It is worth observing that in this case, only packets violating the deadline constraint (99-th percentile of the response time distribution) fail to be correctly delivered.

Loss tolerance. The loss tolerance indicates the capability of an application to tolerate that some of its transmitted packets may be lost. Those applications tagged "High" in the Loss Tolerance column accept losses of up to 5% of transmitted packets; at the other extreme, "Low" identifies applications in which all the transmitted packets must arrive correctly at the destination station. In packet-switching networks losses occur for several reasons

- transmission errors;
- missed deadlines (in the case of time-constrained communications);
- buffer overflows.

As MANs use fiber optics as their physical medium, transmission errors are negligible. Furthermore, in the analysis of a MAN MAC protocol, buffer overflows are generally studied only in the transmitting station, as it can be assumed that the receiver has enough processing capacity to handle the

Table 1.3 Application taxonomy

Application	Offered Load (Mbps)	Burstiness	Response Time	Real time	Loss Tolerance
Voice	0.004 - 0.064	Low	Low	Yes	High
File Transfer	up to 100	High	High	No	Low
Transactions	0.064 - 1.544	High	Low	No	Low
Imaging	0.256 - 25	High	Medium	Yes	Medium
Business video	0.256 - 16	Low	Low	Yes	Medium
Entertainment video	1.5 - 50	Low	Low	Yes	Medium
Isochronous traffic	0.064 - 2.048	Low	Low	Yes	High
LAN interconnection	4 - 100	High	High	No	Low
Server access	4 - 100	Medium	High	No	Low
Hi-Fi audio	0.128 - 1	Low	Low	Yes	Medium
Alarms messaging	less than 0.064	High	Low	Yes	Low

incoming traffic without any congestion. Hence, the packet-loss requirements can be reduced to constraints on buffer overflows in the sending station and on constraints on the missed deadline rate. For these reasons, in the analysis of the QoS provided by MANs, percentiles of the buffer occupancy distribution and percentiles of the response time distribution are investigated. For example, by assuming an infinite buffer in the sending station, a figure which is commonly used to investigate whether a network is suitable to support voice applications is the 99-th percentile of the response time distribution. This figure identifies a value of the response time which is exceeded by only 1% of transmitted packets. Hence, if this value is an acceptable delay for voice applications (see Table 1.3), the network is suitable for this kind of traffic. Note that, in this case, 1% of transmitted packets may violate the deadline constraint, and this is compatible with the voice reliability constraint.

Table 1.3 clearly shows that in LAN networks, whose main target is to support non real-time *EDP* applications (e.g., file transfer, server access and transactions), the quality of service is completely characterized by the average response time and throughput. For MANs, on the other hand, which can support not only EDP data applications but time-constrained applications (e.g., voice, video and alarms) as well, the response-time and packet-loss distributions would also be necessary.

In the next section, each performance figure which is of interest from an application point of view is precisely defined.

1.5 MODELING OF A MAN ENVIRONMENT

Figure 1.8 presents the queueing model used to derive the user-oriented performance figures in a MAN environment. Specifically, the figure focuses on station$\{i\}$, referred to as the *tagged station*, in which a telephone, an EDP and an alarm application generate packets with an aggregate rate (number of packets per time unit) identified by the λ parameter value. Incoming packets are stored in a Local Queue (LQ). This queue may contain a finite or infinite number of packets. In the finite case, M is the maximum number of packets which can be stored in the LQ. When the LQ is full, incoming packets are discarded. P_L indicates the probability that an incoming packet is discarded. Hence, γ is given by $\gamma = \lambda \cdot (1 - P_L)$.

The packet at the head of the local queue waits until its transmission is scheduled according to the MAC protocol. Packet transmission time is denoted by the r.v. B, which is given by the ratio between the packet length (expressed in bits) and the channel speed (expressed in bit/sec); b is the average value of B.

After transmission, the station$\{i\}$-packets experience a propagation delay before arriving at the destination station (say, station$\{j\}$). The propagation delay, $\tau_{i,j}$, depends on the distance between the sender and the receiver, and the speed of the signal in the transmission medium, v.

Finally, packets generated by the synchronous application (telephone) are stored in the jitter absorption buffer before being delivered to the receiver.

With reference to the scenario depicted in Figure 1.8, Figure 1.9 shows the various components of the response time. These are

- *Local Queueing delay*: the time between the arrival of a packet at the LQ and the time at which the packet comes to the head of the LQ, i.e., $t_2 - t_1$;
- *Access delay*: the time between the arrival of a packet at the LQ and the end of packet transmission, i.e., $t_4 - t_1$;
- *MAC delay*: the time between the instant at which the packet comes to the head of the LQ and the end of packet transmission, i.e., $t_4 - t_2$;
- *End-to-End delay*: the time between the arrival of a packet at the LQ and the end of reception of the packet at the receiving side, i.e., $t_6 - t_1$;

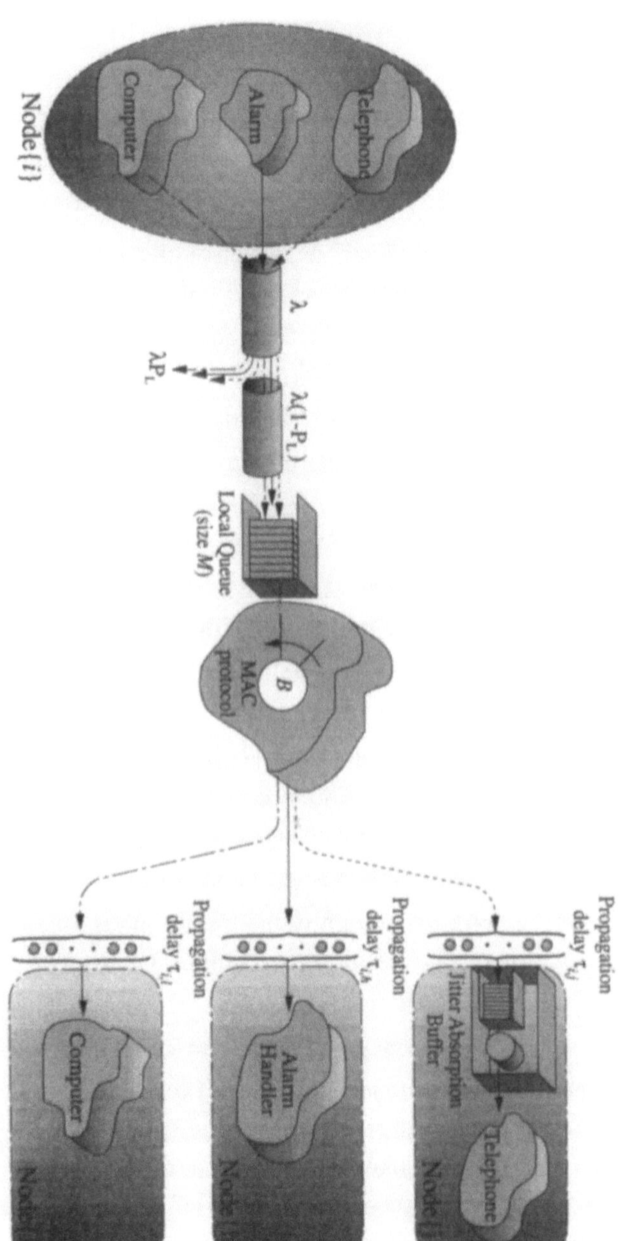

Figure 1.8: Model of a Sender-Receiver communication in a MAN

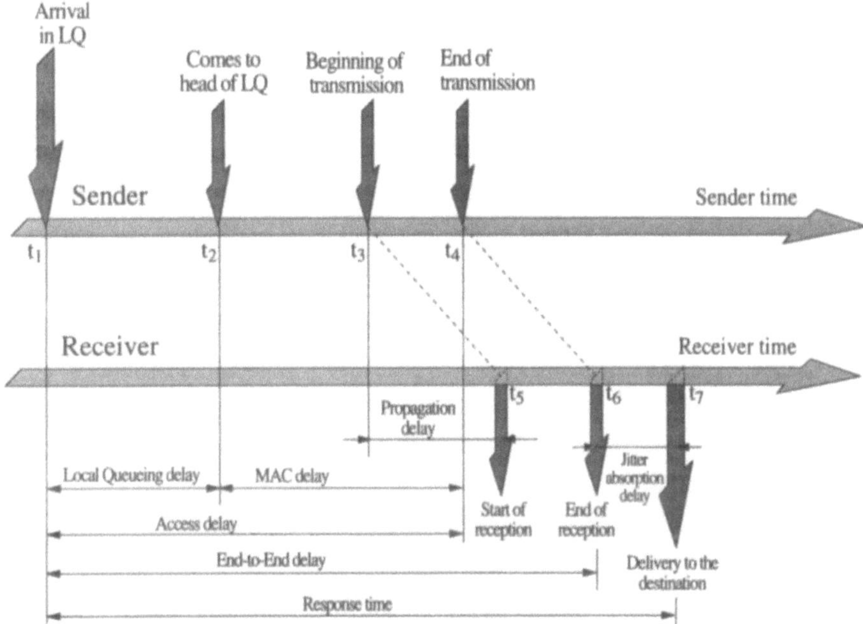

Figure 1.9: Packet delays in a MAN

- *Jitter Absorption delay*: the time added to the packet end-to-end delay to absorb the end-to-end delay jitter, i.e., $t_7 - t_6$;
- *Response time*: The time between the arrival of a packet at the LQ and the delivery of the packet at the destination station, i.e., $t_7 - t_1$. Obviously, for asynchronous applications jitter absorption is not required, and hence, response time and end-to-end delay coincide.

1.6 OUTLINE OF CONTENTS

This section briefly describes the contents of each remaining chapter in this book. Chapters 2, 3 and 4 review the methods and tools for MAN performance analysis. Specifically, Chapter 2 presents the fundamental results of queueing theory. Some of these results are quite recent, and can be found only in very advanced books and papers dealing with stochastic-process theory. This book presents these theoretical results together with illustrative applications.

Chapter 3 presents advanced material on single-server queueing systems with finite or infinite buffers. This chapter focuses on a didactic and self-contained presentation of the well known matrix-analytic solution method for *M/G/*1*-type* systems developed by Marcel Neuts.

Chapter 4 outlines recent results on polling systems with limited service disciplines. It includes analytical methods based on the pseudo-conservation law, and two numerical algorithms proposed by Tran-Gia & Raith and Blanc. The power series algorithm developed by Blanc not only provides solutions for this class of models, but can also be applied to study stochastic processes with a Quasi-birth-and-death (*QBD*) structure.

Chapter 5 describes the FDDI standard. The main focus is on the MAC protocol.

Chapter 6 presents a structured view of the performance modeling activities related to FDDI. A taxonomy of the FDDI models is presented and some relevant models are discussed.

Chapter 7 describes the DQDB standard. The main focus is on DQDB connectionless data services. Simulative results are used extensively to discuss DQDB performance and fairness. Enhancements and variations of the standard are not covered in this book; an overview of these research activities can be found in [120].

Chapter 8 presents a structured view of performance modeling related to DQDB. A taxonomy of the DQDB models is presented and some relevant models are discussed.

Chapter 9 discusses the evolution of local and metropolitan area networks towards gigabit rates.

2 Stochastic Processes For Modeling Metropolitan Area Networks: Basic Results

Two classes of queueing models are generally used to model MAC protocols for MANs: single server queueing system and polling systems. This chapter introduces these models and reviews their basic properties.

Section 2.1 introduces the terminology and the performance figures which are used throughout the book. Section 2.2 outlines some general results on queueing systems: Little, Burke and PASTA theorems. Section 2.3 sketches the results on renewal and regenerative processes relevant for MAN analysis. Conservation laws are a very powerful tool in the analysis of both single server and polling systems. For this reason they are extensively treated in Section 2.4 (*work conservation laws*) and Section 2.6 (*pseudo-conservation laws*). Stochastic decomposition properties, which provide the easiest way to prove the pseudo-conservation laws are presented and applied in several examples in Section 2.5.

Throughout this chapter Laplace-Stieltjes transform (*LST*) and z-transform (or Probability Generating Function, *PGF*) techniques will be extensively used and the reader is assumed to be familiar with them. Introductory on LST and PGF can be found, for example, in [99].

Here, it is worth pointing out that for a continuous r.v. Y with probability distribution function, *PDF*, $Y(\bullet)$ and probability density function, *pdf*, $y(\bullet)$, its LST will be denoted as

$$Y^*(s) = \int_0^\infty e^{-sx} dY(x) \ , \tag{2.1}$$

where s can be either a real or a complex quantity, and, by denoting with $E[Y^i]$ the i-th moment of Y, the following relation holds

$$E[Y^i] = \int_0^\infty x^i dY(x) = i\int_0^\infty x^{i-1}[1-Y(x)]\,dx = (-1)^i Y^{*(i)}(0) , \qquad (2.2)$$

where

$$Y^{*(i)}(0) = \frac{d^i}{ds^i}Y^*(s)\Bigg|_{s=0} \qquad i = 1, 2, 3, ... \quad . \qquad (2.3)$$

Similarly for a discrete r.v. N with probability distribution $N(\bullet)$, its PGF will be denoted as

$$N(z) = \sum_i z^i \cdot P\{N = i\} \qquad (2.4)$$

and, by denoting with $E[N^i]$ the i-th moment of N, the following relations hold

$$E[N] = \sum_i i \cdot P\{N = i\} = N^{(1)}(1) , \qquad (2.5)$$

$$E[N^2] = \sum_i i^2 \cdot P\{N = i\} = N^{(2)}(1) + E[N] , \qquad (2.6)$$

where

$$N^{(i)}(1) = \frac{d^i}{dz^i}N(z)\Bigg|_{z=1} \qquad i = 1, 2, 3, ... \quad . \qquad (2.7)$$

2.1 QUEUEING MODELS FOR MAN MODELING

2.1.1 Single-queue Models

This section presents the notation and results for queueing models made up of one queue (with limited or unlimited length) and c $(c \geq 1)$ servers. Packets arrive at the queueing system, stay there until their service requirements are met, and then leave. An arrival time and a service time process are used to characterize the arrival times of the packets to the system and the amount of service they require, respectively. Each server models a resource. In the context of computer networks, this resource is very likely to be a communication medium or a virtual circuit. If there are multiple servers, they are assumed to be identical and to operate in parallel and independently. A packet can be served by any free server.

Depending upon the action taken by the system when packets are present, the previous models can be further subdivided in two classes. The first one includes models in which a server is never idle if there are packets waiting to be served (*models without vacation*). In the models belonging to the second class there are periods when packets are present but none of them can be served (*vacation models*). In the following, if not explicitly stated, the models are assumed without vacation.

As stated before, to fully specify these models it is necessary to describe the arrival process, the service process, the number and service rate of the servers, and the order in which packets are served. For queueing models with vacation, it is also necessary to specify the rules which govern when vacation periods begin and end. These rules may depend upon either the current state or the past history of the system. However, in the models analyzed hereafter, they do not depend on the future behavior of the arrival process.

ARRIVAL PROCESS. This process is specified by the interval times between successive packet arrivals at the queueing system. These are called interarrival times, and are generally represented by independent and identically distributed (*i.i.d*) random variables (*r.v.*). Throughout the book λ denotes the *arrival rate* of packets to the system.

In several of the models analyzed in this book, the interarrival times are exponential random variables with mean $1/\lambda$. This means that

$$P\{interarrival\ time \leq t\} \ = \ 1 - e^{-\lambda t} \ .$$

Furthermore, if $\{A(t), t \geq 0\}$ is the process that counts the number of packets which arrive at the queueing system in an interval of length t, under the assumption of interarrival times distributed exponentially with rate λ, $\{A(t), t \geq 0\}$ is a Poisson process with rate λ.

$$P\{A(t) = k\} \ = \ \frac{(\lambda t)^k}{k!} e^{-\lambda t}, k = 0, 1, \ldots \ .$$

SERVICE PROCESS. The amount of service required by a packet at the queueing system is described by means of identically distributed (not always statistically independent) random variables which throughout the book, will be denoted by $B_1, B_2, \ldots$. Throughout the r.v. B denotes the generic r.v. of the sequence $\{B_n\}$, and $B(x)$ is its distribution function, i.e.,

$$B(x) = P\{B \le x\} \quad \text{for } x \ge 0.$$

In this book, it is assumed that $B(0) = 0$ and $B(\infty) = 1$. The Laplace-Stieltjes transform $B^*(s)$ of $B(x)$, the mean service time b, and the i-th moment $b^{(i)}$ of the service time distribution $(i = 2, 3, \ldots.)$ are given by

$$B^*(s) = \int_0^\infty e^{-sx} dB(x) \quad , \tag{2.8}$$

$$b = \int_0^\infty x \cdot dB(x) = \int_0^\infty [1 - B(x)] \, dx = -B^{*(1)}(0) \quad , \tag{2.9}$$

and

$$b^{(i)} = \int_0^\infty x^i dB(x) = i \int_0^\infty x^{i-1} [1 - B(x)] \, dx = (-1)^i B^{*(i)}(0) \quad , \tag{2.10}$$

where

$$B^{*(i)}(0) = \frac{d^i}{ds^i} B^*(s) \bigg|_{s=0} \quad \text{for } i = 1, 2, 3, \ldots \quad . \tag{2.11}$$

Several models analyzed in the book have an exponential service time distribution, which is normally expressed as

$$B(x) = 1 - e^{-\mu x} \quad ,$$

where μ is called the *service rate*. μ can be defined as the rate at which a server processes packets when the server is busy. This definition is valid for all service time distributions.

EXAMPLE 2.1   For some models which come up in the networking environment, it is necessary to make a distinction between different types of packets. For example, the protocol-data-units moving along a connection can be classified according to whether or not they carry a user message. If they do not, protocol-data-units are used to control the operation of the connection to which they belong. Think, for example, about acknowledgements or expedited data [137]. Control protocol-data-units are shorter than protocol-data-units which carry user data. From a modeling standpoint this means that the service requirements of a control protocol-data-unit are different from the service requirements of a data protocol-data-unit.

$$\Diamond$$

As shown in the example, the service times for packets of different types may have different probability distribution functions $B_k(x)$ is used to indicate the probability distribution function for k-type packets.

NUMBER OF SERVERS. The single-server queueing system is clearly the simplest one as it can serve only one packet at a time. A multi-server queueing system is characterized by c identical servers and can serve as many as c packets simultaneously. In an infinite-server queueing system, i.e., $c = \infty$, each arriving packet can immediately enter a server without having to wait in the queue.

QUEUEING SYSTEM CAPACITY. In some queueing systems the queue size is assumed to be infinite, while in others it is assumed to be of limited size (denoted by M). In the former systems $(M = \infty)$, an arriving packet is always allowed to join the queue, while in the latter systems, $(M < \infty)$ an incoming packet when the queue is full.

REMARK. According to the *Kendall's notation*, throughout this text, a single queue model is described by $X/Y/c/K$, where X and Y represent the interarrival-time and the service time distribution respectively, c is the number of servers, and K is the queue size. Specifically, the symbol G will denote a stationary distribution. If, in addition, the samples of a stationary distribution are independent from each other the symbol GI will be used. The symbols M and D will be used instead of GI to indicate an exponential and a deterministic distribution, respectively. Hence, the single queue models include the well-known $M/G/1$, $M/G/c$, and $G/GI/c$ queueing systems (see for example [99], [149]).

◊

QUEUEING DISCIPLINE. This is the rule for selecting the next packet to be put in service. First-In-First-Out (*FIFO*) is the most common queueing discipline. This discipline is also referred to as First-Come-First-Serve (*FCFS*). Other queueing disciplines used in this book include: Last-In-First-Out (*LIFO*), Round Robin (*RR*), and Processor Sharing (*PS*) [100]. In addition, in a priority queueing system, packets are divided into priority classes with a priority-associated withheld class. Packets with the highest priority are given preferential treatment and within each class one of the above mentioned service discipline is applied. Priority discipline can be distinguished in *non-preemptive* and *preemptive*. A priority discipline is preemptive if a user in service can be removed from the server (and returns in the queue) because an higher priority user arrives in the system; otherwise it is non-preemptive.

NOTATIONS AND PERFORMANCE MEASURES. The most important performance measures in a queueing system are number of packets in the system, waiting times and throughput.

The process describing the number of packets in the system at time t will hereafter be denoted by $\{N(t), t \geq 0\}$. If $\{N_q(t), t \geq 0\}$ and $\{N_s(t), t \geq 0\}$ are the processes describing the number of packets in the queue and in service, respectively, the following relation holds

$$N(t) = N_q(t) + N_s(t) \quad .$$

Furthermore, if N, N_q, and N_s are the random variables describing the steady-state numbers of packets in the system, in the queue and in service, respectively, then

$$N = N_q + N_s \quad . \tag{2.12}$$

The time a packet spends in a queueing system satisfy the following equation

$$R = W + B \quad , \tag{2.13}$$

where R, W, and B denote the steady-state random variables describing the time a packet spends in the system (*response time*), in the queue waiting for the service (*waiting time*) and in service (*service time*), respectively. Clearly, in (2.12) and (2.13) it is assumed that the queueing system reaches the steady-state condition.

The definition of throughput, offered load, and packet loss introduced in Section 1.4 also apply in the context of queueing systems. From the definition of the offered load it follows that

$$OL = \lambda b \ . \tag{2.14}$$

In single queue models with c servers and an infinite buffer the condition $(OL/c) < 1$ is necessary for system stability [99]. Unless explicitly stated the condition $(OL/c) < 1$ is assumed to hold. Under this hypothesis OL coincides with the average number of busy servers and (OL/c) is the probability of a server being busy. (OL/c) is commonly referred to as utilization factor and denoted with ρ.

When the system is stable the following limiting probabilities exist

$$\lim_{t \to \infty} P\{N(t) = i\} = P\{N = i\} = p_i \ , \quad i \geq 0 \ .$$

2.1.2 Single-queue Models with Server Vacations

A large number of models that will be analyzed in the next chapters can be modeled by means of systems in which the server provides no service during periods called *vacations*. For example, in a token-passing MAC protocol, the server vacation time represents the period between the token's departure from the tagged station and its subsequent return to that station.

A time interval during which a server works continuously is called a *service period* (sometimes referred to also as *busy period*). A server thus alternates between vacations and service periods.

The *M/G/1* queueing systems with server vacation considered in this book have the following property: if the server, on return from a vacation, finds an empty queue, the server immediately starts another vacation (*multiple vacation*), i.e., the length of the corresponding service period is zero. This process continues until the server (on return from a vacation) finds at least one packet in the queue.

The time interval between the end of two successive vacations is called a *service cycle*. Obviously, a service cycle is made up of two components: a service period (which may be of length zero) and the vacation which follows that service period.

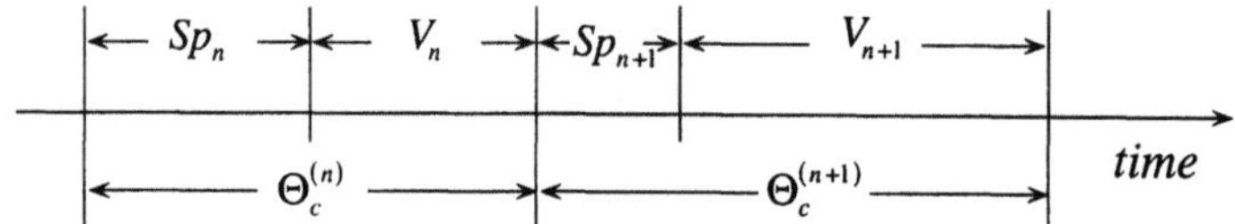

Figure 2.1: Two consecutive service cycles

By denoting with

- V_n the duration of the n-th vacation;
- Sp_n the n-th service period; and
- $\Theta_c^{(n)}$ the n-th service cycle,

in Figure 2.1 a sample of two consecutive service cycles along with their internal structures are shown.

A vacation begins whenever the departing packet leaves the system empty or because other events, depending on the service discipline, occur. For the purpose of this book, vacations begin and end according to the following *visit service disciplines*;

- *Exhaustive service*: the vacation begins as soon as the queue becomes empty;
- *Gated exhaustive service*: only those customers waiting in the queue when the server returns from a vacation are served before the server takes another vacation;
- *l-limited service*: The server serves at most l packets before going on vacation again. Two l-limited service disciplines are defined: exhaustive and gated. In the *exhaustive l-limited service discipline*, the server goes on vacation if either l packets are served or the queue becomes empty (whichever comes first), while in the *gated l-limited service discipline* vacation begins after serving at most l of those packets already present at the end of the previous vacation. When $l = 1$, the 1-limited policy is often referred to as a *Non-Exhaustive* service discipline.

REMARK. Sometimes, rather than the maximum number of packets served in the service period, the visit service discipline specifies the maximum amount of time for each service period. In this case, the service discipline is called *time-limited*.

◊

The *local service discipline* defines how the packets are served. Hence the couple of local and visit service disciplines completely defines the order and the time instants at which packets are served. This couple is referred to as *global service discipline*. Unless explicitly stated, hereafter it is assumed that the local service discipline is FIFO, and therefore the global service discipline is identified by the visit service discipline only.

Let

- V_n be the random variable associated with the duration of the n-th vacation;
- A_n be the number of vacation packets that arrive during the n-th vacation;
- Z_n be the number of packets in the system at the beginning of the n-th vacation.

The sequences $\{V_n\}$, $\{A_n\}$, $\{Z_n\}$ are assumed to be stationary; and V, A, Z denote the generic random variable for $\{V_n\}$, $\{A_n\}$, $\{Z_n\}$, respectively. In the following text, unless otherwise specified, vacations are also assumed to be independent, and the following notation will be used;

- $V(t)$, $v(t)$, $V^*(s)$: distribution function, density function and LST of V, respectively;
- V_+, (V_-) : forward (backward) recurrence time of V (see Section 2.3);
- $V_+(t)$, $V_+^*(s)$, $(V_-(t)$, $V_-^*(s))$ are the distribution function and LST of V_+, (V_-), respectively.

A comprehensive overview of single-server queues with vacations can be found in [60] and [149].

2.1.3 Polling Models

Figure 2.2 shows the queueing model of a polling system. Specifically, a single-server cyclically serves K queues $\{Q_1, Q_2, ..., Q_K\}$ with an infinite buffer size. Without any loss of generality, it is assumed that the queues are indexed by $i = 1, 2, ..., K$ according to the server visit order.

Packets arrive at Q_k (*class-k packets*) according to a Poisson process with rate λ_k. Class-k packets require i.i.d. service times which are distributed as the r.v. B_k (b_k and $b_k^{(2)}$ are respectively, the first and second moments of B_k). The offered load at Q_k is therefore[1]

$$\rho_k = \lambda_k b_k, \quad k = 1, 2, ..., K \ , \tag{2.15}$$

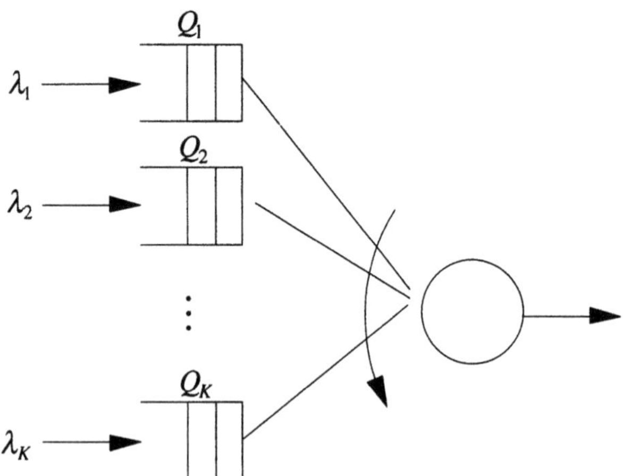

Figure 2.2: Polling model

The total offered load, ρ, is defined as

$$\rho = \sum_{k=1}^{K} \rho_k \ .$$

(2.16)

The time intervals the server needs to switch from Q_k to Q_{k+1} (*switch-over times*) are independent stochastic variables which are distributed as the r.v. S_k (s_k and $s_k^{(2)}$ are the first and second moments, respectively, of S_k). The total switchover time, S, is defined as

$$S = S_1 + S_2 + \ldots + S_K \ .$$

(2.17)

The first and second moments of S are denoted by s and $s^{(2)}$, respectively. Note that s is also

$$s = \sum_{k=1}^{K} s_k \ ,$$

(2.18)

It is assumed below that the interarrival times, service times and switchover times are mutually independent.

---

1. As it is stated below in polling systems the condition $OL < 1$ is necessary for system stability. Under this hypothesis OL coincides with the utilization factor (ρ). To be consistent with the terminology commonly used in polling system analysis, throughout ρ will be used to denote the offered load in polling models.

Similarly to single-queue models with server vacations (see Section 2.1.2), the amount of time the server spends at each queue during a cycle depends on the *global service discipline*. The following visit service disciplines will be considered: exhaustive, gated and *l*-limited.

The concepts of *cycle length*, *intervisit length* and *service period* are used extensively in polling systems. The *cycle lengths* observed by a station Q_k are the lengths of the time intervals between consecutive arrivals of the server at Q_k. The *cycle lengths* observed by Q_k are assumed to be i.i.d. and the r.v. C_k denotes the generic cycle length. The *intervisit lengths* observed by Q_k are the lengths of the time intervals between the departure of the server from Q_k and the next arrival of the server at Q_k. The *intervisit lengths* observed by Q_k are assumed to be identically distributed and the r.v. I_k denotes the generic intervisit length. The *service period* of Q_k is the time interval in which the server is serving packets at Q_k during a given cycle. The *service periods* of Q_k are assumed to be identically distributed and the r.v. Sp_k denotes the generic service period.

The following relationship holds among the quantities defined above

$$C_k = Sp_k + I_k \quad . \tag{2.19}$$

Clearly, $E[C_k]$ does not depend on the queue index k and, in a stationary system, the average cycle length $E[C]$ is

$$E[C] = \frac{s}{1-\rho} \quad . \tag{2.20}$$

By defining the work of a packet as the service time required by it, (2.20) can be derived by noting that, if the system is stable, the average work which arrives at the system during a cycle (2.21) should equal the average work served during a cycle (2.22).

$$E[C] \cdot \sum_{k=1}^{K} \lambda_k \cdot b_k = E[C] \cdot \rho \tag{2.21}$$

$$\sum_{k=1}^{K} E[Sp_k] = E[C] - s \tag{2.22}$$

By applying the same reasoning to the class-*k* packets, it follows that

$$E[Sp_k] = \lambda_k b_k \cdot E[C] = \rho_k \cdot \frac{s}{1-\rho} \ . \qquad (2.23)$$

Finally, from equations (2.19), (2.20) and (2.23) it immediately follows that

$$E[I_k] = E[C_k] - E[Sp_k] = (1-\rho_k) \cdot \frac{s}{1-\rho} \ . \qquad (2.24)$$

The methodologies presented in this book for the analysis of a polling system assume that the system operates in a steady state. Studying stability criteria for multidimensional Markov chains is very difficult [71]. Stability criteria for polling systems have been derived (without a formal proof) by Kuehn [102]. Georgiadis and Szpankowski [71] formally proved the stability condition for polling systems with gated l-limited service discipline.

For polling systems with exhaustive or gated service disciplines, $\rho < 1$ is a necessary and sufficient condition for the system's stability. When the service policy is l-limited, in order to guarantee the stability of all the queues an additional set of constraints should be satisfied

$$\lambda_i \cdot E[C_i] < l_i, \ i = 1, 2, \ldots, K \ , \qquad (2.25)$$

where l_i is the maximum number of class-i packets which can be transmitted in a cycle.

2.2 SYSTEM PROPERTIES

This section focuses on very general properties that a large number of specific models share. For this reason, Heyman and Sobel [84] refer to these properties as *system properties*. "Poisson Arrivals See Time Averages" and the Little theorem are specific examples of such properties.

2.2.1 Little Theorem

Let us now consider a system in which packets arrive at random times to be serviced. In the context of computer networks, this system can be a telecommunication line, a virtual circuit, a segmentation/reassembling process, etc. The Little Theorem provides a general relationship between the average number of users in the system, the system throughout and the average response time.

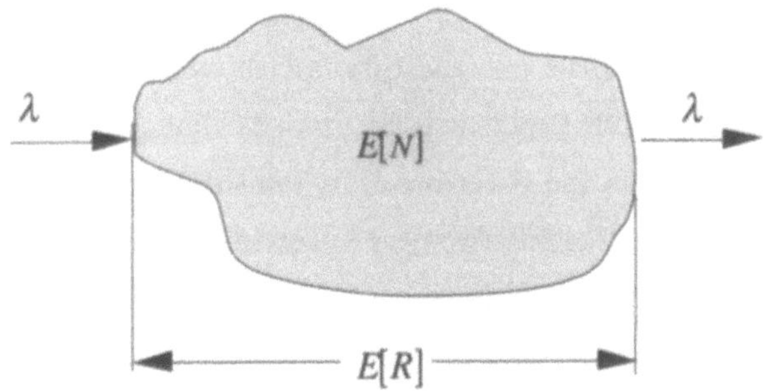

Figure 2.3: The system for Little's formula

THEOREM 2.1   *If* $\lambda < \infty$ *and* $E[R] < \infty$, *then* $E[N] < \infty$ *and* $E[N] = \lambda \cdot E[R]$.

The proof of the theorem can be found in [143].

The Little theorem states that, for stationary systems, the average number of customers in the system is equal to the product of the arrival rate and the average system response time. The Little theorem expresses a very intuitive result: the more crowded the system (large $E[N]$) the longer packets spend in the system ($E[R]$), and viceversa. The system can be viewed as a black box where customers arrive, spend some time and then depart. In fact, no assumptions about service time distribution, arrival distribution, number of servers in the system, nor queueing discipline are required.

REMARK. "System" may refer to a complete queueing system or to a subset of it. For example, the system might only be a queue, or a server or a specific class of customer.

$\lozenge$

EXAMPLE 2.2 The Little Theorem can be used to derive a set of relationships between performance measures in a stable *G/G/c* system with multiple classes of customers (numbered $1, 2, \ldots$). Let λ_i and μ_i ($i = 1, 2, \ldots$) be the arrival rate and the service rate of class-i customers, respectively. If the group of c servers is the system, $\lambda = \sum_i \lambda_i$ is the total arrival rate, and $1/\mu = \sum_i (\lambda_i/\lambda) \cdot (1/\mu_i)$ is the average service time. The quantities in the Little formula have the following interpretation: $E[N]$ is the mean number of packets in service (i.e., the mean number of busy servers) and $E[R]$ is the average time a customer spends in service (i.e., $1/\mu$). In this case, the Little theorem implies that the average number of busy servers is ρ. On the other hand, by considering the same system but only focusing on the class-i customers in service, then $E[N]$ is the average number of class-i customers in service, which is clearly equal to $\rho_i = \lambda_i/\mu_i$.

$$\Diamond$$

The Little Theorem applies to very general systems, but it only provides relationships between average performance figures. By introducing some additional assumptions (in addition to those required by the proof of Theorem 2.1) on system behavior, further relevant relationships can be established among the distributions of the number of packets in the system at arrival, departure epochs, and the distribution of the number of packets at a random point in time (i.e., the steady-state distribution).

2.2.2 Relationships between Arrival and Departure Distributions

This section introduces a result which holds whenever the realizations of the process $N = \{N(t), t \geq 0\}$ are step functions with only unit (upwards and downwards) jumps.

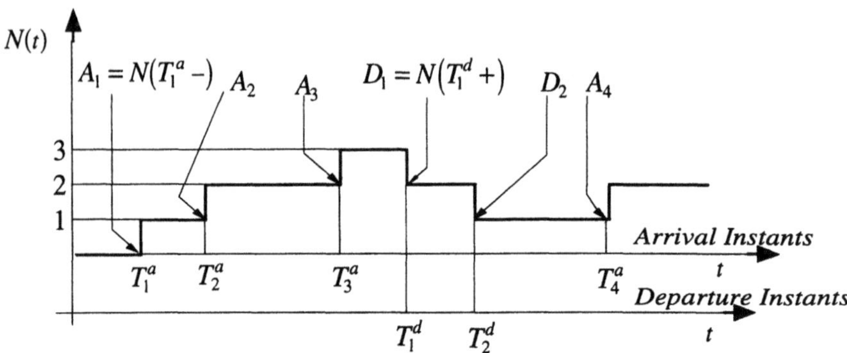

Figure 2.4: A realization of N

Before presenting the main result it is useful to introduce some specific terminology. Let T_i^a (T_i^d) denote the time instant (epoch) at which the i-th arrival (departure) occurs, and $N_i^a = N(T_i^a-)$ $(N_i^d = N(T_i^d+))$ the number of customers in the system just before T_i^a (after T_i^d). As shown in Figure 2.4, which represents a generic realization of the process, A_i is the number of customers found in the system by the i-th arrival, and D_i is the number of customers left in the system by the i-th departure.

The derivation of the following theorem, which was proved by P.J. Burke, proceeds along the lines presented in [54].

THEOREM 2.2   *Let* $\{N(t), t \geq 0\}$ *be a stochastic process with unit realizations that are step functions with only unit jumps. If either*

$$\lim_{i \to \infty} P\{N_i^a \leq k\} \quad \text{or} \quad \lim_{i \to \infty} P\{N_i^d \leq k\}, \quad k = 0, 1, \ldots$$

exists, so does the other, and

$$\lim_{i \to \infty} P\{N_i^a \leq k\} = \lim_{i \to \infty} P\{N_i^d \leq k\} \quad .$$

PROOF

The theorem is proved by showing that, for any k, $P\{N_n^d \leq k\} = P\{N_{n+k+1}^a \leq k\}$, and hence

$$\lim_{n \to \infty} P\{N_n^d \leq k\} = \lim_{n \to \infty} P\{N_{n+k+1}^a \leq k\} = \lim_{n \to \infty} P\{N_n^a \leq k\} \quad .$$

To prove that $P\{N_n^d \leq k\} = P\{N_{n+k+1}^a \leq k\}$ it is sufficient to show that $\{N_n^d \leq k\} \supset \{N_{n+k+1}^a \leq k\}$ and $\{N_{n+k+1}^a \leq k\} \supset \{N_n^d \leq k\}$. To simplify the presentation it is assumed, without any loss of generality, that $N(0) = 0$.

The first step proves that $\{N_n^d \leq k\} \supset \{N_{n+k+1}^a \leq k\}$. Let $N_n^d = j, j \leq k$; i.e., $n + j$ arrivals occur before T_n^d. Hence $T_{n+j}^a < T_n^d < T_{n+j+1}^a$ and in the interval (T_n^d, T_{n+j+1}^a) no arrivals occur, although departures may occur. This implies $\{N(T_n^d+) = j\} \supset \{N(T_{n+j+1}^a-) \leq j\}$. Obviously,

$$\{N(T^a_{n+j+1}-)\le j\} \supset \{N(T^a_{n+j+1+(k-j)}-)\le j+(k-j)\}\quad,$$

and hence $\{N^d_n \le k\} \supset \{N^a_{n+k+1}\le k\}$.

The second step proves that $\{N^a_{n+k+1}\le k\} \supset \{N^d_n \le k\}$. Let $N^a_{n+k+1} = j, j\le k$; i.e., $n+k-j$ departures occur before T^a_{n+k+1}. Hence $T^d_{n+k-j} < T^a_{n+k+1} < T^d_{n+k-j+1}$ and in the interval $(T^d_{n+k-j},T^a_{n+k+1})$ no departures occur, although arrivals may occur. This implies that $\{N(T^a_{n+k+1}-) = j\} \supset \{N(T^d_{n+k-j}+)\le j\}$. Finally, from noting that $T^d_{n+k-j}\ge T^d_n$, it easily follows that

$$\{N(T^d_{n+k-j}+)\le j\} \supset \{N(T^d_n+)\le j+(k-j)\}\quad,$$

and hence $\{N^a_{n+k+1}\le k\} \supset \{N^d_n \le k\}$.

2.2.3 Relationships between Arrival and Steady-state Distributions

This section presents a relevant result which holds when the customer arrival process $\{A(t), t\ge 0\}$ is memoryless, e.g., with exponential or geometric interarrival times.

EXPONENTIAL INTERARRIVAL DISTRIBUTION. Let $\{A(t), t\ge 0\}$ be a Poisson process with rate λ. Arrivals find the system in a given state and interact in some way with it. For example, one arrival can increase by one the number of busy servers in the system. Formally, let $\{\Sigma(t), t\ge 0\}$ be the stochastic process which takes on values in some state space and represents the state of the system. $\{\Sigma(t), t\ge 0\}$ can represent, at time t, the number of customers in the system, the number of busy servers, etc. Because of the interaction between the arrival process and the state of the system, it is expected that $\{\Sigma(t), t\ge 0\}$ will be affected by $\{A(t), t\ge 0\}$. For example, if $\{\Sigma(t), t\ge 0\}$ is the number of packets in the system, this number will increase by one every time a customer arrives, and hence $\{\Sigma(t), t\ge 0\}$ and $\{A(t), t\ge 0\}$ will be dependent processes. However, in general, the system has *no anticipation*, i.e., the future increments of $\{A(t), t\ge 0\}$ are independent of the past of $\{\Sigma(t), t\ge 0\}$. This is the so-called *Lack of Anticipa-*

tion Assumption (LAA), which can be formally expressed by saying that for each $t \geq 0$, $\{A(t+u) - A(t), u \geq 0\}$ and $\{\Sigma(s), s \leq t\}$ are independent. The main result can now be stated.

THEOREM 2.3   *If the LAA holds, the fraction of arrivals which see the process in a given state is equal to the fraction of time the process is in that state.*

This property is called *PASTA*, i.e., "Poisson Arrivals See Time Averages". As implied in the acronym, the PASTA property is primarily concerned with time averages, e.g., the fraction of time a process spends in a given state.

The PASTA property holds both for finite t and for limiting averages. The general proof of PASTA [163] is quite complex and will be omitted. The proof for finite t is presented below, since it is straightforward and highlights the need for the LAA assumption.

Specifically, for finite t, it is proved that for any set Ξ

$$P\{\Sigma(t^-) \in \Xi\} = P\{\Sigma(t^-) \in \Xi | t \text{ is an arrival epoch}\} \quad . \qquad (2.26)$$

In order to overcome probability problems (see [84], page 392), the probability of the right-hand side (*r.h.s.*) of (2.26) is defined as follows

$$P\{\Sigma(t^-) \in \Xi | t \text{ is an arrival epoch}\} =$$

$$\lim_{u \to 0^+} P\{\Sigma(t^-) \in \Xi | A(t+u)-A(t) \geq 1\} \quad ,$$

From this definition the proof easily follows from noting that, for any $u > 0$,

$$P\{\Sigma(t^-) \in \Xi | A(t+u)-A(t) \geq 1\} =$$

$$\frac{P\{A(t+u)-A(t) \geq 1 | \Sigma(t^-) \in \Xi\} \cdot P\{\Sigma(t^-) \in \Xi\}}{P\{A(t+u)-A(t) \geq 1\}} = P\{\Sigma(t^-) \in \Xi\} \quad ,$$

and the last equality holds as

$$P\{A(t+u)-A(t) \geq 1 | \Sigma(t^-) \in \Xi\} = P\{A(t+u)-A(t) \geq 1\}$$

due to the LAA assumption. In the case of t going to infinity, if we define

$$p_\Xi = \lim_{t \to \infty} P\{\Sigma(t) \in \Xi\} \quad,$$

and

$$\pi_\Xi = \lim_{t \to \infty} P\{\Sigma(t^-) \in \Xi \mid t \text{ is an arrival epoch}\} \quad,$$

the PASTA theorem states that $p_\Xi = \pi_\Xi$.

EXAMPLE 2.3   The PASTA theorem simplifies the computation of the loss probability, P_L, in an $M/G/1$ queueing system with finite buffer of size M. In fact the LAA holds for the arrival process of all packets which arrive at the system, including those which will be discarded due to the buffer overflow. Hence the PASTA theorem guarantees that the probability that an arriving packet being discarded, P_L, coincides with the steady-state probability of the buffer being full, p_M. From this observation the relationship between throughput (γ) and arrival rate (λ) for the $M/G/1$ queueing system with finite buffer is

$$\gamma = \lambda(1 - p_M) \quad. \tag{2.27}$$

◊

It can be shown that the PASTA theorem does not hold for systems with an arrival rate which depends on the state of the system [163].

EXAMPLE 2.4   Let us consider a single-server queueing system with an infinite buffer. Let the process which models the system be a birth-and-death process with a rate which depends on the number, j, of packets in the system: $\lambda_j = \lambda/(j+1)$ for $j \geq 0$ and service rate $\mu_j = \mu$ for $j \geq 1$. By indicating with $\tau = 1/\mu$ it can be proved that the probability distribution of the number of packets in the system at any time (i.e., steady-state probability distribution) is given by

$$p_j = \frac{(\lambda\tau)^j}{j!} e^{-\lambda\tau} \text{ for } j \geq 0 \quad,$$

while the probability distribution observed by arrivals is given by

$$\pi_j = \frac{1}{1 - e^{-\lambda\tau}} \cdot p_{j+1} \text{ for } j \geq 0 \quad.$$

◊

The following example, pointed out by Wolff [163], shows a case in which PASTA fails although at first sight the analyzed system appears to be an $M/M/1$ queueing system.

EXAMPLE 2.5   Consider two single-server queues in tandem, modeling for example, two packet-switching nodes connected via a dedicated link. Packet arrivals at the first queue occur according to a Poisson process with rate λ. The packet lengths have an exponential distribution, and thus the service times at both queues are exponentially distributed with rate μ and, obviously, the service time experienced by a packet is the same at both queues. The first queue is $M/M/1$, and therefore, according to Burke's theorem [27], the departure process from this queue (and thus the arrival process at the second queue) is Poisson with rate λ. Thus, the second queue initially seems to behave like an $M/M/1$ queueing system, however, as observed in [163], PASTA does not hold. For an intuitive justification, it is useful to make reference to a characterization of the state of the second queue in which $\{\Sigma(s), s \leq t\}$ is the vector of the service time of the packets in the second queue $\Sigma(s) = [t_1(s), t_2(s), \dots, t_N(s)]$, where $t_j(s)$ is the service time of the j-th packet waiting in the second queue at time s. Obviously, at time s, the probability that the first queue is empty is less than $exp\{-\lambda t_N s\}$. For long packets, this probability comes very close to zero and thus the arrival rate just after a long packet will be μ, instead of λ, and thus the LAA property does not hold.

$\Diamond$

GEOMETRIC INTERARRIVAL DISTRIBUTION. The discrete-time equivalent of the PASTA theorem is reported in [81]. In the discrete-time domain, the geometric distribution is the equivalent of the exponential distribution. In fact the geometric distribution is memoryless, in the sense that at the beginning of each slot, the number of slots until the next arrival does not depend on past history. Let $\{A_n, n = 1, 2, \dots\}$ denote the Bernoulli arrival process (i.e., $A_i \in \{0, 1\}$ denotes the arrivals in the i-th slot time), and $\{\Sigma_n, n = 1, 2, \dots\}$ denote the state of the system. In [81] it is shown that, if the r.v. $\Sigma_1, \Sigma_2, \dots, \Sigma_n$ are independent of the r.v. $A_n, A_{n+1}, \dots$, then the distribution of the state of the system observed by arriving customers and the steady-state distribution are identical.

2.3 SOME RESULTS ON RENEWAL AND REGENERATIVE PROCESSES

In this section are presented some results which hold whenever the processes which characterize a queueing system (e.g., the arrival process, service process, etc.) have a special structure.

RENEWAL PROCESSES. Let S_i be the arrival epoch of the i-th customer in a queueing system, and $X_i = S_i - S_{i-1}$ be the time between successive arrivals. The sequence $S := \{S_n, n = 0, 1, 2, ...\}$ defined as

$$S_0 = 0, \quad S_n = X_1 + X_2 + ... + X_n, \quad n = 1, 2, ...$$

is called a *renewal process* provided that $X_1, X_2, ...$ be i.i.d. non-negative random variables. The arrival epochs S_n are called *renewal epochs*.

Throughout the rest of the book the common distribution function of the r.v. $X_1, X_2, ...$ is $G(x) = P\{X_i \le x\}$ and, according to the Kendall notation, will be indicated by *GI*. β will denote the average of $G(x)$.

EXAMPLE 2.6   Let $S_0, S_1, ...$ be the successive instants at which packets enter a subnetwork. If the packet interarrival times $X_1, X_2, ...$ are i.i.d., then $S := \{S_n, n = 0, 1, 2, ...\}$ is a renewal process. Specifically, if the interarrival times have an exponential distribution, then S is the sequence of the arrival times in a Poisson process.

The following stochastic processes are typically associated with a renewal process

- Number of renewals occurring in $(0, t\,]$; denoted throughout by $N_r(t)$.
- *Renewal function* $(m(t))$: the mean number of renewals occurring in $(0, t\,]$. Thus, $m(t) = E[N_r(t)]$. It can be proved ([29],[84]) that $m(t)$ satisfies the equation

$$m(t) = G(t) + \int_0^t G(t - y) \cdot dm(y) \quad . \tag{2.28}$$

- *Forward Recurrence time* $(X_+(t))$: the remaining time from an arbitrary instant t to the next renewal epoch.

- *Backward Recurrence time* $(X_-(t))$: the elapsed time between an arbitrary instant t and the preceding renewal epoch.

By using the main results of the renewal theory it is possible to show that the limiting distribution of forward and backward recurrence times are

$$X_+(x) = \lim_{t \to \infty} P\{X_+(t) \le x\} = \frac{1}{\beta} \cdot \int_0^x [1 - G(y)]\, dy \ , \tag{2.29}$$

and

$$X_-(x) = \lim_{t \to \infty} P\{X_-(t) \le x\} = \frac{1}{\beta} \cdot \int_0^x [1 - G(y)]\, dy \ . \tag{2.30}$$

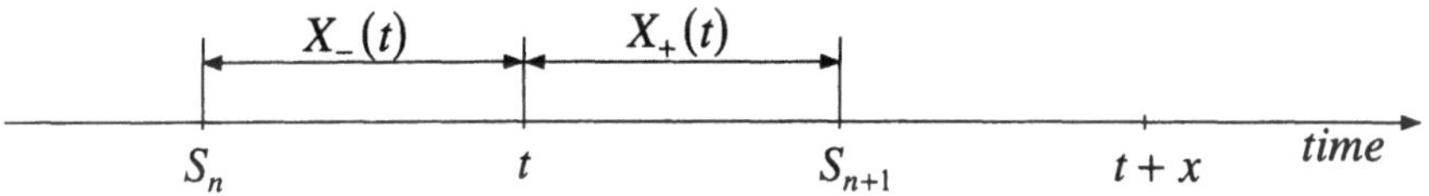

Figure 2.5: Forward and backward recurrence times

An informal proof of these results is given below following the approach presented in [73]. As is clearly shown in Figure 2.5, the following relationship between events holds

$$\{X_+(t) \le x\} = \bigcup_{n=0}^{\infty} \{X_+(t) \le x, S_n < t \le S_{n+1}\} \ ,$$

where $\{S_n < t \le S_{n+1}\}, n = 0, 1, \ldots$ are disjoint events and

$$\sum_{n=0}^{\infty} P\{S_n < t \le S_{n+1}\} = 1 \ .$$

Hence

$$P\{X_+(t) \le x\} = \sum_{n=0}^{\infty} P\{X_+(t) \le x, S_n < t \le S_{n+1}\} \tag{2.31}$$

$$= \sum_{n=0}^{\infty} P\{S_n < t, t < S_{n+1} \le t + x\}$$

$$= \sum_{n=0}^{\infty} \int_0^t P\{t - y < S_{n+1} - S_n \le t + x - y\}\, dP\{S_n \le y\} \ ,$$

By taking into consideration that the random variables $S_{n+1} - S_n = X_{n+1}$ are i.i.d, and by observing that

$$m(y) = E\left[\sum_{n=0}^{\infty} I_{\{S_n \le y\}}\right] = \sum_{n=0}^{\infty} P\{S_n \le y\} \quad,$$

after some algebraic manipulations (2.31) becomes

$$P\{X_+(t) \le x\} = G(t+x) - G(t) + \int_0^t [G(t+x-y) - G(t-y)]\, dm(y). \quad (2.32)$$

Equation (2.32) has the following intuitive stochastic interpretation

1. $G(t+x) - G(t)$ is the probability that the first renewal (S_1) occurs between t and $t+x$; while

2. the integral in the r.h.s. is the probability that a renewal occurs at time $y < t$ and the time elapsing until the next renewal is between $t - y$ and $t - y + x$.

By rewriting the integral in the r.h.s. of (2.32)

$$\int_0^t [G(t+x-y) - G(t-y)]\, dm(y)$$

$$= \int_0^t [1 - G(t-y)]\, dm(y) - \int_0^t [1 - G(t+x-y)]\, dm(y) \quad,$$

and taking the limit as $t \to \infty$, it follows that

$$X_+(x) = \lim_{t \to \infty} P\{X_+(t) \le x\} = \frac{1}{\beta}\int_0^x [1 - G(y)]\, dy \quad. \qquad (2.33)$$

Equation (2.33) is obtained by noting that, as $t \to \infty$, $(G(t+x) - G(t)) \to 0$, and by applying the *Key Renewal Theorem* ([29], [73]) it follows that

$$\int_0^t [1 - G(t-y)]\, dm(y) \to 1 \quad,$$

$$\int_0^t [1 - G(t+x-y)]\, dm(y) \to \frac{1}{\beta}\int_x^{\infty} [1 - G(y)]\, dy \quad.$$

$$\int_0^t [1 - G(t + x - y)]\, dm(y) \rightarrow \frac{1}{\beta}\int_x^\infty [1 - G(y)]\, dy \quad .$$

The limiting distribution of the backward recurrence time can be easily derived by observing that (see Figure 2.6)

$$\{X_+(t) \le x\} \;=\; \{X_-(t + x) \le x\}, \quad \text{for any } t \ge 0 \quad ,$$

from which it results that the limiting distributions of the forward and backward recurrence time coincide.

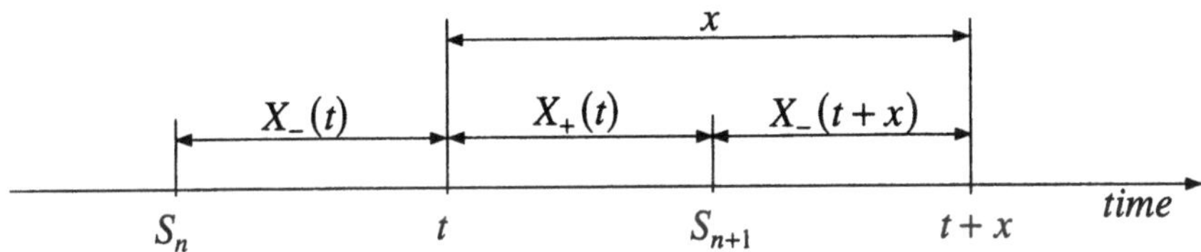

Figure 2.6: Relationship between forward and backward recurrence times

The moments of the limiting distribution of $X_+(x)$ and $X_-(x)$ are obtained with routine computation from their LST, $X_+^*(s)$ and $X_-^*(s)$,

$$X_+^*(s) \;=\; \int_0^\infty e^{-st} dX_+(t) \;=\; X_-^*(s) \;=\; \frac{1 - G^*(s)}{\beta s} \quad , \tag{2.34}$$

where $G^*(s)$ is the LST of $G(x)$. In fact, by denoting the r-th moments about the origin of $X_+(x)$ and $G(x)$ with m_r and μ_r, respectively, then

$$m_r \;=\; \frac{\mu_{r+1}}{(r+1)\,\mu_1} \quad . \tag{2.35}$$

Throughout this text, the formula for the first moment,

$$m_1 \;=\; \frac{\mu_2}{2\mu_1} \quad , \tag{2.36}$$

will be used extensively.

REMARK. Noting that forward and backward recurrence times have the same distribution, it follows that the average length of the interval in which the random point in time falls is μ_2/μ_1. This means that there is a biasing effect on the interval length in which the random point falls. This effect in referred to as *length-biasing phenomenon* (see [56]).

$\Diamond$

REGENERATIVE PROCESSES. A stochastic process $Y := \{Y(t), t \geq 0\}$ is a regenerative process if there exists an epoch, S_1, such that

1. the behavior of the process Y after S_1 is independent of its past behavior; and

2. $\{Y(t + S_1), t \geq 0\}$ is a probabilistic replica of the process behavior starting at time $S_0 = 0$ (i.e., $\{Y(t), t \geq 0\}$).

The epoch S_1 is therefore called a *regenerative point,* and its existence implies the existence of subsequent regeneration points $S_2, S_3, \ldots$ such that $\{Y(t), t \geq 0\}$ regenerates (i.e., satisfies points 1 and 2) at each of these epochs. Thus, $X_n = S_n - S_{n-1}$ is a sequence of i.i.d. random variables and the sequence $S := \{S_n, n = 1, 2, \ldots\}$ is a renewal process embedded in $\{Y(t), t \geq 0\}$.

Hereafter the r.v. X_n are called cycle lengths, and $\{Y(t), S_{n-1} < t \leq S_n\}$, $n \geq 1$ are called *cycles.*

EXAMPLE 2.7   Let $Y := \{Y(t), t \geq 0\}$ be the queue size at time t for an *M/G/*1 queueing system. Suppose that the time S_0 is an instant of departure which left behind j packets. It can be shown that every time a departure occurs leaving behind j packets, the future of Y after such time instants has the same probability law of Y after S_0.

$\Diamond$

The theorem about this class of processes, reported below without any proof, will be used extensively throughout this book. The proof can be found in [84], Theorem 6-7.

THEOREM 2.4   *Let $Y := \{Y(t), t \geq 0\}$ be a regenerative process with a state space S. If the cycle length has a probability density function on some interval and $0 < E[X_n] < \infty$, and*

1. *if $S \subset \{0, 1, 2, \ldots\}$, then $\lim\limits_{t \to \infty} P\{Y(t) = k\} = p_k$ exists and*

$$p_k = \frac{E[T_k]}{E[X_n]} \quad , \tag{2.37}$$

where $E[T_k]$ is the average amount of time the process $\{Y(t), t \geq 0\}$ spends in state k during one cycle.

2. if $S \subset (-\infty, \infty)$, then by denoting with

$$H(x) = \lim_{t \to \infty} P\{Y(t) \leq x\}, x \in S \quad ,$$

then $H(x)$ exists for every real number x, with $H(\infty) = 1$ and

$$H(x) = \frac{E[T_x]}{E[X_n]} \quad , \tag{2.38}$$

where $E[T_x]$ is the average amount of time the event $\{Y(t) \leq x, t \geq 0\}$ is true during a cycle.

REMARK. When Y is a discrete-time regenerative process $Y := \{Y_i, i = 1, 2, \ldots\}$ with $S \subset \{0, 1, 2, \ldots\}$, formula (2.37) becomes

$$p_k = \frac{E\left[\sum_{i=1}^{N} I_{\{Y_i = k\}}\right]}{E[N]} \quad , \tag{2.39}$$

where N is the number of transitions made by process Y during one renewal cycle. Similarly, when Y is a discrete-time regenerative process with $S \subset (-\infty, \infty)$, formula (2.38) becomes

$$H(x) = \frac{E\left[\sum_{i=1}^{N} I_{\{Y_i \leq x\}}\right]}{E[N]} \quad . \tag{2.40}$$

$\Diamond$

The following two examples show relevant applications of Theorem 2.4 in the analysis of the *M/G/1* system.

EXAMPLE 2.8   Let us now come back to the process $Y := \{Y(t), t \geq 0\}$ from Example 2.7. The steady-state probability of an empty system $\lim_{t \to \infty} P\{Y(t) = 0\} = p_0$ can be easily computed by using Theorem 2.4. In fact, by defining S_i as the i-th time instant at which a departure left behind 0 packets, then

$$p_0 = E[T_0]/E[X] \quad ,$$

where $E[X] = [1/\lambda + b/(1-\rho)]$ is the average length of a renewal cycle (delay cycle in [99]) and $E[T_0] = 1/\lambda$ is the average length of the time the system is empty in a renewal cycle (idle period in [99]) in the delay cycle. Hence, $p_0 = 1-\rho$.

$\Diamond$

EXAMPLE 2.9   This example shows a method to derive the *Pollaczek-Kintchine formula* for the average number of packets ($E[N]$) in an *M/G/*1 system [99]. By the PASTA property, an arriving packet observes an average of $E[N]$ packets in the system. It is straightforward to verify that, in average, ρ packets out of $E[N]$ are in service, while $E[N_q]$ are waiting in the queue. Hence, by indicating with $E[W]$ the average waiting time experienced by a packet in the queue, the following relationship holds:

$$E[W] = E[W|empty\ system] (1-\rho) + E[W|not\ empty\ system]\rho \quad .$$

Obviously $E[W|empty\ system] = 0$, hence

$$\begin{aligned} E[W] &= E[W|not\ empty\ system]\rho \\ &= E[B_+]\rho + b \cdot E[N_q|not\ empty\ system]\rho \quad , \end{aligned} \tag{2.41}$$

where $E[B_+]$ is the average residual service time of the packet in service at the arrival epoch. Furthermore,

$$\begin{aligned} E[N_q|not\ empty\ system] &\\ = \frac{\sum_{j=1}^{\infty} (j-1) \cdot p_j}{\rho} &= \frac{E[N_q]}{\rho} = \frac{\lambda E[W]}{\rho} \quad , \end{aligned} \tag{2.42}$$

where the last equality is obtained by applying the Little theorem.

By substituting (2.42) in (2.41), it results

$$E[W] = \frac{\rho}{1-\rho} \cdot E[B_+] \quad . \tag{2.43}$$

By noting that $E[B_+]$ is equal to the average forward recurrence time of the service time process, formula (2.36) gives

$$E[B_+] \;=\; \frac{b^{(2)}}{2b} \;.$$

(2.44)

Hence, by substituting (2.44) in (2.43), the Pollaczek-Kintchine formula is obtained

$$E[W] \;=\; \frac{\lambda b^{(2)}}{2(1-\rho)} \;,$$

from which, by applying the Little theorem,

$$E[N_q] \;=\; \lambda E[W] \;=\; \frac{\lambda^2 b^{(2)}}{2(1-\rho)} \;.$$

$\Diamond$

Formulas for the distribution of the forward and backward recurrence time (equations (2.29) and (2.30)) have been derived for renewal processes. In Theorem 2.5, below, it is shown that these formulas also represent the distribution of the *residual* (i.e., the amount of service to be provided) and *elapsed* (i.e., the service already provided) service times[1] in a queueing system in which service times are identically distributed, but not necessarily independent (e.g., a *GI/G/1* queueing system).

Before proving Theorem 2.5, the following lemma should be proved.

LEMMA 2.1   *Let* $Y := \{Y_i, i = 1, 2, \ldots\}$ *be a discrete-time regenerative process with* $S \subset (0, \infty)$ *, and let* N *be the number of transitions of* Y *during a renewal cycle, then*

$$E[Y] \;=\; \lim_{n \to \infty} E[Y_n] \;=\; \frac{E\left[\sum_{i=1}^{N} Y_i\right]}{E[N]} \;.$$

(2.45)

PROOF

Since Y takes non-negative values

---

1. Please note that forward and backward recurrence times are used in the context of renewal processes, while residual and elapsed service times refer to more general processes.

$$E[Y] = \int_0^\infty P\{Y > x\}\, dx \quad . \tag{2.46}$$

From (2.40), it results

$$P\{Y > x\} = \frac{E\left[\sum_{i=1}^N I_{\{Y_i > x\}}\right]}{E[N]} \quad . \tag{2.47}$$

By substituting (2.47) in (2.46) and exploiting the relationship

$$E\left[I_{\{Y_i > x\}}\right] = P\{Y_i > x\}$$

(2.45) is obtained after routine manipulation.

$$\Diamond$$

THEOREM 2.5   *For a GI/G/1 queueing system with arrival rate λ, in which the arrival times to an empty system are regenerative points for the service time process[1], if one denotes*

- *the service time distribution with $B(x)$;*
- *the average service time b;*
- *the residual (elapsed) service time at time t with $X_{Rs}(t)$ $(X_{El}(t))$;*

then, for every $y \geq 0$, the following relationships hold

$$X_{Rs}(y) = \lim_{t \to \infty} P\{X_{Rs}(t) \leq y \,|\, X_{Rs}(t) > 0\} = \frac{1}{b}\int_0^y (1 - B(u))\, du \quad , \tag{2.48}$$

$$X_{El}(y) = \lim_{t \to \infty} P\{X_{El}(t) \leq y \,|\, X_{El}(t) > 0\} = \frac{1}{b}\int_0^y (1 - B(u))\, du \quad . \tag{2.49}$$

PROOF

According to the hypothesis concerning the service time process, the distribution of the residual (elapsed) service time can be computed by applying (2.38); by denoting the length of a busy period with Θ_b [99] and with N the number of services in a busy period,

---

1. In other words, while the service times within a busy cycle may depend on each other, they are independent of service times belonging to different busy cycles.

$$\lim_{t \to \infty} P\{X_{Rs}(t) > y \,|\, X_{Rs}(t) > 0\} \;=\; \tag{2.50}$$

$$\frac{1}{E[\Theta_b]} E\left[\int_0^{\Theta_b} I_{\{X_{Rs}(t) > y\}} \, dt\right] \;=\; \frac{1}{E\left[\sum_{i=1}^{N} B_i\right]} E\left[\sum_{i=1}^{N} (B_i - y)_+\right] \;,$$

By using Lemma 2.1

$$E\left[\sum_{i=1}^{N} (B_i - y)_+\right] \;=\; E[N] \cdot E[(B_1 - y)_+] \;, \tag{2.51}$$

and

$$E\left[\sum_{i=1}^{N} B_i\right] \;=\; E[N] \cdot E[B_1] \;. \tag{2.52}$$

By substituting (2.51) and (2.52) in (2.50),

$$\lim_{t \to \infty} P\{X_{Rs}(t) > y \,|\, X_{Rs}(t) > 0\} \;=\; \frac{E[(B_1 - y)_+]}{E[B_1]} \;=\; \tag{2.53}$$

$$\frac{1}{E[B_1]} \int_y^{\infty} (u - y) \, dB(u) \;.$$

Formula (2.48) immediately follows from (2.53) with simple algebraic manipulations. The proof of (2.49) can be easily obtained by following the same line of reasoning.

◊

Formulas for the distribution of the forward and backward recurrence times (equations (2.29) and (2.30)) have been derived for renewal processes. Equations (2.48) and (2.49) extend these results to the service time process in a $GI/G/1$ queueing system in which service times belonging to the same busy period are correlated. In [64] it has been proved that equation (2.48) holds true in a much more general setting. Specifically, the authors derived the distribution of the residual service time at time t for a $G/G/1$ queueing system where G is an arbitrary stationary stochastic process without any assumption about the independence between service times. The stochastic process $\{X(t)\}$ is *stationary* if for any $n \in \mathbb{N}^+$, when $t_1 < t_2 < \ldots < t_n$ and $s + t_1, \ldots, s + t_n$ are elements of $\mathbb{R}$ then

$$\{X(t_1), \ldots, X(t_n)\} \Leftrightarrow \{X(s + t_1), \ldots, X(s + t_n)\} \quad .$$

This means that the joint distribution is unaffected by a shift in the time parameter.

THEOREM 2.6   *For a G/G/1 queueing system with arrival rate λ, if one denotes*

- *the service time distribution with $B(x)$;*
- *the average service time with b;*
- *the residual service time at the server at time t with $\beta(t)$;*
- *and λb with ρ,*

for every $y \geq 0$, the following equations hold

$$P\{\beta(0) \leq y\} = (1 - \rho) + \rho \cdot B_R(y) \quad , \tag{2.54}$$

$$P\{\beta(0) \leq y | \beta(0) > 0\} = B_R(y) \quad , \tag{2.55}$$

where $\beta(0)$ denotes the residual service time at an arbitrary point in time and

$$B_R(y) = \frac{1}{b}\int_0^y (1 - B(u))\, du \quad . \tag{2.56}$$

The proof of this theorem is reported in [64] (Theorem 4.5.1, page 126), but as it is beyond the scope of this book is omitted here.

2.4 WORK CONSERVATION LAWS

A MAN is characterized by the following elements:

- a shared transmission medium;
- a finite set of K stations connected to the transmission medium; and,
- a distributed algorithm, called *Medium Access Control* (*MAC*) protocol, which coordinates the stations access to the shared medium.

The basic model which can be employed for describing the behavior of the above system is a single-server queue with K customer classes. The server models the shared medium, while each class of customers represents packets arriving at each station. The arrival processes are mutually independent.

More precisely, a single server serves K queues $(Q_1, Q_2, ..., Q_K)$ with infinite buffer size (see Figure 2.7). Packets arriving at Q_k will be referred to as class-k packets. λ_k is the arrival rate of the class-k packets and each class-k packet requires a service time denoted by the r.v. B_k ($E[B_k]$ and $E[B_k^2]$ are the first and second moments, respectively). Queues are served by the server according to the scheduling discipline implemented by the MAC protocol.[1]

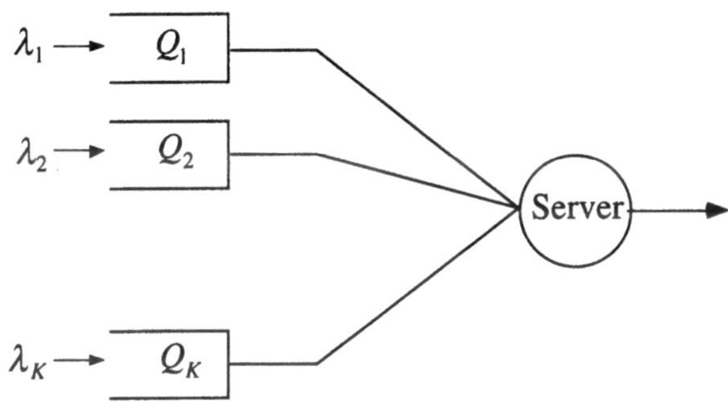

Figure 2.7: MAN model

One of the meaningful performance measures of the system is the *performance vector*

$$E[\mathbf{W}] \;=\; (E[W_1], E[W_2], ..., [W_K]) \quad,$$

where $E[W_k]$ $(k = 1,2,...,K)$ is the steady-state average waiting time for packets of class k (i.e., the time a packet spends in queue k).

Given the network parameters (e.g., arrival and service processes, medium transmission rate, etc.), each *Scheduling Discipline (SD)* will result in a performance vector. In practice, several scheduling disciplines can be employed providing very different performance vectors. However, for a wide class of scheduling disciplines, the performance vectors satisfy an invariant law: the weighted sum of average waiting times is constant. This law is called the *Conservation Law.*

---

1. When a MAC protocol can be modeled with a polling system the scheduling discipline coincides with the global service discipline.

This law is based on the properties of the process $L_{SD} := \{L_{SD}(t), t \geq 0\}$ where $L_{SD}(t)$ is the amount of work in the system at time t (i.e., $L_{SD}(t)$ is the sum of the remaining service times of all customers in the system at time t). Specifically, under a given scheduling discipline, for a given realization of the process L_{SD}, $L_{SD}(t)$ is the sum of the remaining service times of the packets in the system at time t.

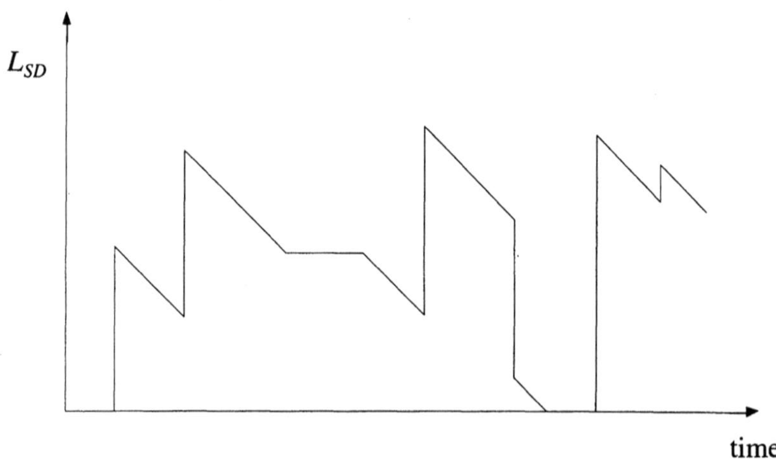

Figure 2.8: One possible realization of the process L_{SD}

Figure 2.8 shows a typical realization of the process L_{SD} by assuming that the service rate is 1. For each packet arrival, L_{SD} jumps upward an amount equal to the packet service time, while it decreases linearly with slope -1 as long as the server is serving a packet. Furthermore, the time interval during which the process has a constant, non-zero value indicates that the scheduling discipline is forcing the server to be idle (i.e., the server is on vacation). Finally, the jump downward represents the departure of a packet from the system before its service has been completed (reneging packet). Scheduling strategies which do not allow server vacations and do not generate reneging packets are termed *work-conserving*.

By assuming that process L_{SD} has a steady-state distribution, with steady state average l_{SD}

$$\lim_{t \to \infty} E\left[L_{SD}(t)\right] = l_{SD} , \qquad (2.57)$$

it is possible to prove that for all work-conserving scheduling strategies, l_{SD} depends only on the arrival and service time processes. Hence, from now on

l_{SD} will be denoted by $E[L]$.

As the average amount of work in the system is strongly correlated to the average number of packets waiting for service, the above property also indicates that there is a relationship among the average waiting times experienced by packets. In the literature there are several theorems which prove this relationship under different assumptions on the system ([98], [136], and [84]). For single-server queueing systems, in [98] it was proved that, for a large class of scheduling disciplines in multiclass $M/G/1$ queueing systems, the weighted sum of the average waiting times of all classes satisfies a simple linear equality constraint.

THEOREM 2.7   *Let $M/G/1$ be a multiclass queueing system in equilibrium (i.e., $\rho < 1$) implementing a non-preemptive, work-conserving scheduling discipline. The following relation holds*

$$\sum_{i=1}^{K} \rho_i \cdot E[W_i] = \frac{\rho}{1-\rho} \cdot \sum_{i=1}^{K} \frac{\lambda_i \cdot E[B_i^2]}{2} \quad , \tag{2.58}$$

where $\rho_k = \lambda_k \cdot E[B_k]$ is the offered load of class-k packets and $\rho = \sum_{k=1}^{K} \rho_k$ is the total offered load.

PROOF

For a single-server queueing system, the average work in the system $E[L]$ is

$$E[L] = E[L_q] + E[L_s] \quad , \tag{2.59}$$

where

- $E[L_s]$ is the average remaining service time for the packet in service at a random point in time; and
- $E[L_q]$ is the average amount of work required by the packets waiting in the queues at a random point in time.

In the first step of the proof $E[L_s]$ is derived. To this end $E[L_s]$ is written as

$$E[L_s] = \sum_{i=1}^{K} E[L_s|\xi = i] \cdot P\{\xi = i\} \quad , \tag{2.60}$$

where $\{\xi = i\}$ denotes the event "server is busy on class-i packets".

Since

- $P\{\xi = i\} = \rho_i = \lambda_i \cdot E[B_i]$ and
- $E[L_s|\xi = i] = E[B_{i+}] = E[B_i^2] / (2 \cdot E[B_i])$, then

$$E[L_s] = \sum_{i=1}^{K} \frac{\lambda_i \cdot E[B_i^2]}{2} \quad . \tag{2.61}$$

Because the system is in equilibrium ($\rho < 1$ by hypothesis), by applying the Little theorem to each class of packets, the following relationship easily follows

$$E[L_q] = \sum_{i=1}^{K} E[L_q(i)] = \sum_{i=1}^{K} \lambda_i E[B_i] E[W_i] = \sum_{i=1}^{K} \rho_i E[W_i] \quad , \tag{2.62}$$

where $E[L_q(i)]$ is the average amount of work required by the class-i packets waiting in the queues at a random point in time.

In the $M/G/1$ queueing system the average waiting time experienced by an incoming packet before being served is equal to $E[L]$, which can be obtained from the Pollaczek-Kintchine formula [99]:

$$E[L] = \frac{\lambda \cdot E[B^2]}{2 \cdot (1 - \rho)} = \frac{\sum_{i=1}^{K} \lambda_i \cdot E[B_i^2]}{2 \cdot (1 - \rho)} \quad . \tag{2.63}$$

By susbstituting (2.61), (2.62) and (2.63) in (2.59), the theorem is proved.

$$\Diamond$$

Theorem 2.7 was generalized by Schrage [136] for a $G/G/1$ queue in which arrival and service processes are stationary arbitrary processes without any independence assumptions. For this queueing system the following theorem holds:

THEOREM 2.8   *Let $G/G/1$ be a multiclass queueing system in equilibrium (i.e., $\rho < 1$) implementing a work-conserving scheduling discipline. Under the assumptions that the scheduling discipline*

- *is not preemptive, and*
- *does not depend on service times,*

the following relation holds

$$\sum_{i=1}^{K} \rho_i \cdot E[W_i] = E[L] - \sum_{i=1}^{K} \frac{(\lambda_i \cdot E[B_i^2])}{2} \quad , \qquad (2.64)$$

where $\rho_k = \lambda_k \cdot E[B_k]$ is the offered load of class-k packets, and $\rho = \sum_{k=1}^{K} \rho_k$ is the total offered load.

The proof of this theorem can be found in [136], but as it is beyond the scope of this book is not included here.

REMARK. It worth noting that in a *G/G/1* queueing system the computation of $E[L]$ is an open problem.

2.5 STOCHASTIC DECOMPOSITION LAWS

2.5.1 Stochastic Decomposition Laws for Single-queue Models

A station of a LAN or MAN is frequently modeled as an *M/G/1* queueing system with server vacation. This system does not possess the work-conserving property, in the sense that the service process may be interrupted although work is present.

A *stochastic decomposition property* has been proved for such systems, i.e., the performance measures for the system with vacation are related to the performance measures of the same system without vacation (*corresponding system*). This decomposition property has been shown to be valid under general conditions. This section focuses on results relating to *M/G/1* systems with exhaustive, gated exhaustive, and *l*-limited visit service disciplines (see Section 2.1.2).

M/G/1 WITH VACATION AND EXHAUSTIVE SERVICE $M/G/1_{VE}$. Several works in the literature have shown the existence of a decomposition property for $M/G/1_{VE}$ queueing system (e.g., [53] and [65]). This property is proven in Theorem 2.9 below. Before proving this theorem, the following terminology must be introduced. *Vacation packets* are the packets which arrive while the server is on vacation, and I_0 is the set of vacation packets. Packets which arrive while members of I_0 are being served belong to set I_1, which is called the *first generation offspring* of I_0. Recursively I_k, $k \geq 2$ (the k-th generation offspring of I_0) denotes the set of packets arriving while members of I_{k-1} are being served. Similarly, for each vacation packet $\hat{A}$, it is possible to define its offspring generations $I_k^{(\hat{A})}$, $k \geq 1$. Finally, the set $I^{(\hat{A})} = \bigcup_{k=0}^{\infty} I_k^{(\hat{A})}$ (where $I_0^{(\hat{A})} = \{\hat{A}\}$) is defined as the ancestral line of $\hat{A}$. Clearly, for any packet $\hat{P}$, there is one and only one vacation packet $\hat{A}$, the ancestor of $\hat{P}$, such that $\hat{P} \in I^{(\hat{A})}$.

THEOREM 2.9   *Let* $M/G/1_{VE}$ *be an* $M/G/1$ *queueing system with vacation and exhaustive service. If*

1. *each service time is independent of the sequence of vacations that preceded it, and*

2. *for any time instant, the future arrivals do not depend on the length of previous vacations,* [1]

then the stationary number of packets present in the $M/G/1_{VE}$ *system at a random point in time is the sum of two independent random variables, one of which is the stationary number of packets in the corresponding* $M/G/1$ *system at a random point in time*

$$\pi_{M/G/1_{VE}}(z) = \chi(z) \cdot \pi_{M/G/1}(z) \quad , \tag{2.65}$$

where

- $\pi_{M/G/1_{VE}}(z)$ *is the PGF for the number of packets left behind by a random departing packet in the* $M/G/1_{VE}$ *system;*
- $\pi_{M/G/1}(z)$ *is the PGF for the number of packets left behind by a random departing packet in the corresponding* $M/G/1$ *system;*
- $\chi(z)$ *is the PGF of the number of Poisson arrivals during a time interval that is distributed as the forward/backward recurrence time of a vacation.*

PROOF

The following proof follows the line of reasoning adopted in [67]. Specifically, it is based on observing the ancestral line of a tagged packet $\hat{P}$. To better clarify the proof, Figure 2.9 shows an example of offspring generations.

---

1. Lack of Anticipation Assumption, see Section 2.2.3.

Without any loss of generality, a LIFO service discipline is assumed. $\hat{A}_1, \hat{A}_2, ..., \hat{A}_n$ are the vacation packets and thus $I_0 = \{\hat{A}_1, \hat{A}_2, ..., \hat{A}_n\}$. Because of the LIFO discipline, the first packet served after the vacation period is $\hat{A}_n$ and

$$I_1^{(\hat{A}_n)} = \{\hat{P}_1^{(\hat{A}_n, 1)}, \hat{P}_2^{(\hat{A}_n, 1)}, ..., \hat{P}_m^{(\hat{A}_n, 1)}\} \ .$$

The next packet to be served is $\hat{P}_m^{(\hat{A}_n, 1)}$. The arrivals during the $\hat{P}_m^{(\hat{A}_n, 1)}$ service,

$$(\hat{P}_1^{(\hat{A}_n, 2)}, \hat{P}_2^{(\hat{A}_n, 2)}, ..., \hat{P}_{l_1}^{(\hat{A}_n, 2)}) \ ,$$

belong to the set $I_2^{(\hat{A}_n)}$ of the second generation offspring of $\hat{A}_n$. The other elements of this set are obviously the packets which arrive during the service of

$$\hat{P}_{m-1}^{(\hat{A}_n, 1)}, \hat{P}_{m-2}^{(\hat{A}_n, 1)}, ..., \hat{P}_1^{(\hat{A}_n, 1)} \ ,$$

which are $(\hat{P}_{l_1+1}^{(\hat{A}_n, 2)}, \hat{P}_{l_1+2}^{(\hat{A}_n, 2)}, ..., \hat{P}_{l_1+l_2}^{(\hat{A}_n, 2)})$, $(\hat{P}_{l_1+l_2+1}^{(\hat{A}_n, 2)}, \hat{P}_{l_1+l_2+2}^{(\hat{A}_n, 2)}, ..., \hat{P}_{l_1+l_2+l_3}^{(\hat{A}_n, 2)})$,

..., $(\hat{P}_{l_1+...+l_{m-1}+1}^{(\hat{A}_n, 2)}, \hat{P}_{l_1+...+l_{m-1}+2}^{(\hat{A}_n, 2)}, ..., \hat{P}_{l_1+...+l_m}^{(\hat{A}_n, 2)})$, respectively.

Let $\hat{P}$ be a random packet, and $N(\hat{P})$ be the number of packets in the system at the instant that $\hat{P}$ departs. As pointed out previously, $\hat{P}$ uniquely identifies an ancestral line $I^{(\hat{A})}$ such that $\hat{P} \in I^{(\hat{A})}$. Hence, $N(\hat{P})$ is the sum of the number of vacation packets which arrived before $\hat{A}$ (denoted by the r.v. χ) and the number of packets belonging to $I^{(\hat{A})}$ already in the system, but not yet served. This last quantity exactly corresponds to the number of packets left behind by a tagged packet in the corresponding $M/G/1$ system.[1] Furthermore, due to hypotheses 1 and 2, these two quantities are independent, and hence $\pi_{M/G/1_{VE}}(z) = \chi(z) \cdot \pi_{M/G/1}(z)$. According to the PASTA and Burke theorems (see Theorems 2.3 and 2.2), the above decomposition result holds for the number of packets in the $M/G/1_{VE}$ system at a random point in time.

$\Diamond$

REMARK. From elementary queueing theory it results that $\chi(z) = V^*(\lambda - \lambda z)$. If vacations are not independent (the sequence $\{A_n\}$ are

---

1. The service period of an ancestral line is equal to the busy period in an $M/G/1$ system.

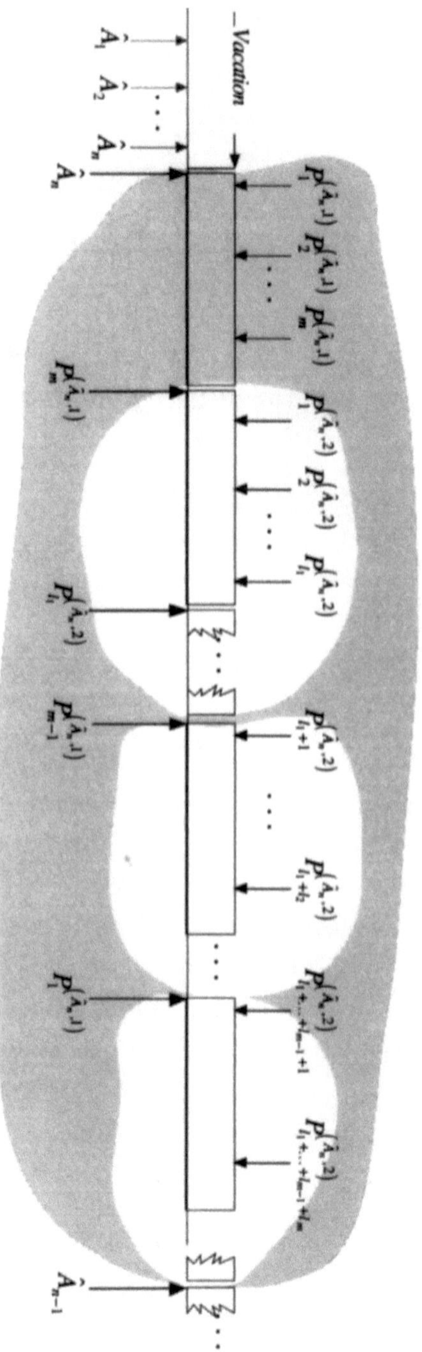

Figure 2.9: An example of offspring generations

only assumed to be stationary), then in [67] it is shown that

$$\chi(z) = \frac{1 - A(z)}{(1 - z) E[A]} \; .$$

(2.66)

◊

$M/G/1$ WITH GENERALIZED VACATION $(M/G/1_{GV})$. The $M/G/1_{VE}$ queueing system described in the previous section assumes that the server goes on vacation only when the queue is empty. This model does not cover important visit service disciplines (e.g., gated and l-limited, see Section 2.1.2) used in computer-network modeling.

This section shows that a decomposition law holds for an $M/G/1$ system with *generalized vacation, M/G/1$_{GV}$* system. Specifically, a system is an $M/G/1_{GV}$ if the number of packets in the system when a vacation begins is independent of the number of packets arriving during that vacation (i.e., A_n and Z_n are independent):

$$\pi_{M/G/1_{GV}}(z) = \zeta(z) \cdot \chi(z) \cdot \pi_{M/G/1}(z) \quad,$$

(2.67)

where

- $\zeta(z)$ is the PGF of Z, and
- $\chi(z)$ is the PGF for the number of arrivals between the start of a vacation and a random point in time during that vacation. This number coincides with the number of packets in the $M/G/1_{VE}$ system at a random point in time when the server is on vacation.

The proof of (2.67) can be found in [67] and is similar to the proof of Theorem 2.9. Here, an intuitive explanation of this result is sketched.

As in the proof of Theorem 2.9, a LIFO service discipline is assumed. Let $\hat{P}$ be a tagged packet and $I^{(\hat{A})}$ be its ancestral line. Due to the LIFO assumption, it is easy to observe that the number of packets in the system at the instant $\hat{P}$ departs, $N(\hat{P})$, can be split up into three components

(i)   the number of packets belonging to $I^{(\hat{A})}$ already in the system but not yet served;

(ii)   the number of vacation packets (χ) which arrived in the same vacation period as $\hat{A}$ but before $\hat{A}$;

(iii)   the number of packets (Z) in the system at the beginning of the vacation period in which $\hat{A}$ arrived.

Due to the hypotheses stated above these three components are independent. Finally, to prove (2.67), it must be shown that the quantity (i) is the number of packets left behind by a tagged packet in the corresponding $M/G/1$ system.

The main difference between $M/G/1_{GV}$ system and an $M/G/1_{VE}$ system is that in the former from the time the service of $\hat{A}$ starts until the departure of $\hat{P}$, a number of vacations may occur before completing the service of the packets belonging to $I^{(\hat{A})}$.

Let *younger ancestral lines* be those ancestral lines whose ancestors arrived during a vacation subsequent to the one in which $\hat{A}$ arrived. In an $M/G/1_{GV}$ system, younger ancestral lines interrupt the service of the packets belonging to $I^{(\hat{A})}$. However, due to the LIFO assumption, when $\hat{P}$ is served the system contains no packets belonging to younger ancestral lines.

The next step of the proof is to show that the number of packets in the system belonging to $I^{(\hat{A})}$ is not influenced by younger ancestral lines. This is proved by observing that arrivals and departures of $I^{(\hat{A})}$ packets occur *iff* the server is serving an $I^{(\hat{A})}$ packet. Hence, the number of packets in the system belonging to this ancestral line remains constant whenever the server goes on vacation or it serves younger ancestral lines. Thus, as shown in Figure 2.10, at any departure epoch of an $I^{(\hat{A})}$ packet the distribution of the number of $I^{(\hat{A})}$ packets in the system is exactly the same as the distribution of the number of packets in the corresponding $M/G/1$ system.

REMARK. The decomposition property for the $M/G/1_{VE}$ system can be easily derived from the decomposition property of the systems with generalized vacation by noting that in the $M/G/1_{VE}$ system $Z = 0$ with probability 1 (*w. p.* 1), which means $\zeta(z) = 1$.

2.5.2 From the Decomposition Law to the Average Response Time

For an $M/G/1_{GV}$ system with FIFO service discipline, it is easy to observe that the number of packets in the system when the tagged packet $\hat{P}$ departs are those which arrive during its response time in the system, hence

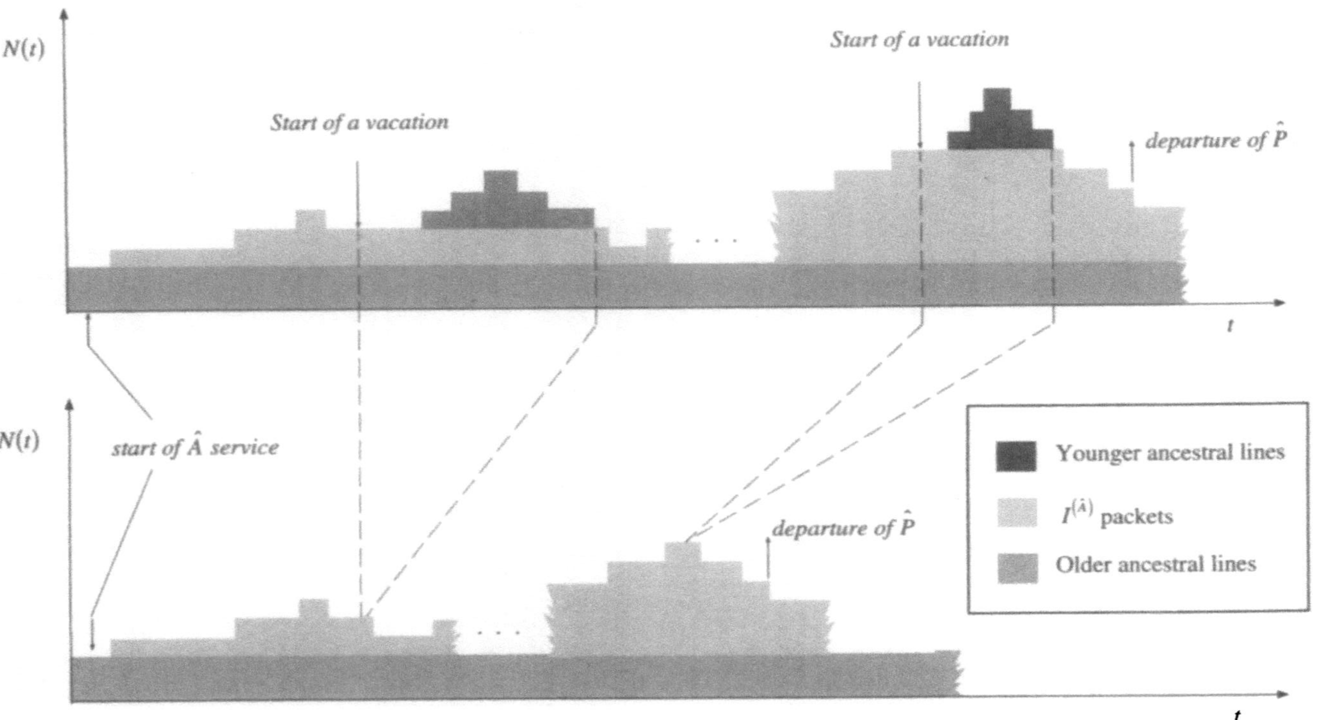

Figure 2.10: $I^{(\hat{A})}$ evolution in $M/G/1_{GV}$ and in $M/G/1_{VE}$ systems

$$R^*_{M/G/1_{GV}}(\lambda - \lambda z) = \zeta(z) \cdot \chi(z) \cdot \pi_{M/G/1}(z) \quad , \tag{2.68}$$

where $R^*_{M/G/1_{GV}}(\cdot)$ is the LST of the response time in the system.

By denoting $\lambda - \lambda z$ with s, equation (2.68) can be written as

$$R^*_{M/G/1_{GV}}(s) = \zeta(1 - s/\lambda) \cdot \chi(1 - s/\lambda) \cdot \pi_{M/G/1}(1 - s/\lambda) \quad . \tag{2.69}$$

By noting that

$$R^*_{M/G/1}(s) = \pi_{M/G/1}(1 - s/\lambda) \quad , \tag{2.70}$$

the following decomposition result holds also for the waiting time distribution

$$R^*_{M/G/1_{GV}}(s) = \zeta(1 - s/\lambda) \cdot \chi(1 - s/\lambda) \cdot R^*_{M/G/1}(s) \quad . \tag{2.71}$$

Using equation (2.71), which is valid for the FIFO service discipline, the derivation of the average response time in the system is straightforward

$$E[R_{M/G/1_{GV}}] = -\frac{d}{ds}(R^*_{M/G/1_{GV}}(s))\Big|_{s=0} \quad .$$

A more general formula for the average response time (i.e., not limited to the FIFO service discipline) can be derived using the Little theorem. In fact, the Burke and PASTA theorems guarantee that the distribution of the number of packets in the system at departure epochs $(N_{M/G/1_{GV}})$ coincides with the distribution at a random instant, and as a consequence

$$E[R_{M/G/1_{GV}}] = \frac{E[N_{M/G/1_{GV}}]}{\lambda} \quad , \tag{2.72}$$

where

$$E[N_{M/G/1_{GV}}] = \frac{d}{dz}\pi_{M/G/1_{GV}}(z)\Big|_{z=1} \quad . \tag{2.73}$$

From (2.67), after some routine computation, the following expression is obtained

$$E[R_{M/G/1_{GV}}] = \frac{E[Z]}{\lambda} + \frac{A^{(2)}(1)}{2\lambda E[A]} + \frac{\lambda b^{(2)}}{2(1-\rho)} + b \quad . \tag{2.74}$$

Considering that $A(z) = V^*(\lambda - \lambda z)$, the above formula can be written as

$$E\left[R_{M/G/1_{GV}}\right] = \frac{E\left[Z\right]}{\lambda} + \frac{V^{(2)}(0)}{2E\left[V\right]} + \frac{\lambda b^{(2)}}{2\left(1-\rho\right)} + b \quad . \tag{2.75}$$

It is easy to see that $E\left[Z\right]$ depends upon the specific vacation discipline being used. Some relevant examples are given below.

$M/G/1_{VE}$ SYSTEM. As noted previously, $Z = 0$ (w.p.1) in an $M/G/1_{VE}$ system. Thus, formula (2.75) reduces to

$$E\left[R_{M/G/1_{VE}}\right] = \frac{V^{(2)}(0)}{2E\left[V\right]} + E\left[R_{M/G/1}\right] \quad ,$$

which has a simple interpretation. The average response time is the sum of two components: the average forward recurrence time of the vacation and the average response time in the corresponding $M/G/1$ system.

$M/G/1_{gated}$ SYSTEM. In the $M/G/1_{GV}$ system with gated visit service discipline, $M/G/1_{gated}$, the r.v. Z represents the number of packets which arrived during the previous service cycle.

Let $\Theta_c^{(n)}$ denote the length of the n-th service cycle (i.e., as shown in the time interval between the end of the $(n-1)$ th vacation and the beginning of the n-th vacation), hence by denoting with Z_n the number of packets in the system at the beginning of the n-th vacation (see Section 2.1.2)

$$E\left[Z_n\right] = \lambda E\left[\Theta_c^{(n)}\right] \quad , \tag{2.76}$$

and since, during a service cycle, all the customers which arrived during the previous service cycle and vacation (i.e., previous delay cycle) are served

$$E\left[\Theta_c^{(n)}\right] = b\lambda\left(E\left[V_{n-1}\right] + E\left[\Theta_c^{(n-1)}\right]\right) \quad . \tag{2.77}$$

By substituting (2.77) in (2.76) and taking the limit for $n \to \infty$, after some algebraic manipulation it results that

$$E\left[Z\right] = \lambda E\left[V\right] \cdot \frac{\rho}{1-\rho} \quad . \tag{2.78}$$

Finally, by substituting (2.78) in (2.75), the average response time is obtained

$$E[R_{M/G/1_{VE}}] = E[V] \cdot \frac{\rho}{1-\rho} + \frac{V^{(2)}(0)}{2E[V]} + E[R_{M/G/1}]$$

$$= E[V] \cdot \frac{\rho}{1-\rho} + E[R_{M/G/1_{VE}}] \quad . \tag{2.79}$$

$M/G/1_{1-limited}$ SYSTEM. In the $M/G/1$ system with 1-limited visit service discipline, $M/G/1_{1-limited}$, $E[Z]$ can be derived by exploiting the following relationship

$$E[Z] = E[Z|1\,service\ in\ the\ last\ service\ cycle] \\ \cdot P\{1\,service\ in\ the\ last\ service\ cycle\} \quad ,$$

By observing that Z, under the condition that a packet was served in the previous service cycle, coincides with the number of packets at a departure epoch, the following relationship holds

$$E[Z|1\ service\ in\ the\ last\ service\ cycle] = \frac{d}{dz}\pi_{M/G/1_{1-limited}}(z)\Big|_{z=1} .$$

$P\{1\ service\ in\ the\ last\ service\ cycle\}$ is computed by observing that, in the stationary region, the throughput of this system is

$$\lambda = \frac{E[number\ of\ transmissions\ per\ service\ cycle]}{E[V] + E[service\ cycle]} \quad ,$$

where

- $E[number\ of\ transmissions\ per\ service\ cycle]$
  $= P\{1\,service\ in\ the\ last\ service\ cycle\}$,

- $E[service\ cycle] = E[B]\,P\{1\,service\ in\ the\ last\ service\ cycle\}$.

After some algebraic manipulations, it results

$$P\{1\,service\ in\ the\ last\ service\ cycle\} = \frac{\lambda E[V]}{1-\rho} \quad .$$

Finally, substituting $E[Z]$ in (2.75), and after some routine manipulation, the average response time is obtained

$$E[R_{M/G/1_{1-limited}}] = \frac{1-\rho}{1-\rho-\lambda E[V]} \cdot \left(\frac{V^{(2)}(0)}{2E[V]} + E[R_{M/G/1}]\right) \quad .$$

2.5.3 Stochastic Decomposition Laws for Polling Models

The extension to polling models of the decomposition result for the number of packets in the system was presented by Fuhrmann [66] under the assumption of symmetric queues. To generalize the decomposition result to asymmetric polling models, Boxma and Groenendijk [24] consider the amount of work in the system rather than the number of packets. In fact, for each service discipline which does not create or destroy work while the server is visiting a queue, the work in the system is equal to the amount of work in the *equivalent M/G/1 system with vacation*, in which

- vacations have a one-to-one correspondence with switchover times in the polling system: *i*) the beginning of a switchover time coincides with the beginning of the corresponding vacation, and *ii*) the switchover time and its corresponding vacation have the same length;
- the arrival process is Poisson with rate Λ, $\Lambda = \sum_{i=1}^{K} \lambda_i$;
- the service time distribution is

$$B(x) = \sum_{i=1}^{K} \frac{\lambda_i}{\Lambda} B_i(x) \quad .$$

The decomposition result for polling models is developed in [24] and is based on Lemma 2.2, below.

LEMMA 2.2   *The amount of work in the polling system at an arbitrary epoch in a service period ($L_{Polling}^{Sp}$) is the sum of two independent random variables*

$$L_{Polling}^{Sp} \Rightarrow L_{M/G/1}^{Sp} + L_Y \quad ,$$

where

- $\Rightarrow$ *means convergence in distribution;*
- $L_{M/G/1}^{Sp}$ *is the work in the system in the corresponding M/G/1 system (see Section 2.5.1 on page 55) at an arbitrary epoch in a service period;*
- L_Y *is the amount of work in the polling system at an arbitrary time during a switchover time.*

PROOF

The proof of this lemma involves the concepts of "ancestral line" and "offspring of a packet" introduced in Section 2.5.1, and the definition of "equivalent *M/G/1* system with vacation".

According to the above definitions, $L_{Polling}^{Sp}$ is equal to the amount of work at an arbitrary time during the service period of the corresponding *M/G/1* system with vacation. Hence the decomposition result can be proved by focusing on the latter. The proof moves along the same line of reasoning used to prove the decomposition property for $M/G/1_{GV}$ systems (see Section 2.5.1).

Without any loss of generality, a LIFO service discipline for the equivalent $M/G/1$ system with vacation is assumed. Furthermore, as the vacations in the equivalent $M/G/1$ system are forced by the beginning of switchover times, it may happen that a service is interrupted by a vacation. An interrupted service is resumed when all the packets belonging to younger ancestral lines (see Section 2.5.1) have been served.

Let $\hat{P}$ be the packet in service at a random point in time and $I^{(\hat{A})}$ be its ancestral line. By the PASTA property, the amount of work observed in the system by $\hat{A}$ on its arrival is distributed as L_Y. Furthermore, when the $\hat{P}$ packet is in service, all packets belonging to younger ancestral lines have already been served; at that time, the amount of work in the system is L_Y plus the amount of work carried by the packets belonging to $I^{(\hat{A})}$ already in the system but not yet served. The latter quantity has the same distribution as the work in the system in the corresponding $M/G/1$ system $(L^{Sp}_{M/G/1})$. Thus

$$L^{Sp}_{Polling} \Rightarrow L^{Sp}_{M/G/1} + L_Y \quad . \tag{2.80}$$

$\Diamond$

The stochastic decomposition of the work in the system at an arbitrary point in time is now obtained by using Lemma 2.2 and observing that in a polling system,

$$P\{server\ is\ serving\} \ = \ \rho, \ P\{server\ is\ switchng\} \ = \ 1-\rho \quad .$$

Specifically, since

$$E\left[e^{-sL_{Polling}}\right] = E\left[e^{-sL_{Polling}}\,\middle|\,server\ is\ serving\right] \cdot \rho \tag{2.81}$$
$$+ E\left[e^{-sL_{Polling}}\,\middle|\,server\ is\ switchng\right] \cdot (1-\rho) \quad ,$$

and

$$E\left[e^{-sL_{Polling}}\,\middle|\,server\ is\ switchng\right] = E\left[e^{-sL_Y}\right] \quad , \tag{2.82}$$

$$E\left[e^{-sL_{Polling}}\Big|\,server\ is\ serving\right] \;=\; E\left[e^{-sL_{Polling}^{Sp}}\right] \;=\; \tag{2.83}$$

$$E\left[e^{-sL_{M/G/1}^{Sp}}\right]\cdot E\left[e^{-sL_{Y}}\right]\;,$$

after some algebraic manipulation it is easy to verify that (see Theorem 1 in [24])

$$E\left[e^{-sL_{Polling}}\right] \;=\; E\left[e^{-sL_{M/G/1}}\right]\cdot E\left[e^{-sL_{Y}}\right]\;. \tag{2.84}$$

2.6 PSEUDO-CONSERVATION LAWS

*M/G/*1 systems with vacation and polling systems are not work-conserving, as they violate the constraint that no work is created within the system. The vacation periods or the switchover times can be interpreted as additional work created within the system. Hence, in these systems, the amount of work in the system is not equal to the amount of work in the corresponding *M/G/*1 system.

In this section, by exploiting the decomposition property (2.84), it is shown that the *M/G/*1 conservation law can be extended to polling models with non-zero switchover times.

From (2.84) it follows that

$$E\,[L_{Polling}] \;=\; E\,[L_{M/G/1}] + E\,[L_{Y}]\;. \tag{2.85}$$

$E\,[L_{Y}]$ in (2.85) implies that for this class of system, the amount of work in the system is no longer independent of the service discipline, as will be shown below.

From (2.85) and (2.63),

$$E\,[L_{Polling}] \;=\; E\,[L_{M/G/1}] + E\,[L_{Y}] \;=\; \frac{\sum_{i=1}^{K}\lambda_{i}\cdot b_{i}^{(2)}}{2\,(1-\rho)} + E\,[L_{Y}]\;. \tag{2.86}$$

On the other hand,

$$E\,[L_{Polling}] \;=\; \sum_{i=1}^{K} b_{i}E\,[Nq_{i}] + \sum_{i=1}^{K}\rho_{i}\frac{b_{i}^{(2)}}{2b_{i}}\;, \tag{2.87}$$

where the second term on the right-hand side keeps count of the residual service time of the packet in service: ρ_i is the probability that a class-i packet is in service, while $b_i^{(2)}/2b_i$ is its mean residual service time.

By applying the Little theorem to the first term on the right-hand of (2.87),

$$E[L_{Polling}] = \sum_{i=1}^{K} \rho_i E[W_i] + \frac{1}{2}\sum_{i=1}^{K} \lambda_i b_i^{(2)} \quad . \tag{2.88}$$

From (2.86) and (2.88),

$$\sum_{i=1}^{K} \rho_i E[W_i] = \rho\frac{\sum_{i=1}^{K} \lambda_i \cdot b_i^{(2)}}{2(1-\rho)} + E[L_Y] \quad . \tag{2.89}$$

Formula (2.89) is referred to as the *pseudo-conservation law*. The term *pseudo* was introduced because for this class of systems, the weighted sum of the average waiting times depends on the service discipline. To show this dependency, $E[L_Y]$ is expressed as

$$E[L_Y] = \sum_{i=1}^{K} \frac{s_i}{s} \cdot E[L_{Y,i}] \quad ,$$

where $E[L_{Y,i}]$ is the amount of work in the polling system during the i-th switchover time (i.e., the time the server is switching between station $\{i\}$ and station $\{i+1\}$). Furthermore,

$$E[L_{Y,i}] = E[M_i^{(1)}] + E[M_i^{(2)}] + \rho \cdot \frac{s_i^{(2)}}{2s_i} \quad ,$$

where

- $E[M_i^{(1)}]$ is the mean amount of work in Q_i when the server departs from this queue;
- $E[M_i^{(2)}]$ is the mean amount of work in the rest of the system when the server departs from Q_i; and
- $\rho \cdot (s_i^{(2)}/2s_i)$ is the mean amount of work already arrived during the i-th switchover time (i.e., the work arrived during the backward recurrence time of the i-th switchover time).

After some algebraic manipulations, it can be shown that (see [24])

$$\sum_{i=1}^{K} \rho_i E[W_i] = \rho \frac{\sum_{i=1}^{K} \lambda_i \cdot b_i^{(2)}}{2(1-\rho)} + \rho \frac{s^{(2)}}{2s} \qquad (2.90)$$

$$+ \frac{s}{2(1-\rho)} \cdot \left[\rho^2 - \sum_{i=1}^{K} \rho_i^2 \right] + \sum_{i=1}^{K} E[M_i^{(1)}] \quad .$$

In Formula (2.90) there are K unknowns,

$$E[M_i^{(1)}], \quad i = 1, 2, \ldots, K \quad ,$$

representing the amount of work at Q_i when the server departs from this queue. $E[M_i^{(1)}]$ depends on the service discipline at Q_i, and thus (2.90) can be completely derived by specifying the service discipline at each queue.

In the following cases, (2.90) will be computed by assuming the same service discipline for all the queues in the system.

2.6.1 Exhaustive Service Discipline

The server departs from Q_i only when the queue is empty, therefore

$$E[M_i^{(1)}] = 0 \quad , \qquad (2.91)$$

and the pseudo-conservation law for a polling system with an exhaustive service discipline is

$$\sum_{i=1}^{K} \rho_i E[W_i] = \qquad (2.92)$$

$$\rho \frac{\sum_{i=1}^{K} \lambda_i \cdot b_i^{(2)}}{2(1-\rho)} + \rho \frac{s^{(2)}}{2s} + \frac{s}{2(1-\rho)} \cdot \left[\rho^2 - \sum_{i=1}^{K} \rho_i^2 \right] \quad .$$

2.6.2 Gated Service Discipline

When the server departs from Q_i, it leaves in the queue only those packets which arrived during the last service period, therefore

$$E[M_i^{(1)}] = \lambda_i b_i \cdot E[Sp_i] \quad . \qquad (2.93)$$

By substituting (2.15) and (2.23) in (2.93), we get

$$E[M_i^{(1)}] = \rho_i^2 \frac{s}{1-\rho} \quad . \tag{2.94}$$

The pseudo-conservation law for polling systems with a gated service discipline is obtained by substituting (2.94) in (2.90):

$$\sum_{i=1}^{K} \rho_i E[W_i] = \tag{2.95}$$

$$\rho \frac{\sum_{i=1}^{K} \lambda_i \cdot b_i^{(2)}}{2(1-\rho)} + \rho \frac{s^{(2)}}{2s} + \frac{s}{2(1-\rho)} \cdot \left[\rho^2 + \sum_{i=1}^{K} \rho_i^2 \right] \quad .$$

2.6.3 1-limited Service Discipline

In a polling system with a 1-limited (non-exhaustive) service discipline, the average number of packets removed from Q_i during a cycle is (see (2.25))

$$\lambda_i \cdot \frac{s}{1-\rho} < 1 \quad , \tag{2.96}$$

and it is equal to the probability that the queue is not empty when the server visits the queue. Hence, as

$$E[M_i^{(1)} | no\ Q_i\ packet\ removed] = 0 \quad ,$$

it follows that

$$E[M_i^{(1)}] = E[M_i^{(1)} | one\ Q_i\ packet\ removed] \frac{\lambda_i s}{1-\rho} \quad , \tag{2.97}$$

and $E[M_i^{(1)} | one\ Q_i\ packet\ removed] = E[M_i^{(1)} | Q_i\ depature\ epoch]$.

The work in Q_i at a packet departure epoch corresponds to the work which arrives at Q_i while this packet is waiting in the queue; thus

$$E[M_i^{(1)} | Q_i\ depature\ epoch] = \lambda_i b_i (E[W_i] + b_i) \quad . \tag{2.98}$$

By substituting (2.98) in (2.97) we get

$$E[M_i^{(1)}] = \rho_i \cdot E[W_i] \cdot \frac{\lambda_i \cdot s}{1-\rho} + \rho_i^2 \cdot \frac{s}{1-\rho} \quad . \tag{2.99}$$

The pseudo-conservation law for polling systems with a non-exhaustive service discipline is finally obtained by substituting (2.99) in (2.90):

$$\sum_{i=1}^{K} \rho_i E[W_i] \cdot \left(1 - \frac{\lambda_i \cdot s}{1-\rho}\right) = \tag{2.100}$$

$$\rho \frac{\sum_{i=1}^{K} \lambda_i \cdot b_i^{(2)}}{2(1-\rho)} + \rho \frac{s^{(2)}}{2s} + \frac{s}{2(1-\rho)} \cdot \left[\rho^2 + \sum_{i=1}^{K} \rho_i^2\right] \quad .$$

2.6.4 *l*-limited Service Disciplines

In a polling system with *l*-limited service disciplines the amount of work in Q_i when the server departs from this queue depends on

- N_i^{sa}: the number of packets in Q_i when the server arrives at this queue;
- N_i^{arr}: the number of packets which arrive during the service period; and
- N_i^{tr}: the number of packets transmitted during the service period.

Specifically, the following relationship holds:

$$E[M_i^{(1)}] = E[N_i^{sa} - N_i^{tr} + N_i^{arr}] \cdot b_i = \tag{2.101}$$
$$(E[N_i^{sa}] - E[N_i^{tr}] + E[N_i^{arr}]) \cdot b_i \quad .$$

By using flow balancing arguments (see (2.21)), one can deduce that

$$E[N_i^{tr}] = \lambda_i \cdot \frac{s}{1-\rho} \quad . \tag{2.102}$$

Furthermore,

$$E[N_i^{arr}] = \lambda_i E[Sp_i] = \lambda_i b_i E[N_i^{tr}] \quad . \tag{2.103}$$

Hence, by substituting (2.102) and (2.103) in (2.101)

$$E[M_i^{(1)}] = (E[N_i^{sa}] - E[N_i^{tr}] + E[N_i^{tr}] \rho_i) b_i \quad , \tag{2.104}$$

where the only unknown is $E[N_i^{sa}]$. In order to compute this quantity one must first specify whether the *l*-limited service discipline is exhaustive or gated.

EXHAUSTIVE L-LIMITED SERVICE DISCIPLINE. To compute $E[N_i^{sa}]$, the queue length of Q_i ($i = 1, 2, ..., K$) is analyzed with the embedding Markov-chain technique. Specifically, the embedded chain describes the queue length both at the time instant at which the server arrives at Q_i (*intervisit termination instant*) and after each transmission. The state of Q_i at the embedding points is described by the couple $\{\xi, N\}_{i,n}$. $\xi = 0$ indicates the intervisit termination instant just before the n-th service period of Q_i, while $\xi = j$ $(j > 0)$ is the embedding point just after the j-th transmission during the n-th service period. Finally, the r.v. N denotes the queue length of Q_i at the embedding point.

Under the assumption that the stability criteria are satisfied (see (2.25)) the following steady-state joint probabilities can be defined

$$p_{i,k}^{(m)} = \lim_{n \to \infty} P\{\xi = m, N = k\}_{i,n}, \quad m = 0, ..., l_i, \quad k = 0, 1, ... \quad , \quad (2.105)$$

where $\sum_{m=0}^{l_i} \sum_{k=0}^{\infty} p_{i,k}^{(m)} = 1$.

By denoting with $a_{i,n}$ the probability that n packets arrive at Q_i during a service time

$$a_{i,n} = \int_0^\infty \frac{(\lambda_i t)^n \cdot e^{-\lambda_i t}}{n!} dB_i(t) \quad ,$$

the following relationship exists among the steady-state probabilities at the m-th and $(m+1)$th embedding points

$$p_{i,k}^{(m+1)} = \sum_{j=1}^{k+1} p_{i,j}^{(m)} \cdot a_{i,k-j+1} \quad .$$

By defining

$$P_{i,m}(z) = \sum_{j=0}^{\infty} p_{i,j}^{(m)} \cdot z^j, \quad m = 0, 1, ..., l_i \quad ,$$

and taking into consideration that

$$B_i^*(\lambda_i - \lambda_i z) = \sum_{j=0}^{\infty} a_{i,j} \cdot z^j \quad ,$$

after some algebraic manipulation, the following relationship is obtained:

$$P_{i,m}(z) = P_{i,0}(z) \left[\frac{B_i^*(\lambda_i - \lambda_i z)}{z} \right]^m - \qquad (2.106)$$

$$\sum_{j=0}^{m-1} p_{i,0}^{(j)} \cdot \left[\frac{B_i^*(\lambda_i - \lambda_i z)}{z} \right]^{m-j} , \quad m = 1, ..., l_i \quad .$$

From (2.105), it follows that the probability of k packets being present in the system at any departure instant is

$$P\{\xi > 0,\ N=k\}_i = \sum_{m=1}^{l_i} p_{i,k}^{(m)} \quad,$$

and hence

$$p_{i,k}^{d} = P\{N = k \mid \xi > 0\}_i = \frac{\sum_{m=1}^{l_i} p_{i,k}^{(m)}}{1 - P_{i,0}(1)} \quad.$$

The PGF of the number of packets in Q_i just after a departure instant is therefore

$$P_i^{d}(z) = \sum_{k=0}^{\infty} p_{i,k}^{d} \cdot z^{k} = \frac{\sum_{k=0}^{\infty} \sum_{m=1}^{l_i} p_{i,k}^{(m)} \cdot z^{k}}{1 - P_{i,0}(1)} = \frac{\sum_{m=1}^{l_i} P_{i,m}(z)}{1 - P_{i,0}(1)} \quad,\qquad (2.107)$$

and hence

$$E_i[N \mid \xi > 0] = \frac{d}{dz} P_i^{d}(z) \bigg|_{z=1} = \frac{1}{1 - P_{i,0}(1)} \sum_{m=1}^{l_i} \frac{d}{dz} P_{i,m}(z) \bigg|_{z=1}, \qquad (2.108)$$

where

$$\frac{d}{dz} P_{i,m}(z) \bigg|_{z=1} = \frac{d}{dz} P_{i,0}(z) \bigg|_{z=1} + (1-\rho_i) \left[\sum_{j=0}^{m-1} (m-j)\, p_{i,0}^{(j)} - P_{i,0}(1) \cdot m \right] \quad.$$

It can readily be verified that

$$P\{N_i^{tr} = h\} = p_{i,0}^{(h)} / P_{i,0}(1), \qquad h = 0, 1, \ldots, l_i - 1 \quad, \qquad (2.109)$$

and hence

$$P\{N_i^{tr} = l_i\} = 1 - \left(\sum_{h=0}^{l_i - 1} p_{i,0}^{(h)} / P_{i,0}(1) \right) \quad. \qquad (2.110)$$

From (2.106) and the normalization condition $\sum_{m=0}^{l_i} P_{i,m}(1) = 1$, after some algebraic manipulation, it follows that

$$\sum_{j=0}^{l_i - 1} (l_i - j)\, p_{i,0}^{(j)} = (l_i + 1)\, P_{i,0}(1) - 1 \quad, \qquad (2.111)$$

which, together with (2.109) and (2.110), gives

$$P_{i,0}(1) = 1/(1 + E[N_i^{tr}]) \quad . \tag{2.112}$$

Equation (2.112) has a simple stochastic interpretation: $(1 + E[N_i^{tr}])$ is the average number of transitions between two consecutive visits at state $\{\xi = 0, N \geq 0\}_i$.

Finally, since

$$\frac{1}{P_{i,0}(1)} \cdot \frac{d}{dz} P_{i,0}(z) \Big|_{z=1} = E[N_i^{sa}] \quad ,$$

by exploiting (2.108)-(2.112), it can be shown that

$$E_i[N|\xi > 0] = \frac{1}{E[N_i^{tr}]} \cdot \left(l_i \cdot E[N_i^{sa}] + \frac{(1 - \rho_i)}{2} \cdot E[N_i^{tr}(N_i^{tr} - 1)] - (1 - \rho_i) \cdot l_i \cdot E[N_i^{tr}] \right) \quad . \tag{2.113}$$

Since the local service discipline at Q_i is FIFO, the average number of packets in Q_i just after a packet departure corresponds to the average number of packets which arrive during a time interval equal to the packet sojourn time (i.e., waiting time plus service time) in Q_i:

$$\lambda_i (E[W_i] + b_i) = E_i[N|\xi > 0] \quad . \tag{2.114}$$

Finally, by exploiting (2.113) and (2.114), an expression for $E[N_i^{sa}]$ is obtained:

$$E[N_i^{sa}] = (\lambda_i E[W_i] + \rho_i) \frac{E[N_i^{tr}]}{l_i} - \frac{(1 - \rho_i)}{2l_i} E[N_i^{tr}(N_i^{tr} - 1)] + (1 - \rho_i) E[N_i^{tr}] \quad . \tag{2.115}$$

By substituting (2.115) in (2.104), the following relationship is derived:

$$E[M_i^{(1)}] = \rho_i E[W_i] \cdot \frac{E[N_i^{tr}]}{l_i} - \frac{\rho_i \cdot (1 - \rho_i)}{2\lambda_i l_i} \cdot E[N_i^{tr}(N_i^{tr} - 1)] + \frac{\rho_i^2 \cdot E[N_i^{tr}]}{\lambda_i l_i} \quad , \tag{2.116}$$

from which the pseudo-conservation law for polling systems with an exhaustive l-limited service discipline is immediately obtained:

$$\sum_{i=1}^{K} \rho_i E[W_i] \cdot \left(1 - \frac{E[N_i'']}{l_i}\right) = \rho \frac{\sum_{i=1}^{K} \lambda_i \cdot b_i^{(2)}}{2(1-\rho)} + \rho \frac{s^{(2)}}{2s} + \qquad (2.117)$$

$$\frac{s}{2(1-\rho)} \cdot \left[\rho^2 - \sum_{i=1}^{K} \rho_i^2\right] + \frac{s}{(1-\rho)} \cdot \sum_{i=1}^{K} \frac{\rho_i^2}{l_i} -$$

$$\sum_{i=1}^{K} \frac{\rho_i \cdot (1-\rho_i)}{2\lambda_i l_i} E[N_i''(N_i''-1)] \quad .$$

In order to apply this pseudo-conservation law, the second factorial moments $(E[N_i''(N_i''-1)])$ in (2.117) must be estimated.

REMARK. When $l_i = 1$ $(i = 1, 2, ..., K)$, the second factorial moments are all equal to zero. In this case, Formula (2.117) reduces to Formula (2.100), i.e., the pseudo-conservation law for the non-exhaustive service discipline.

REMARK. In Fuhrmann & Wang [68], the second factorial moments are set equal to zero. This value is, in fact, the lower bound for the second factorial moments. In this case the last term on the right-hand side of (2.117) is zero, and the relationship becomes an inequality between a weighted sum of the average waiting times and the system parameters.

REMARK. Chang & Shandu [31] have derived an expression for the amount of work at Q_i when the server departs from this queue (denoted in the following as $E[M_i^{(1)}]_{cs}$):

$$E[M_i^{(1)}]_{cs} = \rho_i E[W_i] \cdot \frac{E[N_i'']}{l_i} + \frac{\rho_i E[N_i'']}{2\lambda_i l_i} \cdot \qquad (2.118)$$

$$\left[(1-\rho_i) \cdot \sum_{j=1}^{l_i-1} j(l_i-j) p_{i,0,j} - [l_i - 1 - (l_i+1)\rho_i]\right] ,$$

where $p_{i,0,j}$, $i = 1, 2, ..., K$, $j = 1, 2, ..., l_i$ are the (unknown) steady-state joint probabilities that Q_i becomes empty after transmitting j packets in a service period. Formulas (2.116) and (2.118) are the same, although they are expressed in different forms. The differences are due to a slight difference in the definition of the embedded Markov chain used to analyze the number of packets queued at Q_i. In fact, the embedded Markov chain used in Chang & Shandu [31] describes the state of the system only at the departure instants. By describing (as before) the state of the system at the embedding points by the couple of random variables $\{\xi, N\}_{i,n}$, and under the assumption that the stability criteria are satisfied, the following steady-state joint probabilities are defined:

$$p_{i,k,m} = \lim_{n \to \infty} P\{\xi = m, N = k\}_{i,n}, \quad m = 1, ..., l_i, \quad k = 0, 1, ... \quad (2.119)$$

where $\sum_{m=1}^{l_i} \sum_{k=0}^{\infty} p_{i,k,m} = 1.$ \hfill (2.120)

By defining $P_{i,m}(z) = \sum_{k=0}^{\infty} p_{i,k,m} \cdot z^k$, it can be observed that

$$P\{N_i'' = k \mid N_i'' > 0\} = p_{i,0,k}/P_{i,1}(1), \quad k = 0, 1, ..., l_i - 1, \quad (2.121)$$

and

$$P\{N_i'' = l_i \mid N_i'' > 0\} = 1 - \left(\sum_{k=0}^{l_i-1} p_{i,0,k}/P_{i,1}(1) \right). \quad (2.122)$$

Furthermore,

$$P\{N_i'' > 0\} = E[N_i'']/E[N_i'' \mid N_i'' > 0], \quad (2.123)$$

where

$$E[N_i'' \mid N_i'' > 0] = \quad (2.124)$$

$$\left[\sum_{j=1}^{l_i-1} j \cdot \frac{p_{i,0,j}}{P_{i,1}(1)} + l_i \cdot \left(1 - \sum_{j=1}^{l_i-1} \frac{p_{i,0,j}}{P_{i,1}(1)} \right) \right] = \frac{1}{P_{i,1}(1)}.$$

The last equality in (2.124) was derived by exploiting equation (2.120).

From (2.121), (2.123) and (2.124), it follows that

$$p_{i,0,k} = P\{N_i^{tr} = k\}/E[N_i^{tr}], \qquad k = 0, 1, ..., l_i - 1 \quad . \tag{2.125}$$

By exploiting relationship (2.125), after some algebraic manipulations it can be shown that equations (2.116) and (2.118) are equal.

GATED L-LIMITED SERVICE DISCIPLINE. Equation (2.104) holds also when the service discipline is gated. Therefore, to derive the pseudo-conservation law of polling systems, expressions for $E[N_i^{sa}]$, the only unknown quantities, must be derived.

An expression for $E[N_i^{sa}]$ can be derived by following the procedure used (in the previous subsection) for deriving $E[N_i^{sa}]$ in a polling system with an exhaustive l-limited service discipline. Specifically, the queue length of Q_i $(i = 1, 2, ..., K)$ is still analyzed with a Markov chain embedded at the intervisit termination instants and after each transmission. The state of the system at the embedding points is described by the couple $\{\xi, N\}_{i,n}$, and for a system which satisfies the stability criteria the following steady-state probabilities can be defined:

$$p_{i,k}^{(m)} = \lim_{n \to \infty} P\{\xi = m, N = k\}_{i,n}, \quad m = 0, ..., l_i, \ k = 0, 1, ... \quad .$$

The main difference from the previous analysis is due to the gated transmission discipline, which makes eligible for transmission only those packets which are in the queue when the server arrives at Q_i. In fact, in this case, a relationship between the steady-state probabilities at the intervisit termination instant and at the m-th embedding point can be immediately derived

$$p_{i,k}^{(m)} = \sum_{j=m}^{m+k} p_{i,j}^{(0)} \cdot a_{i,k-j+m}^{(m)} \quad , \tag{2.126}$$

where $a_{i,k}^{(m)}$ $(k = 0, 1, ...)$ is the probability that k packets arrive at Q_i in a time interval which is distributed as the sum of m service times.

By applying the z-transform technique to relationship (2.126), after routine algebraic manipulation, it can be shown that

$$P_{i,m}(z) = \frac{P_{i,0}(z) - \sum_{j=0}^{m-1} p_{i,j}^{(0)} \cdot z^j}{z^m} [B_i^*(\lambda - \lambda z)]^m, \quad m = 1, 2, ..., l_i \quad . \tag{2.127}$$

Equation (2.107) also holds in this case, and hence

$$E[N_i|\xi > 0] = \frac{1}{1 - P_{i,0}(1)} \sum_{m=1}^{l_i} \frac{d}{dz} P_{i,m}(z) \Big|_{z=1}, \qquad (2.128)$$

where

$$\frac{d}{dz} P_{i,m}(z) \Big|_{z=1} = \frac{d}{dz} P_{i,0}(z) \Big|_{z=1} - \sum_{j=0}^{m-1} j p_{i,j}^{(0)} - m(1 - \rho_i) \left[P_{i,0}(1) - \sum_{j=0}^{m-1} p_{i,j}^{(0)} \right].$$

By noting that

$$P\{N_i^{tr} = k\} = p_{i,k}^{(0)} / P_{i,0}(1), \qquad k = 0, 1, ..., l_i - 1, \qquad (2.129)$$

$$P\{N_i^{tr} = l_i\} = 1 - \sum_{k=0}^{l_i-1} p_{i,k}^{(0)} / P_0(1), \qquad (2.130)$$

equation (2.112) still holds, and then

$$E[N_i|\xi > 0] = \frac{1}{E[N_i^{tr}]} \left(l_i E[N_i^{sa}] - l_i E[N_i^{tr}] + \right. \qquad (2.131)$$
$$\left. E[N_i^{tr}(N_i^{tr} - 1)] \frac{(1 + \rho_i)}{2} + \rho_i E[N_i^{tr}] \right).$$

Finally, by exploiting (2.129) and (2.114), it follows that

$$E[N_i^{sa}] = \lambda_i E[W_i] \frac{E[N_i^{tr}]}{l_i} - \qquad (2.132)$$
$$\frac{(1 + \rho_i)}{2l_i} E[N_i^{tr}(N_i^{tr} - 1)] + E[N_i^{tr}].$$

By substituting (2.132) in (2.104) the following relationship is obtained:

$$E[M_i^{(1)}] = \rho_i E[W_i] \frac{E[N_i^{tr}]}{l_i} - \qquad (2.133)$$
$$\frac{\rho_i(1 + \rho_i)}{2\lambda_i l_i} E[N_i^{tr}(N_i^{tr} - 1)] + \frac{\rho_i^2 E[N_i^{tr}]}{\lambda_i},$$

from which the pseudo-conservation law for polling systems with a gated l-limited service discipline is immediately obtained

$$\sum_{i=1}^{K} \rho_i E\left[W_i\right]\left(1 - \frac{E\left[N_i''\right]}{l_i}\right) = \rho \frac{\sum_{i=1}^{K} \lambda_i \cdot b_i^{(2)}}{2(1-\rho)} + \rho \frac{s^{(2)}}{2s} + \tag{2.134}$$

$$\frac{s}{2(1-\rho)}\left[\rho^2 + \sum_{i=1}^{K}\rho_i^2\right] - \sum_{i=1}^{K} \frac{\rho_i(1+\rho_i)}{2\lambda_i l_i} \cdot E\left[N_i''\left(N_i''-1\right)\right] \quad .$$

REMARK. In previous subsections, pseudo-conservation laws have been derived under the assumption that the same service discipline is applied to all the queues in the polling model. However, pseudo-conservation laws also exist for polling models with mixed service disciplines, i.e., some groups of queues are served according to an exhaustive service discipline, another set of nodes with a gated service discipline, and the remaining nodes with an l-limited service discipline. The pseudo-conservation law for polling systems with mixed service strategies is obtained by substituting in (2.90), for each Q_i, the expression of $E\left[M_i^{(1)}\right]$ which corresponds to the service discipline at Q_i. Specifically, $E\left[M_i^{(1)}\right]$ is given by

- equation (2.91) for all queues with an exhaustive service discipline;
- equation (2.94) for all queues with a gated service discipline;
- equation (2.99) for all queues with a non-exhaustive service discipline;
- equation (2.116) for all queues with an exhaustive l-limited service discipline; and,
- equation (2.133) for all queues with a gated l-limited service discipline.

Hence, the pseudo-conservation law for mixed service strategies reduces to

$$\sum_{i \in e} \rho_i E\left[W_i\right] + \sum_{i \in g} \rho_i E\left[W_i\right] + \sum_{i \in 1l} \rho_i E\left[W_i\right]\left(1 - \frac{\lambda_i s}{1-\rho}\right) = \tag{2.135}$$

$$\rho \frac{\sum_{i=1}^{K} \lambda_i b_i^{(2)}}{2(1-\rho)} + \rho \frac{s^{(2)}}{2s} + \frac{s}{2(1-\rho)}\left[\rho^2 - \sum_{i=1}^{K}\rho_i^2\right] + \frac{s}{1-\rho} \sum_{i \in (g \cup 1l)} \rho_i^2 \quad ,$$

where e, g, $1l$ denote the sets of the station indices with exhaustive, gated and 1-limited service disciplines, respectively.

BIBLIOGRAPHIC NOTES. The pseudo-conservation laws for polling systems with exhaustive and gated service disciplines were first derived by Watson [159] and Ferguson & Aminetzah [63] by a tough algebraic manipulation of the PGFs of the steady-state probabilities of the number of packets queued in $\{Q_1, Q_2, ..., Q_K\}$. With the same technique, Watson [159] also derived the pseudo-conservation law for polling systems with a non-exhaustive service discipline. The approach presented in this section is based on the work decomposition property proved in Groenendijk [76] and Boxma & Groenendijk [24], through which a general expression for the pseudo-conservation law is obtained (see Equation (2.90)) which includes the previous derivation as a special case. Expressions for $E[M_i^{(1)}]$ in exhaustive (2.92), gated (2.95) and non-exhaustive (2.100) polling systems are derived by following the approach in [24]. For the l-limited service disciplines ((2.116) and (2.133)), expressions for $E[M_i^{(1)}]$ are derived by following the method presented in Everitt [62]. In the case of a polling model with an l-limited service discipline, the term $E[M_i^{(1)}]$ was also derived by Chang & Shandu [31] (see (2.118)).

3 Methods for the Analysis of Node-in-isolation Models

As pointed out in Chapter 2, a LAN or a MAN station is frequently analyzed in isolation and hence each station is modeled as an *M/G/1* queueing system with or without server vacation (*node-in-isolation models*). This chapter presents methods extensively used in the literature to analyze the performance figures of these systems. Specifically, it focuses on the study of the stochastic process $\{N(t), t \geq 0\}$ which represents the number of packets queued in the system (e.g., a station of a MAN, a packet-switching node, etc.) at a random point in time.

M/G/1 queueing system with and without server vacation are presented in Section 3.1 and in Section 3.2, respectively. For these systems the finite and the infinite buffer size are considered. It worth noting that, according to the classical Kendall notation [99], in this chapter the symbol K is used to denote the size of the buffer, rather than the number of stations in a MAN. This should not cause any problem since node-in-isolation models focus on a tagged station only.

For systems with vacation the book focuses only on the exhaustive and *l*-limited (global) service disciplines as they have practical relevance in MAN modeling.

The classical solution methods used for the analysis of these systems include PGFs for the infinite buffer cases, and numerical methods together with supplementary variables for the finite buffer cases.

In the analysis of the *M/G/1* queueing system (with or without server vacation) a special attention is devoted to the structure of their embedded Markov chains. All these chains belong to the class of the so called *M/G/1-type* Markov chains. Section 3.3 introduces the basic concepts of the matrix analytic techniques developed by M.F. Neuts [28] to provide the exact solution for *M/G/1-type* Markov chains.

The methodologies presented in this chapter are applied in Section 3.4 to derive QoS figures of a tagged station in a MAN.

3.1 *M/G/1* SYSTEMS

An output link of a packet switching node can be modeled with a single server and a queue (see Figure 3.1). Packets arrive according to a Poisson process, with rate λ, and are stored in the queue that models the output-link buffer. The server models the packet transmission over the output link. In a real packet-switching node, the buffer is finite and thus the resulting model is an *M/G/1/K* queueing system. However, to simplify the analysis the queue it is often assumed to be infinite, and the model is reduced to an *M/G/1* queueing system. Both models have been extensively studied in the literature ([54], [99], [149]) thus in the following only relevant results for the purposes of this book will be presented.

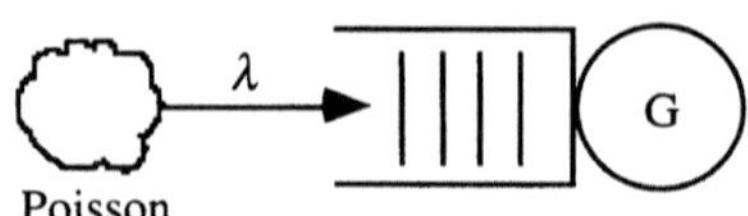

Figure 3.1: *M/G/1* queueing system

3.1.1 Infinite Buffer Systems

The stochastic process $\{N(t), t \geq 0\}$, associated to an *M/G/1* system, is a non-Markovian continuous-time process. A Markovian process is obtained by observing the system at particular time instants, named *embedding points*. This process is called *embedded Markov chain*. In the *M/G/1* systems the embedding points are the service termination instants.

By denoting with N_j the number of packets queued in the system at the end of the *j*-th service time, it is easy to verify that the process $\{N_j, j = 0, 1, ...\}$ is a homogeneous Markov chain with the following transition matrix

$$P = \begin{bmatrix} a_0 & a_1 & a_2 & a_3 & \dots \\ a_0 & a_1 & a_2 & a_3 & \dots \\ 0 & a_0 & a_1 & a_2 & \dots \\ \dots & 0 & a_0 & a_1 & \dots \\ \dots & \dots & \dots & \dots & \dots \end{bmatrix} , \qquad (3.1)$$

where

$$a_i = \int_0^\infty e^{-\lambda x} \frac{(\lambda x)^i}{i!} dB(x), \qquad i = 0, 1, \dots \quad , \qquad (3.2)$$

is the probability of i arrivals in a service time.

By defining $\pi_k = \lim_{j \to \infty} P\{N_j = k\}$, in [29] it is formally proved that if the offered load is less than one (i.e., $\lambda b < 1$) the steady-state probabilities $\{\pi_k, k = 0, 1, \dots\}$ exist and are the unique solution of the following system of linear equations

$$\pi_k = \sum_{j=1}^{k+1} \pi_j \cdot a_{k-j+1} + \pi_0 \cdot a_k, \qquad k = 0, 1, \dots \quad ,$$

$$\sum_{k=0}^\infty \pi_k = 1 \quad .$$

By exploiting the special structure of matrix P the following (numerically stable) recursive scheme can be applied to derive the steady state probabilities [126]

$$\pi_0 = 1 - \lambda b \quad ,$$

$$\pi_k = \frac{1}{a_0} \cdot \left[\pi_0 \cdot \hat{a}_{k-1} + \sum_{j=1}^{k-1} \pi_j \cdot \hat{a}_{k-j} \right], \qquad k \geq 1 \quad ,$$

where $\hat{a}_j = 1 - \sum_{h=0}^j a_h$.

Moments of the $\{\pi_k, k = 0, 1, \dots\}$ distribution can be easily derived via the probability generating function, $\Pi_{M/G/1}(z) = \sum_{k=0}^\infty \pi_k \cdot z^k$, for which a closed formula exists ([54], [99], [149])

$$\Pi_{M/G/1}(z) = B^*(\lambda - \lambda z) \cdot \frac{(1 - \lambda b)(1 - z)}{B^*(\lambda - \lambda z) - z} \quad . \qquad (3.3)$$

For the $M/G/1$ queueing system the Burke (see Theorem 2.2) and PASTA (see Theorem 2.3) theorems hold, hence the steady-state distribution at the embedding points is equal to the steady-state distribution at a random point in time, i.e., $p_k = \lim_{t \to \infty} P\{N(t) = k\}$.

If the queueing discipline is FIFO, the number of arrivals during the waiting time of a tagged packet coincides with the number of packets in the system at a departure instant, and thus

$$\Pi_{M/G/1}(z) = W^*(\lambda - \lambda z) \cdot B^*(\lambda - \lambda z) \ . \tag{3.4}$$

By substituting (3.3) in (3.4) it follows that

$$W^*(s) = \frac{s \cdot (1 - \lambda b)}{s - \lambda + \lambda B^*(s)} \ , \tag{3.5}$$

and by exploiting the LST and PGF properties, the Pollaczek-Kintchine formulae are obtained

$$E[W] = \frac{\lambda b^{(2)}}{2(1 - \rho)} \ , \tag{3.6}$$

$$E[N_q] = \lambda E[W] = \frac{\lambda^2 b^{(2)}}{2(1 - \rho)} \ . \tag{3.7}$$

3.1.2 Finite Buffer Systems

In the $M/G/1/K$ queueing system an embedded Markov chain is still used to derive the steady-state probabilities of i packets in the system at departure instants, $\{\pi_i , 0 \le i \le K - 1\}$.   For   this   system   the   process $\{N_j , j = 0, 1, ..., K - 1\}$ is a homogeneous Markov chain with the following transition matrix.

$$P = \begin{bmatrix} a_0 & a_1 & a_2 & \cdots & a_{K-2} & \tilde{a}_{K-1} \\ a_0 & a_1 & a_2 & \cdots & a_{K-2} & \tilde{a}_{K-1} \\ 0 & a_0 & a_1 & \cdots & a_{K-3} & \tilde{a}_{K-2} \\ \cdots & \cdots & \cdots & \cdots & \cdots & \cdots \\ \cdots & \cdots & \cdots & \cdots & a_0 & \tilde{a}_1 \end{bmatrix} \ , \tag{3.8}$$

where $\tilde{a}_j = \sum_{i \ge j} a_i$.

Since this Markov chain is finite, irreducible and aperiodic, its steady-state probabilities exist and are generally calculated by numerically solving the following linear system

$$\begin{cases} \pi_k = \sum_{j=1}^{k+1} \pi_j \cdot a_{k-j+1} + \pi_0 \cdot a_k & k = 0, 1, \ldots, K-2 \\ \pi_{K-1} = 1 - \sum_{j=0}^{K-2} \pi_j \end{cases} \tag{3.9}$$

REMARK. Obviously, $(\pi_i , 0 \le i \le K-1)$ are different from the steady-state probabilities at a random point in time $(p_i , 0 \le i \le K)$. When the buffer is finite, the distribution at departure instants is still equal to the distribution at arrival instants (see Theorem 2.2) but the PASTA property (see Theorem 2.3) does not hold as the number of packets in the system $\{N(t), t \ge 0\}$ does not meet the LAA requirements.

$\Diamond$

When the steady-state distribution at a random point in time and the steady-state distribution at the embedding point differ, the former can generally be derived from the latter using the method of *supplementary variables* ([55], [149])[1]. This method first investigates the behavior of the embedded Markov chain and then adds a new variable (called *supplementary variable*) to describe the system state at a random point in time.

Since the distribution function of the service time $B(x)$ is not exponential, the probability of a packet service being completed in the interval $(t, t + \Delta t]$ depends on the service time elapsed. Therefore, the state of the system at time t is defined by the following couple $\{N(t), X_+(t)\}$, where $X_+(t)$ is the r.v. describing the remaining service time of the packet in service at time t. Hence the state space contains the following points

- 0, if there are zero packets in the system;

- (n, y), if there are n packets in the system $(n = 1, 2, \ldots, K)$, and the remaining service time of the packet in service at time t is equal to y.

---

1. For an *M/G/1/K* system ad hoc techniques can also be applied to derive the steady-state probabilities at a random point in time [77].

Let $P_n(t, y)$ be the probability density function of the state (n, y) at time t. More explicitly, if $X_+(t)$ is the r.v. describing the remaining service time of the packet in service at time t,

$$P_n(t, y)dy = P\{N(t) = n, y < X_+(t) \le y + dy\}, \quad 1 \le n \le K \ . \qquad (3.10)$$

Throughout this chapter it is assumed that the process $\{N(t), X_+(t), t \ge 0\}$ can reach the steady state, and hence its steady-state probabilities exist. Below $p_n(y) = \lim_{t \to \infty} P_n(t, y)$ denotes the steady-state probability density function of the process $\{N(t), X_+(t), t \ge 0\}$, and $P_n^*(s)$ denotes its LST, i.e.,

$$P_n^*(s) = \int_0^\infty e^{-sy} P\{N = n, y < X_+ \le y + dy\} \quad 1 \le n \le K \ , \qquad (3.11)$$

and $p_n = P_n^*(0)$.

In ([149], [109]) $P_n^*(s)$ is expressed in terms of the steady-state probabilities at the embedding points $(\pi_i, 0 \le i \le K - 1)$ by taking into consideration that the event

$$\{N = n, y < X_+ \le y + dy\}$$

can be partitioned into the following mutually exclusive events:

1. there were no packets at the last service completion time and $n - 1$ packets arrived during the elapsed service time $X_- = x$ of the first packet in a busy period;
2. there were j $(j \ge 1)$ packets at the last service completion time and $n - j$ packets arrived during the elapsed service time $X_- = x$ of the packet in service.

By exploiting 1 and 2, after some algebraic manipulations, it can be shown that

$$P_n^*(s) = \rho' \pi_0 \psi_{n-1}^*(s) + \rho' \sum_{j=1}^{n} \pi_j \psi_{n-j}^*(s), \quad 1 \le n \le K - 1 \ , \qquad (3.12)$$

and

$$P_K^*(s) = \rho' \pi_0 \sum_{n=K-1}^{\infty} \psi_n^*(s) + \rho' \sum_{j=1}^{n} \pi_j \sum_{n=K-j}^{\infty} \psi_{n-j}^*(s) \ , \qquad (3.13)$$

where

- $\psi_n^*(s) = E[e^{-sX_+}|A(X_-) = n] P\{A(X_-) = n\}, \quad n = 0, 1, 2, \dots$
- ρ' is the probability that the server is busy at a random point in time; and
- $A(X_-)$ denotes the number of packets that join the system during the elapsed service time.

By exploiting the following relations

$$\psi_n^*(s) = \int_0^\infty e^{-sy} \int_0^\infty P\{A(X_-) = n, y < X_+ \le y+dy, x < X_- \le x+dx\} \quad , \quad (3.14)$$

$$P\{x < X_- \le x+dx\} = \frac{1 - B(x)}{b} \quad , \quad (3.15)$$

$$P\{y < X_+ \le y+dy | X > x\} = \frac{b(x+y)}{1 - B(x)} \quad , \quad (3.16)$$

and by conditioning on X_-, (3.14) becomes

$$\psi_n^*(s) = \int_0^\infty \frac{(\lambda x)^n}{n!} e^{-\lambda x} \frac{1 - B(x)}{b} dx \int_0^\infty e^{-sy} \frac{b(x+y)}{1 - B(x)} dy \quad , \quad (3.17)$$

which, after some algebraic manipulations, is rewritten as

$$\psi_n^*(s) = \frac{1}{\rho}\left[B^*(s)\left(\frac{\lambda}{\lambda - s}\right)^{n+1} - \sum_{j=0}^{n} a_j\left(\frac{\lambda}{\lambda - s}\right)^{n-j+1}\right] \quad . \quad (3.18)$$

Substituting (3.18) in (3.12) and (3.13) the following formulas are obtained

$$P_n^*(s) = \frac{1}{\pi_0 - \rho}\left\{ B^*(s)\left[\pi_0\left(\frac{\lambda}{\lambda - s}\right)^{n} + \sum_{j=1}^{n} \pi_j\left(\frac{\lambda}{\lambda - s}\right)^{n-j+1}\right] - \sum_{j=0}^{n-1} \pi_j\left(\frac{\lambda}{\lambda - s}\right)^{n-j}\right\} \quad 1 \le n \le K - 1 \quad , \quad (3.19)$$

$$P_K^*(s) = \frac{\lambda}{(\pi_0 + \rho) s}\left\{ B^*(s)\left[\pi_0\left(\frac{\lambda}{\lambda - s}\right)^{K-1} + \sum_{j=1}^{K-1} \pi_j\left(\frac{\lambda}{\lambda - s}\right)^{K-j}\right] - \sum_{j=0}^{K-1} \pi_j\left(\frac{\lambda}{\lambda - s}\right)^{K-j-1}\right\} \quad . \quad (3.20)$$

From (3.19) and (3.20), by exploiting the LST properties, it is possible to derive:

- the steady state probabilities of the number of packets in the system at a random point in time;
- the moments of any order of the number of packets in the system at a random point in time.

REMARK. In accordance with PASTA, $p_K = P_K^*(0)$ is the probability that an arriving packet cannot join the system because the buffer is full. As introduced in Chapter 1, this probability is called the *packet loss probability,* and is denoted by P_L.

$\Diamond$

If the service discipline is FCFS, from the joint distribution of the number of packet in the system and the remaining service time, the waiting time of a tagged packet is obtained by noting that

- it is equal to zero if the tagged packet finds an empty system, otherwise
- it is the sum of the remaining service time and the service times of the packets ahead (i.e., if the tagged packet finds $N = n$ packets in the system, its waiting time is X_+ plus $n - 1$ service times).

Hence,

$$W^*(s) = \frac{1}{1 - P_L} \left\{ p_0 + \sum_{n=0}^{K-1} P_n^*(s) \left[B^*(s) \right]^{n-1} \right\} . \tag{3.21}$$

Finally, by substituting (3.20) in (3.21)

$$W^*(s) = \frac{\pi_0 s \left(1 - \left[\frac{\lambda B^*(s)}{\lambda - s} \right]^K \right)}{s - \lambda + \lambda B^*(s)} + \left[B^*(s) \right]^{K-1} \cdot \sum_{j=0}^{K-1} \pi_j \left(\frac{\lambda}{\lambda - s} \right)^{K-j} . \tag{3.22}$$

3.2 *M/G/*1 SYSTEMS WITH VACATION

As pointed out in Chapter 2, a LAN or a MAN station is frequently modeled as an *M/G/*1 queueing system with server vacation.

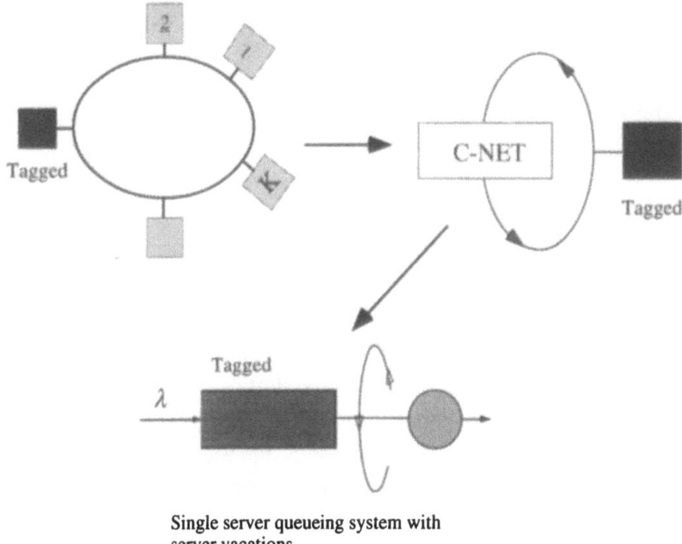

Single server queueing system with
server vacations

Figure 3.2:   Tagged station modeling

For example, by focusing on a tagged station of a Token Ring network [82]
the tagged station is modeled as a single server queueing system with server
vacations (see Figure 3.2). Specifically, with respect to the tagged station the
network is partitioned into the tagged station itself, and the complementary
part of the network (*C-NET*) which aggregates all the other stations. The
server vacation time represents the period between the end of a service
period at the tagged station and the subsequent arrival of a (free) token at
this station, i.e., the time interval during which the token is in the C-NET
part of the network.

*M/G/*1 queueing systems with server vacation can be classified according to the buffer capacity and the service discipline (see Section 2.1.2). As
stated at the beginning of this chapter only exhaustive and *l*-limited service
disciplines are considered in the following.

*M/G/*1 WITH VACATION, EXHAUSTIVE SERVICE AND INFINITE BUFFER $(M/G/1_{VE})$.
As in the *M/G/*1 queueing system, the stochastic process which describes
the number of packets in the system, $\{N(t), t \geq 0\}$, is a non Markovian continuous time process, and hence the embedded Markov chain technique is
employed. The embedded Markov chain $\{N_j, j = 0, 1, \ldots\}$, where N_j is

the number of packets queued in the system at the end of the j-th service time, is a homogeneous Markov chain. To define its transition matrix the following probabilities must be introduced

- α_j is the probability that j packets arrive during a vacation given that at least one packet arrives

$$\alpha_j = \frac{v_j}{1 - v_0} \quad , \quad j \geq 1 \quad , \tag{3.23}$$

where, recalling that A is a random variable denoting the number of arrivals during a vacation (see Section 2.1.2), $v_j = P\{A = j\}$;

- c_i is the probability that there are i packets in the queue at the end of the first service after a vacation. c_i is computed from the convolution between the probability α_j that j $(j \geq 1)$ packets arrive during a vacation[1], and the probability a_{i-j+1} that $i-j+1$ packets arrive during a service time

$$c_i = \sum_{j=1}^{i+1} \alpha_j \cdot a_{i-j+1} \qquad i = 0, 1, \ldots \quad . \tag{3.24}$$

From (3.24) it can be verified, e.g., [149], that if the offered load is less than one (i.e., $\lambda b < 1$) the steady-state probabilities of $\{N_j , j = 0, 1, \ldots\}$ exist and satisfy the following system of linear equations

$$\pi_k = \sum_{j=1}^{k+1} \pi_j \cdot a_{k-j+1} + \pi_0 \cdot c_k \qquad k = 0, 1, \ldots \quad ,$$

$$\sum_{k=0}^{\infty} \pi_k = 1 \quad .$$

Hence, the transition matrix of the embedded Markov chain has the following structure

$$P = \begin{bmatrix} c_0 & c_1 & c_2 & c_3 & \cdots \\ a_0 & a_1 & a_2 & a_3 & \cdots \\ 0 & a_0 & a_1 & a_2 & \cdots \\ \cdots & 0 & a_0 & a_1 & \cdots \\ \cdots & \cdots & \cdots & \cdots & \cdots \end{bmatrix} \quad . \tag{3.25}$$

---

1. Several consecutive vacations with zero arrivals may precede the vacation in which at least one arrival occurs.

By exploiting the special structure of matrix P, with   algebraic manipulations, a closed formula for the generating function of the number of packets in the system at departure instants $(\Pi_{M/G/1_{VE}}(z))$ can be obtained

$$\Pi_{M/G/1_{VE}}(z) = B^*(\lambda - \lambda z) \cdot \frac{\pi_0 (1 - \alpha(z))}{B^*(\lambda - \lambda z) - z} \quad , \tag{3.26}$$

where $\alpha(z) = \sum_{j=1}^{\infty} \alpha_j \cdot z^j$.

π_0 can be computed by the normalization condition, i.e., $\Pi_{M/G/1_{VE}}(1) = 1$, from which

$$\pi_0 = (1 - \lambda b) \cdot \frac{1 - v_0}{E[A]} \quad . \tag{3.27}$$

By substituting (3.27) in (3.26) and keeping into consideration (3.3) the following *decomposition property* is obtained

$$\Pi_{M/G/1_{VE}}(z) = \Pi_{M/G/1}(z) \cdot \frac{1 - A(z)}{(1 - z) \cdot E[A]} \quad . \tag{3.28}$$

REMARK. It is interesting to note that (3.28), which was obtained by using the z-transform technique, coincides with (2.65) in Theorem 2.9 where a completely different line of reasoning was used. The $M/G/1_{VE}$ decomposition property was first derived in [53] by exploiting the z-transform technique. The stochastic explanation of this result, used in Theorem 2.9, was then provided by Furhmann [65].

$\Diamond$

As for the $M/G/1_{VE}$ queueing system, the Burke (see Theorem 2.2) and PASTA (see Theorem 2.3) theorems hold, hence (3.28) also defines the PGF of the steady-state distribution of $\{N(t), t \geq 0\}$.

Formula (3.27) defines the probability of there being an empty system at a random point in time. To obtain the complete distribution, the following (numerically stable) recursive scheme can be applied (see Section 1.4 in [126])

$$\pi_1 = \frac{1}{a_0} \cdot (1 - c_0) \cdot \pi_0 \quad , \tag{3.29}$$

and

$$\pi_k = \frac{1}{a_0} \cdot \left[\pi_0 \cdot \hat{c}_{k-1} + \sum_{j=1}^{k-1} \pi_j \cdot \hat{a}_{k-j} \right], \qquad k \geq 2 \;, \tag{3.30}$$

where $\hat{c}_j = 1 - \sum_{h=0}^{j} c_h$.

$M/G/1$ WITH VACATION, EXHAUSTIVE SERVICE AND FINITE BUFFER ($M/G/1/K_{VE}$). Going back to the analysis of a real system, e.g. a station in a token ring network, an $M/G/1/K_{VE}$ queueing system can be used to model a tagged station with a finite buffer. This system can be studied by using the embedded Markov chain technique used for the $M/G/1_{VE}$ system. For this model the transition matrix of the embedded Markov chain is

$$P = \begin{bmatrix} c_0 & c_1 & c_2 & \ldots & c_{K-2} & \tilde{c}_{K-1} \\ a_0 & a_1 & a_2 & \ldots & a_{K-2} & \tilde{a}_{K-1} \\ 0 & a_0 & a_1 & \ldots & a_{K-3} & \tilde{a}_{K-2} \\ \ldots & \ldots & \ldots & \ldots & \ldots & \ldots \\ \ldots & \ldots & \ldots & \ldots & a_0 & \tilde{a}_1 \end{bmatrix}, \tag{3.31}$$

where $\tilde{c}_{K-1} = \sum_{i \geq K-1} c_i$ and $\tilde{a}_j = \sum_{i \geq j} a_i$.

The PASTA theorem does not hold for this system since the buffer is finite. Hence, the steady-state probabilities at a random point in time can be derived by using the supplementary variable technique as proposed by T. T. Lee in [109]. Specifically, he derived the steady-state probabilities at a random point in time starting from the steady-state probabilities of an embedded Markov chain. The embedding points are both at the service completion instants and at the vacation termination instants. The state of the system at the n-th embedding point is described by a couple $\{\xi, N\}_n$, where ξ indicates the type of the n-th embedding point and the r.v. N denotes the number of packets in the system at that embedding point. Specifically, $\xi = 0$ indicates a vacation termination instant, while $\xi = 1$ denotes a transmission termination instant. The transition matrix of the embedded Markov chain is

$$
P = \begin{bmatrix}
B_0 & B_1 & B_2 & \dots & B_{K-2} & \tilde{B}_{K-1} \\
A_0 & A_1 & A_2 & \dots & A_{K-2} & \tilde{A}_{K-1} \\
0 & A_0 & A_1 & \dots & A_{K-3} & \tilde{A}_{K-2} \\
\dots & \dots & \dots & \dots & \dots & \dots \\
\dots & \dots & \dots & \dots & A_0 & \tilde{A}_1
\end{bmatrix} ,
\tag{3.32}
$$

where $\tilde{B}_{K-1} = \sum_{i \geq K-1} B_i$, $\tilde{A}_j = \sum_{i \geq j} A_i$, and B_k, A_k are 2×2 matrices, whose (i, j) element has the following meaning

$$
B_k(i, j) = \begin{cases}
P\{\xi = 0, N = k | \xi = 0, N = 0\} & \text{if } (i, j) = (0, 0) \\
P\{\xi = 0, N = k | \xi = 1, N = 0\} & \text{if } (i, j) = (1, 0) \\
0 & \text{otherwise}
\end{cases}
$$

$$
A_k(i, j) = \begin{cases}
P\{\xi = 1, N = n + k - 1 | \xi = 0, N = n\} & \text{if } (i, j) = (0, 1) \\
P\{\xi = 1, N = n + k - 1 | \xi = 1, N = n\} & \text{if } (i, j) = (1, 1) \\
0 & \text{otherwise}
\end{cases}
$$

Hence, B_k and A_k have the following structures

$$
B_k = \begin{bmatrix} v_k & 0 \\ v_k & 0 \end{bmatrix} , \qquad
A_k = \begin{bmatrix} 0 & a_k \\ 0 & a_k \end{bmatrix} , \qquad k \geq 0 .
$$

Details on Lee's method for deriving the steady-state distribution at a random point in time can be found in [150] and [109].

3.2.1 $M/G/1$ Queue with Vacation and E-limited Service Discipline

The $M/G/1$ system with vacation and Exhaustive-limited, $M/G/1_{E-limited}$, is a queueing system in which the number of packets served in a service period is limited to l. The server continues to serve packets until it has served l packets or the system empties, whichever occurs first.

Like the $M/G/1_{VE}$ system, the $M/G/1_{E-limited}$ system is analyzed with the embedding point technique. The system is observed at the vacation termination instants and at the service completion instants. The state of the system at the n-th embedding point is described by the couple $\{\xi, N\}_n$, where ξ indicates the type of embedding point and N denotes the number of packets in the system. $\xi = 0$ indicates the vacation termination instant, while

$\xi = j$ is the embedding point just after the j-th transmission during a service period.

Under the assumption that the stability criteria are satisfied (see Section 2.1.3) the steady state joint probabilities

$$\pi_k^{(m)} = \lim_{n \to \infty} P\{\xi = m, N = k\}_n, \quad m = 0, 1, ..., l, \quad k = 0, 1, ...$$

exist and satisfy the following system of linear equations

$$\pi_k^{(0)} = \sum_{j=0}^{l-1} \pi_0^{(j)} v_k + \sum_{j=0}^{k} \pi_j^{(l)} v_{k-j}, \tag{3.33}$$

$$\pi_k^{(m)} = \sum_{j=1}^{k+1} \pi_j^{(m-1)} a_{k-j+1}, \quad m = 1, ..., l, \tag{3.34}$$

where a_i (v_i), for $i \geq 0$, is the probability that i packets arrive during a service (vacation) time.

The transition matrix P of the embedded Markov chain has the following structure

$$P = \begin{bmatrix} B_0 & B_1 & B_2 & B_3 & \cdots \\ A_0 & A_1 & A_2 & A_3 & \cdots \\ 0 & A_0 & A_1 & A_2 & \cdots \\ \cdots & \cdots & A_0 & A_1 & \cdots \\ \cdots & \cdots & \cdots & \cdots & \cdots \end{bmatrix}, \tag{3.35}$$

where B_k and A_k are $(l+1) \times (l+1)$ matrices, whose (i, j) element has the following meaning

$$B_k(i, j) = \begin{cases} P\{\xi = 0, N = k | \xi = i, N = 0\} & if \quad (j = 0) \\ 0 & otherwise \end{cases}$$

$$A_k(i, j) =$$

$$\begin{cases} P\{\xi = i+1, N = n+k-1 | \xi = i, N = n\} & \text{if } (i < l \wedge j = i+1 \wedge n > 0) \\ P\{\xi = 0, N = n+k | \xi = l, N = n\} & \text{if } (i = l \wedge j = 0) \\ 0 & \text{otherwise} \end{cases}$$

Hence B_k and A_k have the following structure

$$B_k = \begin{bmatrix} v_k & 0 & \cdots & 0 & 0 \\ v_k & 0 & \cdots & 0 & 0 \\ \cdots & \cdots & \cdots & \cdots & \cdots \\ v_k & 0 & \cdots & 0 & 0 \\ v_k & 0 & \cdots & 0 & 0 \end{bmatrix}, \qquad A_k = \begin{bmatrix} 0 & a_k & 0 & \cdots & 0 \\ 0 & 0 & a_k & \cdots & 0 \\ \cdots & \cdots & \cdots & \cdots & 0 \\ 0 & 0 & 0 & \cdots & a_k \\ v_k & 0 & 0 & \cdots & 0 \end{bmatrix}, \qquad k \geq 0 .$$

By defining $\Pi_m(z) = \sum_{j=0}^{\infty} \pi_j^{(m)} \cdot z^j$, $m = 0, 1, \ldots, l$, after routine manipulations (see [110], [149])[1]

$$\Pi_0(z) = V^*(\lambda - \lambda z) \cdot \left[\Pi_l(z) + \sum_{i=0}^{l-1} \pi_0^{(i)} \right] , \tag{3.36}$$

and

$$\Pi_m(z) = \frac{B^*(\lambda - \lambda z)^m}{z} \Pi_0(z) - \sum_{i=0}^{m-1} \frac{B^*(\lambda - \lambda z)^{m-i}}{z} \pi_0^{(i)}, \quad m > 0 . \tag{3.37}$$

Finally by substituting (3.37) with $m = l$ in (3.36), the following expression for $\Pi_0(z)$ is obtained

$$\Pi_0(z) = \frac{V^*(\lambda - \lambda z) \cdot z^l}{z^l - V^*(\lambda - \lambda z) \cdot B^*(\lambda - \lambda z)^l} \sum_{i=0}^{l-1} \pi_0^{(i)} \left[1 - \left(\frac{B^*(\lambda - \lambda z)}{z} \right)^{l-i} \right] . \tag{3.38}$$

In (3.38) there are l unknowns $\{ \pi_0^{(m)}, m = 0, 1, \ldots, l-1 \}$, which are called *boundary probabilities*. A widespread method to compute them exploits Rouche's Theorem and the analyticity of PGFs in the unit circle, i.e., $|z| \leq 1$. Specifically, in [110], first it is shown that, if the stability criteria are satis-

---

1. In (3.37), for $m = 1$ the summation is zero.

fied, the denominator of (3.38) has l zeros in the unit disk (one of which is $z = 1$). Then, by imposing the constraint that these zeros are roots of the numerator it is possible to obtain $l - 1$ equations among the l unknowns. The root $z = 1$ does not provide any meaningful equation since for $z = 1$ the numerator is trivially zero as well. An additional equation is obtained by utilizing the normalization condition $\sum_{m=0}^{l} \Pi_m(1) = 1$.

For this queueing system the Burke and PASTA theorems guarantee that the distribution of the number of packets in the system at a departure epoch coincides with the distribution at a random point in time $\{p_k, k = 0, 1, ...\}$, hence

$$P(z) = \frac{\sum_{m=1}^{l} \Pi_m(z)}{1 - \Pi_0(1)} \, , \tag{3.39}$$

where $P(z) = \sum_{k=0}^{\infty} p_k \cdot z^k$.

From (3.39) and using Little's theorem the average response time can be evaluated. When the queueing discipline is FIFO, the LST of the waiting time can also easily be obtained from (3.39) by observing that the number of arrivals during a tagged-packet response time coincides with the number of packets in the system at the tagged-packet departure instant, thus

$$P(z) = W^*(\lambda - \lambda z) \cdot B^*(\lambda - \lambda z) \, , \tag{3.40}$$

from which

$$W^*(s) = \frac{P(1 - s/\lambda)}{B^*(s)} \, . \tag{3.41}$$

3.2.2 *M/G/1/K* Queue with Vacation and E-limited Service Disciplines

The only difference between the queueing system presented in this section and the queueing system studied in the previous section is the queue size, which is now limited.

Again the system is analyzed at the vacation termination instants and the service completion instants (*embedding points*); and its state at the n-th embedding point is described by the couple

$$\{\xi = m, N = k\}_n \, , \quad m = 0, 1, ..., l \, , \quad k = 0, 1, ..., K \, .$$

The embedded Markov chain is irreducible and aperiodic, hence its steady-state probabilities

$$\pi_k^{(m)} = \lim_{n \to \infty} P\{\xi = m, N = k\}_n$$

exist and satisfy the following relationships

- when $\{\xi = m, N = k\}$ with $m > 0$ and $k < K - 1$ equation (3.34) holds;
- when $\{\xi = 0, N = k\}$ with $k < K$ equation (3.33) holds;
- when $\{\xi = m, N = K - 1\}$ the following equations hold

$$\pi_{K-1}^{(1)} = \sum_{j=1}^{K} \pi_j^{(0)} \sum_{h=K-j}^{\infty} a_h \ , \tag{3.42}$$

and

$$\pi_{K-1}^{(m)} = \sum_{j=1}^{K-1} \pi_j^{(m-1)} \sum_{h=K-j}^{\infty} a_h \ , \quad m = 2, \ldots, l \ ; \tag{3.43}$$

- when $\{\xi = 0, N = K\}$ the following equations hold

$$\pi_K^{(0)} = \sum_{j=0}^{l-1} \pi_0^{(j)} \cdot \sum_{j=K}^{\infty} v_j + \sum_{j=0}^{K-1} \pi_j^{(l)} \cdot \sum_{h=K-j}^{\infty} v_h \ . \tag{3.44}$$

In addition, the steady-state probabilities satisfy the normalization condition $\sum_{j=0}^{l} \sum_{k=0}^{K} \pi_k^{(j)} = 1$.

REMARK. Equations (3.42)-(3.44) take into account that the buffer is of limited size K, and hence the maximum number of packets at a service is $K-1$, while the maximum number of packets at a vacation termination instant is K.

$\Diamond$

The steady-state probabilities of the embedded Markov chain are usually numerically calculated by solving the system of linear equations defined above. The computation of the distribution of the number of packets in the system at an arbitrary point in time can be performed using the supplementary variable technique [110].

An important performance figure for this class of queueing system is the loss probability P_L, i.e., the probability that the buffer is full. This probability has been shown to be [110]

$$P_L = \frac{\rho - \rho'}{\rho} \; ,$$

where $\rho = \lambda b$, and ρ' which is the fraction of time the server is working. Specifically,

$$\rho' = \frac{(1 - P_V) \cdot b}{P_V \cdot E[V] + (1 - P_V) \cdot b} \; , \tag{3.45}$$

where $P_V = \sum_{j=1}^{K} \pi_j^{(0)}$ is the probability that an embedding point is a vacation termination point.

3.3 *M/G/1*-TYPE MODELS

In Section 3.1.1 it is shown that the transition matrix of the embedded Markov chain for an *M/G/1* system has the following structure

$$P = \begin{bmatrix} a_0 & a_1 & a_2 & a_3 & \dots \\ a_0 & a_1 & a_2 & a_3 & \dots \\ 0 & a_0 & a_1 & a_2 & \dots \\ \dots & 0 & a_0 & a_1 & \dots \\ \dots & \dots & \dots & \dots & \dots \end{bmatrix} \; , \tag{3.46}$$

where the elements are scalars.

Furthermore, Section 3.2.1 introduced a system whose behavior is described by an embedded Markov chain with the following transition matrix

$$P = \begin{bmatrix} B_0 & B_1 & B_2 & B_3 & \dots \\ A_0 & A_1 & A_2 & A_3 & \dots \\ 0 & A_0 & A_1 & A_2 & \dots \\ \dots & \dots & A_0 & A_1 & \dots \\ \dots & \dots & \dots & \dots & \dots \end{bmatrix} \; , \tag{3.47}$$

where the elements are matrices. Hence (3.47) is a generalization of the structure of (3.46). For this reason, in the literature, Markov chains whose transition matrices have the structure (3.47) are called *M/G/1-type* Markov chains, and they frequently occur in modeling packet-switching networks.

This section introduces the basic concepts of the matrix analytic techniques developed by M.F. Neuts [28] to provide the exact solution for *M/G/1-type* Markov chains.

The most general structure of the transition matrix for this class of Markov chains is shown in Figure 3.3.

$$
P = \begin{matrix} level\ 0 \\ level\ 1 \\ level\ 2 \\ \cdots \\ \cdots \end{matrix}
\begin{bmatrix}
B_0 & B_1 & B_2 & B_3 & \cdots \\
C_0 & A_1 & A_2 & A_3 & \cdots \\
0 & A_0 & A_1 & A_2 & \cdots \\
 & \cdots & 0 & A_0 & A_1 & \cdots \\
\cdots & \cdots & \cdots & \cdots & \cdots
\end{bmatrix}
$$

Figure 3.3: Structure of *M/G/1*-type Markov chains

The elements of P are matrices with a finite size, specifically $B_0 \in \mathbb{R}^{m \times m}$, $B_i \in \mathbb{R}^{m \times n}$, $C_0 \in \mathbb{R}^{n \times m}$, $A_j \in \mathbb{R}^{n \times n}$ where $i > 0$ and $j \geq 0$.

The state space of the Markov chain is the set of couples (l, ph), where the first component is called *level*, and the second *phase*. This set is generally made up of m couples $(0, 1)$, $(0, 2)$, ..., $(0, m)$ in level 0, and n couples $(i, 1)$, $(i, 2)$, ..., (i, n) in level i, for each $i > 0$.

The structure of the matrix reflects the fact that to go from a state in level i to a lower level, the chain has to visit each intermediate level at least once. Furthermore, for levels greater than 1 the transition probabilities (described by matrices A_j) are independent of the level, i.e., the chain is *spatially homogeneous*. For levels 0 and 1, which describe the behaviour of the chain on the border, there may be ad hoc transitions that are described by B_i and C_0.

3.3.1 Solution Method

The solution of an *M/G/1-type* stochastic model consists in deriving the steady-state probability distribution or moments of the distribution. The existence of these steady-state probabilities can be traced back to the verification that P is positive recurrent, assuming that it is stochastic and irreducible [29]. Whether the moments or the whole set of the steady-state probabilities have to be computed, the vectors of the steady-state probabilities of levels 0 and 1 have to be obtained. Hereafter, $\mathbf{x}_0$ and $\mathbf{x}_1$ indicate these vectors.

Vector $\mathbf{x}_0$ is obtained by an embedded Markov chain, with transition matrix $K \in \mathbb{R}^{m \times m}$. This Markov chain is built by observing the system only at the instants in which the chain hits level 0. Hence,

$$K_{st} = P\{(0, s) \rightarrow (0, t)\}, \quad 1 \le s, t \le m ,$$

where t is the *first passage phase in level* 0, i.e., the state $(0, t)$ is the first state in level 0 reached by a Markov chain after leaving state $(0, s)$. In general, the *first passage phase* of level i indicates the first phase reached by the Markov chain when going back to level i.

The main step in the computation of matrix K (and hence of vector $\mathbf{x}_0$) is the study of two special processes:

1. the first passage process from a generic level i $(i > 1)$ to level $i - 1$;

2. the first passage process from level 1 to level 0.

THE FIRST PASSAGE PROCESS $i \rightarrow i - 1$. The first passage process $i \rightarrow i - 1$ is a fundamental part of Neuts' theory. For each level i $(i > 1)$ the first passage process is obtained by observing the state of the *M/G/1-type* chain in the return instants to the level preceding the departure level. The transition matrix of this process is denoted by $G \in \Re^{n \times n}$. Specifically, $G_{\varphi, \phi}$ is the probability of the transition $(i, \varphi) \rightarrow (i - 1, \phi)$, $i \ge 2$, $1 \le \varphi, \phi \le n$, where ϕ is the phase of the *M/G/1-type* chain when it reaches for the first time level $i - 1$ starting from the state (i, φ).

The process described by G is built by fixing a level $i > 1$ and taking "photos" of the instants in which the *M/G/1-type* chain hits the preceding level for the first time in any number of steps (Figure 3.4).

The figure shows a possible sequence of steps that intervenes in the transitions of G. From the starting state the chain can generally go up by several levels and go down by one (since P has an *M/G/1-type* structure), before going back to the preceding level.

When describing G it is useful to introduce matrix $G(k)$ that represent the first passage process in a number k of steps; where the generic element of the matrix $G(k)$ represents

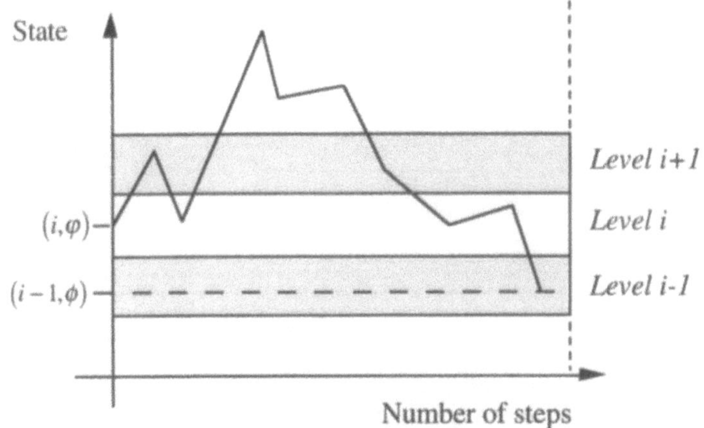

Figure 3.4: Relationship between a realization of the $M/G/1$-*type* chain and a transition in the first passage process

$$G_{st}(k) = P\{ (i, s) \to (i-1, t) \ in \ k \ steps\}, \quad 1 \le s, t \le m, i > 1, k \ge 1 \quad .$$

The relationship between G and $G(k)$ is given by

$$G = \sum_{k=1}^{\infty} G(k) \quad , \tag{3.48}$$

and bearing in mind the definition of $G(k)$, some possible transitions are examined.

Figure 3.5 shows some first-passage-process transitions in the case of $G(1)$, $G(2)$ and $G(3)$. Case (i) represents a transition of $G(1)$. From the structure of P it can easily be deduced that $G(1) = A_0$. Type (ii) transitions, which are part of matrix $G(2)$, are those that describe the transitions in two steps. In the first one the chain remains in the same level (A_1), and in the second step goes down one level, i.e., it behaves as in case (i); therefore $G(2) = A_1 A_0 = A_1 G(1)$. The last two cases give the possible transitions of $G(3)$. In case (iii), in the first two steps the $M/G/1$-*type* chain stays at the same level, whilst in the last it goes back $(A_1^2 A_0)$. In case (iv) the $M/G/1$-*type* chain goes up one level (A_2) and in the next two steps it goes down, hence

$$G(3) = A_1 A_1 A_0 + A_2 A_0 A_0 = A_1 G(2) + A_2 G(1) G(1) \quad .$$

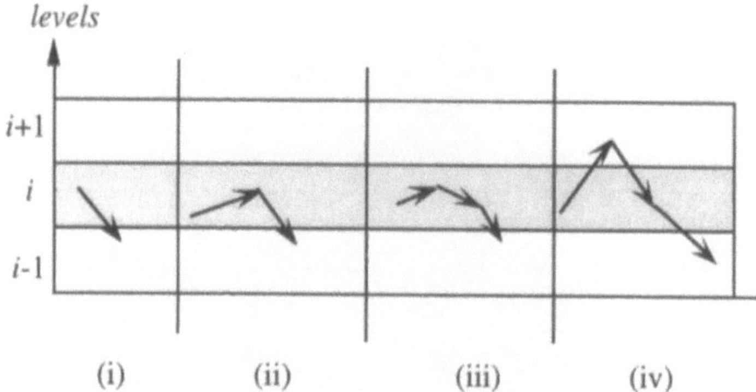

LEGEND: (i) return in one step; (ii) return in two steps; (iii) return in three steps; (iv) return in three steps.

Figure 3.5: Transitions that intervene in the first passage process

In order to get the formula $G(k)$, the case $k{=}4$ is analyzed. The possible ways to go back to the preceding level in four steps can be partitioned into three groups, depending upon the behavior of the *M/G/1-type* chain in the first step:

1. in the first step the *M/G/1-type* chain remains at the same level and goes down to the proceeding one in the other three steps;

2. in the first step it goes up one level and in the other three steps goes down two levels;

3. in the first step it goes up two levels and in the other three steps it goes down three levels.

If the process goes up more than two levels, it will never be able to go back in the other three steps to the level before the departure one. Therefore events (1), (2) and (3) are all possible and, in addition, are mutually exclusive. By denoting with $G_{(1)}(4)$, $G_{(2)}(4)$, and $G_{(3)}(4)$ the matrices describing the transitions outlined in cases (1), (2) and (3), then $G(4) = G_{(1)}(4) + G_{(2)}(4) + G_{(3)}(4)$.

The first and third matrices can easily be expressed as follows

$$G_{(1)}(4) = A_1 G(3) \ ,$$

and

$$G_{(3)}(4) = A_3 [G(1)G(1)G(1)] = A_3 A_0^3 \ .$$

To derive matrix $G_{(2)}(4)$ Figure 3.6 is helpful. Cases (b.1) and (b.2) are the two possible ways to go down two levels in three steps, and (b.3) summarizes the previous two cases. In both cases the process goes up one level in the first step, but they differ in the next three steps. In (b.1) the process goes down one level in one step $(G(1))$, and another level down in the next two steps $(G(2))$, whilst in (b.2) the opposite happens.

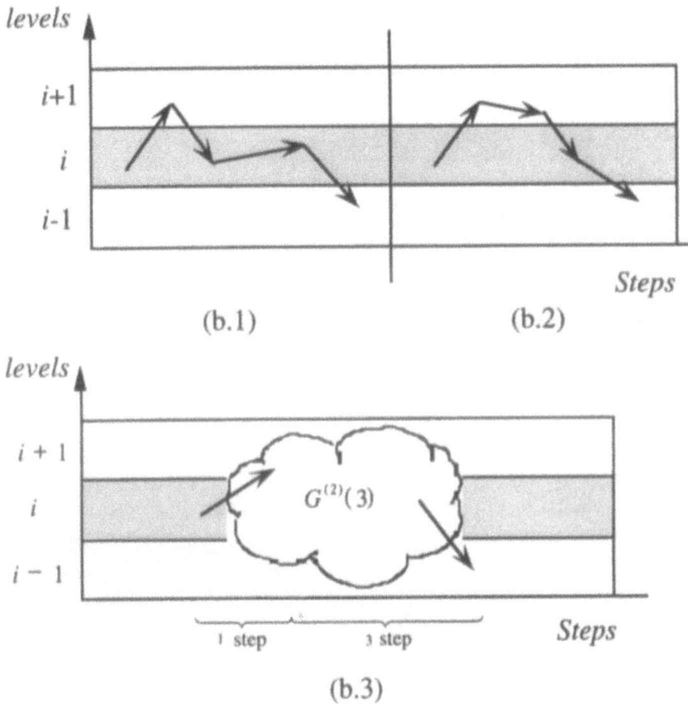

Figure 3.6:   Transitions of $G_{(2)}(4)$

The above considerations lead to

$$G_{(2)}(4) \;=\; A_2 G(1)G(2) + A_2 G(2)G(1) = A_1 \cdot G^{(2)}(3) \quad ,$$

where $G^{(2)}(3)$ represents all the possible terns of steps that make the process go down two levels in three steps (see case b.3 in Figure 3.6).

Hereafter $G^{(v)}(k)$ indicates the matrix that describes all the possible transitions to go down v levels in k steps. For ease of representation $G(k) = G^{(1)}(k)$.

To sum up, the results obtained are

$$G(1) = A_0, \; G(2) = A_1 G(1), \; G(3) = A_1 G(2) + A_2 G^{(2)}(2) \; ,$$

and

$$G(4) = A_1 G(3) + A_2 G^{(2)}(3) + A_3 G^{(3)}(3) \; . \tag{3.49}$$

Formula (3.49) can easily be extended to any v and k $(k \geq 1)$

$$G(k) = \sum_{v=1}^{k-1} A_v G^{(v)}(k-1) \; . \tag{3.50}$$

Via (3.48) and (3.50)

$$G = A_0 + \sum_{k=2}^{\infty} \sum_{v=1}^{\infty} A_v G^{(v)}(k-1) \tag{3.51}$$

$$= A_0 + \sum_{v=1}^{\infty} A_v \sum_{k=v+1}^{\infty} G^{(v)}(k-1) \; .$$

Bearing in mind the stochastic meaning of $G^{(v)}(k)$, $G^{(v)}(k) = 0, \; \forall k < v$,

$$\sum_{k=v+1}^{\infty} G^{(v)}(k-1) = \sum_{k=1}^{\infty} G^{(v)}(k-1) = G^{(v)} = G^v \; , \tag{3.52}$$

where the last equality can easily be proved (see [126]). By substituting (3.52) in (3.51), the following fundamental relationship is obtained

$$G = \sum_{v=0}^{\infty} A_v G^v \; . \tag{3.53}$$

Formula (3.53) which describes the first passage process is formally proved in [126]. Formula (3.53) describes that to make a transition from level i to the preceding one, given that in the first step the $M/G/1$-type chain moves to level $i + v - 1$, v transitions in the first-passage process $i \to i - 1$, $(i > 1)$ are needed.

FIRST PASSAGE PROCESS TO LEVEL 0. The transitions responsible for the first passage process from level 0 to level 0 can be partitioned into the following three classes:

(i)   direct transitions within level 0;
(ii)   direct transitions from level 0 to 1 and return to level 0;
(iii) direct transitions from level 0 to i $(i > 1)$ and return to level 0.

These three classes need to be taken into consideration in relation to the structure of P. This entails distinguishing between the first passage process $i \rightarrow i - 1$, $(i > 1)$ and the first passage process from level 1 to level 0, which in turn are different from a transition in one step from 0 to 0.

Due to matrix C_0, the first passage process from level 1 to level 0 differs from the first passage process $i \rightarrow i - 1$, $(i > 1)$. L denotes the transition matrix of the former process. By exploiting matrices L and G, the transition probabilities related to (i), (ii) and (iii) are shown in Figure 3.7.

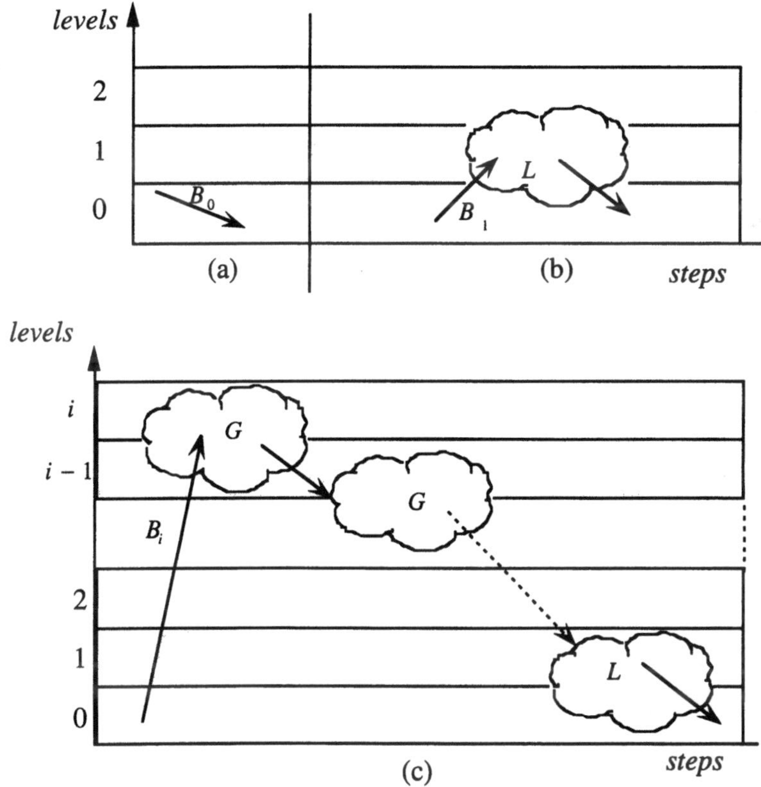

Figure 3.7: The transitions that define the process K

The transition probabilities related to class (i) (see Figure 3.7a) are trivially described by matrix B_0. In the second class (see Figure 3.7b), matrix B_1 describes the probabilities of going to level 1, and matrix L those of going down to level 0. Case (iii) (see Figure 3.7c) involves matrix G, since the first step leads the process to a level i which is greater than 1 (via B_i). To get back to level 0, $(i-1)$ returns to the previous level have to occur until the process is back in level 1. At this point an analogous situation to case (ii) is reached.

The above considerations lead to

$$K = B_0 + B_1 L + \sum_{i=2}^{\infty} B_i G^{i-1} L \ .$$
(3.54)

MATRIX L. As with G and K, matrix L is derived by analyzing all the possible transitions from level 1 to 0. These can be partitioned into the following

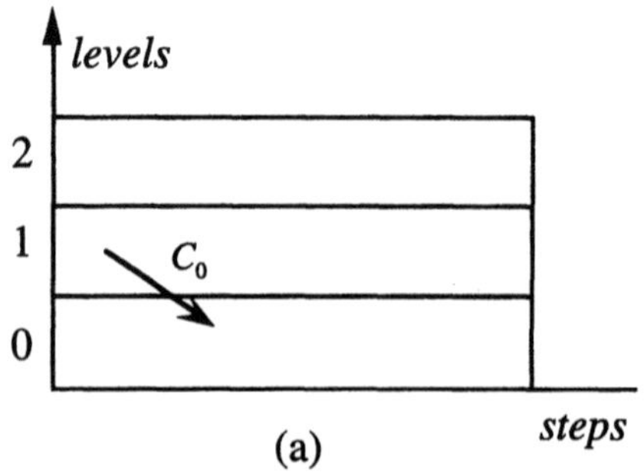

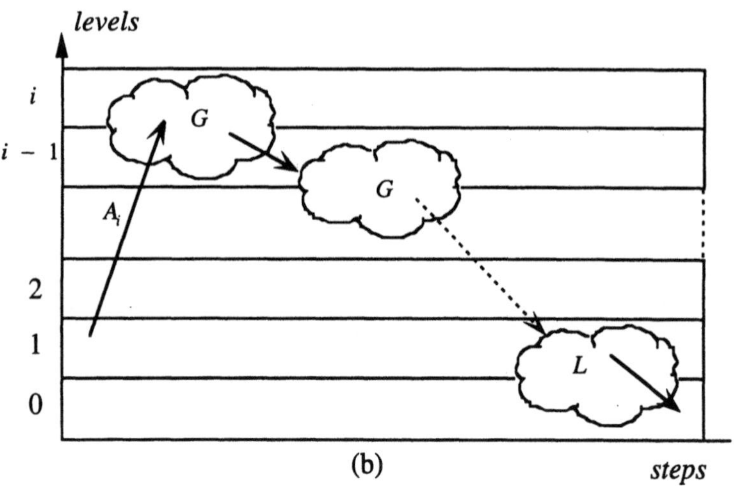

Figure 3.8: The transitions that define process L

three (mutually exclusive) classes

(i)   direct return from level 1 to 0;

(ii)   transition from level 1 to 1 and then return to level 0;

(iii)   visiting levels $i \geq 1$ and subsequently returning to level 1 and then going to level 0.

Figure 3.8a shows the transition probabilities (C_0) of the events in class (i). Classes (ii) and (iii) (Figure 3.8b) take into account possible changes in level caused by i arrivals. Events belonging to class (ii) occur with probabilities $A_1 L$. For the same reasons as outlined in the analysis of matrix K, in case (iii), given that in the first transition i $(i > 1)$ arrivals occur, the first passage process moves from level i to level 1 with probability G^{i-1}, and then moves from level 1 to level 0 with probability L. Hence

$$L = C_0 + A_1 L + \sum_{i=2}^{\infty} A_i G^{i-1} L \quad . \tag{3.55}$$

3.3.2 Implementing the Solution Method

In this subsection the implementation of the solution method for an $M/G/1$-*type* Markov chains with transition matrix P is presented.

$$P = \begin{bmatrix} B_0 & B_1 & B_2 & B_3 & \cdots \\ C_0 & A_1 & A_2 & A_3 & \cdots \\ 0 & A_0 & A_1 & A_2 & \cdots \\ \cdots & 0 & A_0 & A_1 & \cdots \\ \cdots & \cdots & \cdots & \cdots & \cdots \end{bmatrix} \quad .$$

As matrix P is stochastic, it follows that matrix A $(A = \sum_{v=0}^{\infty} A_v)$ is stochastic and

$$B_0 \mathbf{e} + \sum_{v=1}^{\infty} B_v \cdot \mathbf{e} = \mathbf{e} \quad , \tag{3.56}$$

where $\mathbf{e}^T = (1, 1, \ldots, 1)$. It also results that

- $C_0 \mathbf{e} + \sum_{v=1}^{\infty} A_v \mathbf{e} = \mathbf{e}$,
- $\sum_{v=0}^{\infty} A_v \mathbf{e} = \mathbf{e}$,

and hence

$$C_0 \mathbf{e} = A_0 \mathbf{e} \quad . \tag{3.57}$$

The general goal of the analysis is to obtain the steady-state probabilities of the Markov chain. When only the moments of the steady-state distribution are required the analysis simplifies. Both types of analysis, entail calculating first the vectors of the steady-state probabilities of levels 0 ($\mathbf{x}_0$) and 1 ($\mathbf{x}_1$). As mentioned before, the existence of these vectors can be traced back to the verification that P is positive recurrent, assuming that it is stochastic and irreducible.

COMPUTATION OF MATRIX G. Before presenting the computation method for G it is useful to introduce some quantities that will be useful for developing this method

- $\boldsymbol{\pi}$ is the left eigenvector of matrix A corresponding to the eigenvalue 1, i.e., $\boldsymbol{\pi}A = \boldsymbol{\pi}, \boldsymbol{\pi} \cdot \mathbf{e} = 1$;
- $\boldsymbol{\beta}$ is the vector defined as $\boldsymbol{\beta} = \sum_{v=0}^{\infty} v A_v \mathbf{e}$.

Hereafter, A is always assumed to be irreducible. In addition the matrix G is initially assumed to be irreducible as well. In Section 3.3.3 a particular case where matrix G is reducible, is analyzed.

When matrix A is irreducible there is only one vector $\boldsymbol{\pi}$ that satisfies $\boldsymbol{\pi}A = \boldsymbol{\pi}$ and $\boldsymbol{\pi} \cdot \mathbf{e} = 1$. When A is irreducible, matrix G is the unique stochastic matrix which satisfies equation (3.53) [126]. It has also been proven that matrix G is stochastic only if the following inequality holds ([126] Theorem 2.3.1)

$$\rho = \boldsymbol{\pi} \cdot \boldsymbol{\beta} \leq 1 \ . \tag{3.58}$$

When (3.58) holds, matrix G can be computed according to the following iterative scheme [126]

$$G_n = \begin{cases} A_0 & n = 0 \\ \displaystyle\sum_{j=0}^{\infty} A_j G_{n-1}^j & n > 0 \end{cases} \tag{3.59}$$

and $G = \lim_{n \to \infty} G_n$.

Inequality (3.58) is a necessary condition for the recurrence of the $M/G/1$-type Markov chain. To have necessary and sufficient conditions for the positive recurrence of the Markov chain, the following conditions must hold

1. $\rho = \pi \cdot \beta < 1$, and

2. matrix $\sum_{v=0}^{\infty} v B_v$ is finite.

COMPUTING MATRIX K. This is the matrix that describes the first passage process to level 0.

The irreducibility of K depends directly on the irreducibility of matrix P. In fact, if K were reducible, there would be at least two states $(0, j)$ and $(0, s)$ such that, from state $(0, j)$ it would be impossible to reach state $(0, s)$. This also holds for matrix P which would therefore also be reducible. Consequently, if P is irreducible then so is K.

As already shown, the formula for K is

$$K = B_0 + \sum_{v=1}^{\infty} B_v G^{v-1} L \ , \tag{3.60}$$

where $L = C_0 + \sum_{i=1}^{\infty} A_i G^{i-1} L$.

Furthermore, L can be written as

$$L = \left[I - \sum_{i=1}^{\infty} A_v G^{i-1} \right]^{-1} C_0 \ . \tag{3.61}$$

where the existence of the inverse of $[I - \sum_{i=1}^{\infty} A_v G^{i-1}]$ is guaranteed by the irreducibility of G [126].

By substituting (3.61) into (3.60), matrix K can be rewritten as follows

$$K = B_0 + \sum_{v=1}^{\infty} B_v G^{v-1} \cdot \left[I - \sum_{i=1}^{\infty} A_i G^{i-1} \right]^{-1} C_0 \ . \tag{3.62}$$

REMARK. From formula (3.62) it is easy to prove that if G is stochastic, then so is K. This follows by first observing that

$$G = A_0 + \sum_{i=1}^{\infty} A_i G^i = A_0 + \left[\sum_{i=1}^{\infty} A_i G^{i-1} \right] G \ ,$$

and hence, after some algebraic manipulations

$$G = \left[I - \sum_{i=1}^{\infty} A_i G^{i-1} \right]^{-1} A_0 \ . \tag{3.63}$$

Furthermore, by bearing in mind that $C_0 \mathbf{e} = A_0 \mathbf{e}$, it follows $K \mathbf{e} = G \mathbf{e}$.

$\Diamond$

Since K is irreducible and stochastic, there exists only one vector $\mathbf{k}$ such that $\mathbf{k} K = \mathbf{k}$ and $\mathbf{k} \mathbf{e} = 1$, where k_j indicates the steady-state probability of being in $(0, j)$ given that the Markov chain is in level 0.

COMPUTING VECTOR $\mathbf{x}_0$. To compute the boundary probabilities (the unconditional probability of being in level 0) it is useful to define vector $\tilde{\mathbf{k}}_1$ whose j-th component is the average number of steps to go back to level 0 starting from the state $(0, j)$. Therefore, $\mathbf{k} \cdot \tilde{\mathbf{k}}_1$ is the average number of transitions between two consecutive visits to level 0 and hence, $(\mathbf{k} \cdot \tilde{\mathbf{k}}_1)^{-1}$ is the steady-state probability of level 0.

In [126] it is shown that

$$\tilde{\mathbf{k}}_1 = \mathbf{f}_2 + \sum_{v=1}^{\infty} B_v G^{v-1} \cdot \left[I - \sum_{i=1}^{\infty} A_i G^{i-1} \right]^{-1} \mathbf{f}_1 \ , \tag{3.64}$$

where vectors $\mathbf{f}_1$ and $\mathbf{f}_2$ are

$$\mathbf{f}_1 = \left[I - A_0 - \sum_{i=1}^{\infty} A_i G^{i-1} \right] \cdot [I - A + (\mathbf{e} - \boldsymbol{\beta}) \mathbf{g}]^{-1} \cdot \mathbf{e} \tag{3.65}$$

$$+ (1 - \rho)^{-1} \cdot A_0 \mathbf{e} \ ,$$

$$\mathbf{f}_2 = \mathbf{e} + \left[\sum_{v=1}^{\infty} B_v - \sum_{v=1}^{\infty} B_v G^{v-1} \right] [I - A + (\mathbf{e} - \boldsymbol{\beta}) \mathbf{g}]^{-1} \mathbf{e} + (1 - \rho)^{-1} \tag{3.66}$$

$$\cdot \sum_{v=1}^{\infty} (v - 1) B_v \mathbf{e} \ .$$

To make the computation easier, the following matrices and vectors are defined

$$\bar{B}_1 = \sum_{v=1}^{\infty} B_v G^{v-1} \ , \tag{3.67}$$

$$\bar{A}_1 = \sum_{i=1}^{\infty} B_i G^{i-1} \ , \tag{3.68}$$

and

$$\mathbf{w} = [I - A + (\mathbf{e} - \boldsymbol{\beta})\mathbf{g}]^{-1}\mathbf{e} \ . \tag{3.69}$$

Since G is stochastic $G^v \cdot \mathbf{e} = \mathbf{e}$, hence

$$\bar{B}_1\mathbf{e} = \sum_{v=1}^{\infty} B_v G^{v-1}\mathbf{e} = \sum_{v=1}^{\infty} B_v\mathbf{e} \ , \tag{3.70}$$

and after some algebraic manipulations it follows that

$$\bar{B}_1 (I - \bar{A}_1)^{-1} \cdot \mathbf{f}_1 = \bar{B}_1\mathbf{w} + (1 - \rho)^{-1}\sum_{v=1}^{\infty} B_v\mathbf{e} - \bar{B}_1 G\mathbf{w} \ . \tag{3.71}$$

Using (3.67), (3.68), and (3.71), after some algebraic manipulations (3.64) can be rewritten as

$$\tilde{\mathbf{k}}_1 = \mathbf{e} + \sum_{v=1}^{\infty} B_v\mathbf{w} + (1 - \rho)^{-1}\sum_{v=1}^{\infty} vB_v\mathbf{e} - \bar{B}_1 G\mathbf{w} = \tag{3.72}$$

$$\mathbf{e} + \left[\sum_{v=1}^{\infty} B_v - \bar{B}_1 G\right]\mathbf{w} + (1 - \rho)^{-1}\sum_{v=1}^{\infty} vB_v\mathbf{e} \ .$$

$\mathbf{x}_0$ can now be computed by exploiting vectors $\mathbf{k}$ and $\tilde{\mathbf{k}}_1$ according to the following theorem.

THEOREM 3.1   *When matrix G is irreducible, then chain P is positive recurrent iff $\rho < 1$ and $\sum_{v=1}^{\infty} vB_v$ is finite. In this case vector $\mathbf{x}_0$ is given by* $\mathbf{x}_0 = (\mathbf{k}\tilde{\mathbf{k}}_1)^{-1}\mathbf{k}$.

Leaving aside the proof of the theorem, an interpretation of $\mathbf{x}_0$ can be given. In fact, the j-th component of $\mathbf{x}_0$

$$(\mathbf{x}_0)_j = k_j \cdot (\mathbf{k}\tilde{\mathbf{k}}_1)^{-1}, \tag{3.73}$$

represents the steady state probability of being in state $(0, j)$. It is given by the product of the probability of being in level 0 $(\mathbf{k}\tilde{\mathbf{k}}_1)^{-1}$ and the probability of being in $(0, j)$ conditioned to the event "being in level 0" (k_j).

COMPUTING VECTOR $\mathbf{x}_1$. To compute vector $\mathbf{x}_1$ it is useful to introduce the first passage process from level 1 to level 1 whose transition matrix will be denoted by H. The $H_{i,j}$ element is the first passage probability from $(1, i)$ to $(1, j)$. As in the first passage process from level 0 to level 0, there exists only one vector $\mathbf{h}$ such that $\mathbf{h}H = \mathbf{h}$ and $\mathbf{h}\mathbf{e} = 1$, where h_j indicates the steady-state probability of being in $(1, j)$ given that the Markov chain is in level 1. Again to compute the unconditional probability of being in level 1, it is useful to define a vector $\tilde{\mathbf{h}}_1$ whose j-th component is the average number of steps to go back to level 1 starting from the state $(1, j)$.

Hence vector $\mathbf{x}_1$ can be expressed as $\mathbf{x}_1 = \mathbf{h}(\mathbf{h} \cdot \tilde{\mathbf{h}}_1)^{-1}$, where $\mathbf{h}$ is the steady state solution of the Markov chain with transition matrix H. Specifically, in [126] it is shown that

$$H = C_0 [I - B_0]^{-1} \sum_{v=1}^{\infty} B_v G^{v-1} + \sum_{v=1}^{\infty} A_v G^{v-1}, \tag{3.74}$$

and

$$\tilde{\mathbf{h}}_1 = \mathbf{f}_1 + C_0 [I - B_0]^{-1} \mathbf{f}_2, \tag{3.75}$$

where $\mathbf{f}_1$ and $\mathbf{f}_2$ are defined in (3.65) and (3.66), respectively.

COMPUTING THE DISTRIBUTION. On the basis of $\mathbf{x}_0$ and $\mathbf{x}_1$, the other steady-state probabilities vectors $\mathbf{x}_i$, $i > 1$ of the Markov chain can be computed using Ramaswami's theorem [131], according to which

$$\mathbf{x}_i = \left[\mathbf{x}_0 \overline{B}_i + \sum_{j=1}^{i-1} \mathbf{x}_j \overline{A}_{i+1-j} \right] (I - \overline{A}_1)^{-1}, \quad i > 0, \tag{3.76}$$

where

$$\overline{A}_v = \sum_{i=v}^{\infty} A_i G^{i-v} \, , \quad \overline{B}_v = \sum_{i=v}^{\infty} B_i G^{i-v} \, , \quad v > 1 \, .$$

$$(3.77)$$

Formula (3.76) can be efficiently computed by observing that $\lim_{i \to \infty} \overline{B}_i = \lim_{i \to \infty} \overline{A}_i = 0$. It is thus convenient to choose an index i such that vectors

$$\sum_{k=i+1}^{\infty} \overline{B}_k \mathbf{e} \, , \text{ and } \sum_{k=i+1}^{\infty} \overline{A}_k \mathbf{e}$$

have negligibly small components. Then, by setting $\overline{A}_i = \overline{B}_i = 0$, matrices $\overline{A}_k$ and $\overline{B}_k$ can be computed by using, instead of (3.77), the following backward recursion scheme

$$\overline{B}_k = B_k + \overline{B}_{k+1} G \, , \qquad (3.78)$$

$$\overline{A}_k = A_k + \overline{A}_{k+1} G \, . \qquad (3.79)$$

COMPUTING THE MOMENTS. Although it is possible to calculate the moments once the steady-state distribution $\mathbf{x}_i$ $(i \geq 0)$ has been obtained, due to the high computational cost of Ramaswami's method, another approach is preferable with which the first and second moments[1] can be directly obtained.

The PGF of the steady state probability vectors is

$$\mathbf{X}(z) = \sum_{i=1}^{\infty} \mathbf{x}_i \cdot z^i \, . \qquad (3.80)$$

Using formula (3.80) both the first and second moments can be obtained by computing the derivatives of $\mathbf{X}(z)$ in $z = 1$

$$\mathbf{X}'(1\text{-}) = L_1 \boldsymbol{\pi} + [\mathbf{U}'(1) + \mathbf{X}(1\text{-}) \cdot A^{*'}(1\text{-}) - \mathbf{X}(1\text{-})] Z \, , \qquad (3.81)$$

and

$$\mathbf{X}''(1\text{-}) = L_2 \boldsymbol{\pi} + [\mathbf{U}''(1) + \qquad (3.82)$$
$$2\mathbf{X}'(1\text{-}) \cdot A^{*'}(1\text{-}) + \mathbf{X}(1\text{-}) \cdot A^{*''}(1\text{-}) - 2\mathbf{X}'(1\text{-})] Z \, ,$$

---

1. In theory, moments above the second order could be computed. However, the formulae needed would quickly become unmanageable.

where

- $L_1 = [2(1-\rho)]^{-1} \cdot [2\theta_1 + \mathbf{U}''(1)\mathbf{e} + \mathbf{X}(1\text{-})]\,Z$
- $\theta_1 = [\mathbf{U}'(1) + \mathbf{X}(1\text{-}) \cdot A^{*'}(1\text{-}) - \mathbf{X}(1\text{-})]\,Z\beta$
- $L_2 = [3(1-\rho)]^{-1} \cdot [2\theta_2 + \mathbf{U}'''(1\text{-})\mathbf{e} + 3\mathbf{X}'(1\text{-})\mathbf{a}_2 + \mathbf{X}'(1\text{-})\mathbf{a}_3]\,Z$
- $\theta_2 = [\mathbf{U}''(1\text{-}) + 2\mathbf{X}'(1\text{-}) \cdot A^{*'}(1\text{-}) + \mathbf{X}(1\text{-}) \cdot A^{*''}(1\text{-}) - 2\mathbf{X}'(1\text{-})]\,Z\beta$
- $\mathbf{X}(1\text{-}) = (1 - \mathbf{x}_0\mathbf{e}) \cdot \boldsymbol{\pi} + \mathbf{U}(1\text{-})Z$
- $Z = (\mathbf{I} - A + \mathbf{e} \cdot \boldsymbol{\pi})^{-1}$ with $\boldsymbol{\pi}Z = \boldsymbol{\pi}$ and $Z\mathbf{e} = \mathbf{e}$
- $\mathbf{U}(1\text{-}) = \mathbf{x}_0 B(1\text{-}) - \mathbf{x}_1 A_0$ with $\mathbf{U}(1\text{-})\mathbf{e} = 0$
- $\mathbf{U}'(1\text{-}) = \mathbf{x}_0 B'(1\text{-}) + \mathbf{x}_0 B(1\text{-}) - \mathbf{x}_1 A_0$ with $\mathbf{U}'(1\text{-})\mathbf{e} = \mathbf{x}_0\mathbf{b}_1$
- $\mathbf{U}''(1\text{-}) = \mathbf{x}_0 B''(1\text{-}) + 2\mathbf{x}_0 B'(1\text{-})$ with $\mathbf{U}''(1\text{-})\mathbf{e} = \mathbf{x}_0\mathbf{b}_2 + 2\mathbf{x}_0\mathbf{b}_1$
- $\mathbf{U}'''(1\text{-})\mathbf{e} = \mathbf{x}_0\mathbf{b}_3 + 3\mathbf{x}_0\mathbf{b}_2$
- $A^{*}(1\text{-}) = A = \sum_{v=0}^{\infty} A_v$
- $A^{*(n)}(1\text{-}) = \sum_{v=1}^{\infty} v(v-1) \ldots (v-n+1)\,A_v$
- $\mathbf{a}_n = A^{*(n)}(1\text{-})\mathbf{e}$
- $B^{*(n)}(1\text{-}) = \sum_{k=1}^{\infty} k(k-1) \ldots (k-n+1)\,B_k$
- $\mathbf{b}_n = B^{*(n)}(1\text{-})\mathbf{e}$

This section has shown how to calculate both the moments and the distribution of a general *M/G/1-type* process. There are, however, special cases in which, by exploiting the particular structures of matrices G and P, these performance figures can be obtained more efficiently.

3.3.3 Special Cases

For some models of practical interest, further simplification can be made which reduce the complexity of the formulae and thereby the computational cost of the algorithm. The cases analyzed in this section are those in which:

(i)   matrix P has a special structure;
(ii) G is reducible.

In the first case it is assumed that $A_0 = C_0$, thus the system only behaves differently in level 0 states. The P matrix has the following structure

$$P = \begin{bmatrix} B_0 & B_1 & B_2 & B_3 & \ldots \\ A_0 & A_1 & A_2 & A_3 & \ldots \\ 0 & A_0 & A_1 & A_2 & \ldots \\ \ldots & 0 & A_0 & A_1 & \ldots \\ \ldots & \ldots & \ldots & \ldots & \ldots \end{bmatrix}, \qquad (3.83)$$

and the sub-matrices are all squared.

This leads to a simplification in the computation of matrix K and vector $\tilde{\mathbf{k}}_1$. Recalling (3.62) and (3.63), we have

$$K = B_0 + \sum_{v=1}^{\infty} B_v G^{v-1} \cdot \left[I - \sum_{v=1}^{\infty} A_v G^{v-1} \right]^{-1} A_0 = \tag{3.84}$$

$$B_0 + \left(\sum_{v=1}^{\infty} B_v G^{v-1} \right) G = \sum_{v=0}^{\infty} B_v G^v \quad ,$$

by virtue of the fact that matrices B_i now have the same size.

For the same reason $\tilde{\mathbf{k}}_1$ becomes

$$\tilde{\mathbf{k}}_1 = \mathbf{e} + \left[\sum_{v=1}^{\infty} B_v - \bar{B}_1 G \right] \mathbf{w} + (1-\rho)^{-1} \sum_{v=1}^{\infty} v B_v \mathbf{e} \tag{3.85}$$

$$= \mathbf{e} + \left[\sum_{v=0}^{\infty} B_v - K \right] \mathbf{w} + (1-\rho)^{-1} \sum_{v=1}^{\infty} v B_v \mathbf{e} \quad .$$

Another quite frequent particular structure of P is the one that describes a system with a behavior that is homogeneous for all the states, i.e. when matrices A_i are identical to B_i. In fact, in this case $K = G$, and $\tilde{\mathbf{k}}_1$ becomes

$$\tilde{\mathbf{k}}_1 = \mathbf{e} + \left[\sum_{v=0}^{\infty} B_v - K \right] \mathbf{w} + (1-\rho)^{-1} \sum_{v=1}^{\infty} v B_v \mathbf{e} = \tag{3.86}$$

$$\mathbf{e} + [A - G] \mathbf{w} + (1-\rho)^{-1} \beta \quad .$$

The formula for the moments cannot be simplified, but the computation of some of its elements is reduced since $A_i = B_i$.

So far the working hypotheses have been based on the irreducibility of matrix G. Below a particular case where G is reducible will be examined. This case occurs when A_0 has some empty columns.

As can readily be seen from (3.59), matrix G inherits all the empty columns from matrix A_0. Therefore, by permuting the phases in levels $i, i > 0$, the structure of A_0, and thus of G and the remaining A_i, may be expressed as

$$A_0 = \begin{bmatrix} A_0(1) & 0 \\ A_0(3) & 0 \end{bmatrix}, \quad G = \begin{bmatrix} G(1) & 0 \\ G(3) & 0 \end{bmatrix}, \quad A_i = \begin{bmatrix} A_i(1) & A_i(2) \\ A_i(3) & A_i(4) \end{bmatrix}, \quad i \geq 1 \quad .$$

The complexity of equation (3.59) is simplified because the two smaller matrices, namely $G(1)$ and $G(3)$, have to be computed

$$G(1) = \sum_{v=0}^{\infty} A_v(1) G^v(1) + \sum_{v=1}^{\infty} A_v(2) G(3) G^{v-1}(1) \quad , \tag{3.87}$$

$$G(3) = \sum_{v=0}^{\infty} A_v(3) G^v(1) + \sum_{v=1}^{\infty} A_v(4) G(3) G^{v-1}(1) \quad . \qquad (3.88)$$

If the square matrix $G(1)$ is irreducible, vector $\mathbf{g}_1 = \mathbf{g}_1 G(1)$, $\mathbf{g}_1 \mathbf{e} = 1$, can be computed. Then after routine computation [126], vector $\mathbf{w}$, defined in (3.69) for the computation of $\tilde{\mathbf{k}}_1$ is in this case

$$\mathbf{w} = \left[I - A + G^* - \sum_{v=1}^{\infty} v A_v G^* \right]^{-1} \mathbf{e} \quad ,$$

where

$$G^* = \begin{bmatrix} \mathbf{e}\mathbf{g}_1 & 0 \\ \mathbf{e}\mathbf{g}_1 & 0 \end{bmatrix} \quad .$$

After computing vector $\mathbf{w}$, the computation of the steady-state distribution or of the moments follows the procedure defined in the previous subsection, for the general case. Obviously, there are computational gains since the sub-matrices of G that are not empty are being worked on.

3.3.4 Case Study: Analysis of a Packet-switching Node

M/G/1-type models frequently occur when the system's behavior (service times and/or scheduling algorithms) depends on the past, and the phase is a useful mean of inserting into the state variable the past history that still affects the dynamics of the process. In this section an *M/G/1-type* model is used to study the performance figures related to the gateways in packet-switching networks, when the incoming traffic is described by correlated processes. Specifically, both the incoming links and the outgoing link are slotted. The resulting queueing model of the packet-switching node is depicted in Figure 3.9. The process which describes the arrivals at the gateway is made up of k independent sources each representing an input line, and incoming packets are buffered in an infinite buffer. Packets are of constant length and thus the service time is deterministic. Input lines are assumed to be synchronized both each other and with the output line. Throughout it is assumed that packets which arrive at the same time instant are randomly inserted into the queue.

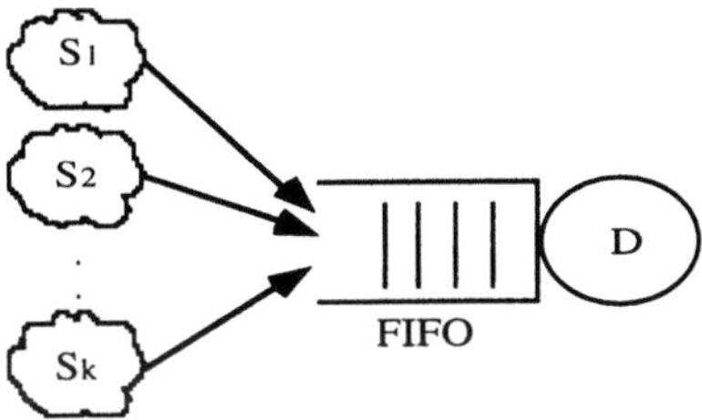

Figure 3.9: Gateway model

As the state of the system can only change at discrete time instants, its behavior can easily be described via a Markov chain derived by observing the system at the time instants $\{t_s\}_{s \in \mathbb{N}}$, $t_s > 0$; where t_s (see Figure 3.10) is the time instant immediately after the reception of the header of the s-th slot on each input line (i.e., at this time instant it is already known whether an incoming slot is occupied or not).[1]

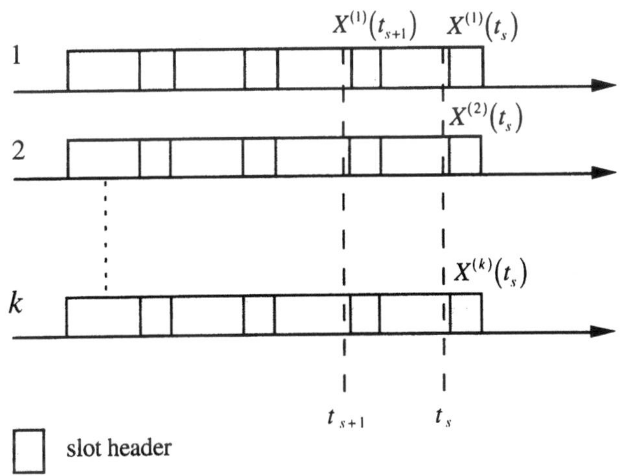

Figure 3.10: Relationship between embedding points and input lines

Traffic on an input line of a packet-switching network is typically highly correlated. In order to capture some of these dependencies, arrival processes are modeled with a homogeneous C-th order Markov chain. Specifically, with reference to the i-th source, the r.v. $X^{(i)}(t_v)$ indicates the state of the slot arriving at the gateway at time t_v (see Figure 3.10)

---

1. It is assumed that a packet should be completely received before its transmission on the output line can start.

$$X^{(i)}(t_v) = \begin{cases} 1 & \textit{busy slot} \\ 0 & \textit{empty slot} \end{cases},$$

and according to the above assumption $X^{(i)}(t_v)$ depends on the r.v. $X^{(i)}(t_{v-j})$, $j = 1, \ldots, C$ (i.e., the state of the last C slots generated by the i-th source). In addition, the Markov chains are homogeneous, i.e., the transition probabilities

$$P\{X^{(i)}(t_v)\,|\,X^{(i)}(t_{v-1}), \ldots, X^{(i)}(t_{v-c})\}$$

do not depend on the time instants t_v. Figure 3.11 shows generic transitions in which the source index i is omitted to simplify the notation. For the homogeneity hypothesis index t_{v-j} is reduced to j, i.e.,

$$P\{X^{(i)}(t_v)\,|\,X^{(i)}(t_{v-1}), \ldots, X^{(i)}(t_{v-c})\} = P\{X\,|\,X_1, \ldots, X_C\} \quad .$$

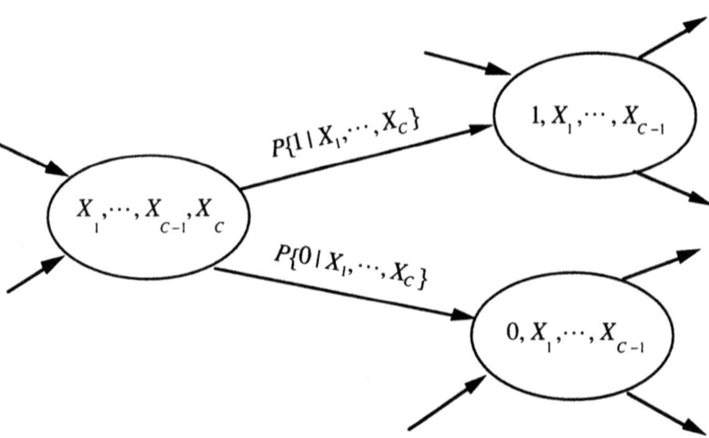

Figure 3.11:   Generic transitions of a C-th order Markov chain

The state space of the Markov chain which models the gateway at the embedding points is

$$E = \{(Q, \mathbf{SL}_1, \ldots, \mathbf{SL}_k) : Q \in \mathbb{N} \cup \{0\}, \mathbf{SL}_i \in \{0, 1\}^C\} \quad .$$

In fact, to satisfy the Markov property the state variable must include: both the number of packets in the queue (Q), and, for each source S_i, a vector

$\mathbf{SL}_i = (X_1^{(i)}, X_2^{(i)}, ..., X_C^{(i)})$ representing the state of the last C slots generated by that source.

By observing the state of the Markov chain at two consecutive embedding points it is possible to note that it behaves like an *M/G/1-type* Markov chain, in which level i corresponds to the states with $Q = i$. Whenever $Q > 0$, after a transition its value becomes $Q = Q - 1 + N_{Pack}$, where N_{Pack} is the number of arrivals: $N_{Pack} = \sum_{j=1}^{k} X_1^{(i)}$. The embedded Markov chain is also spatially homogeneous, since the number of arrivals at the gateway, given $\mathbf{SL}_1, ..., \mathbf{SL}_k$, does not depend on Q. Specifically, the Markov chain transition matrix has the following structure

$$P = \begin{bmatrix} A_0 & \cdots & A_k & 0 & \cdots \\ A_0 & \cdots & A_k & 0 & \cdots \\ 0 & A_0 & \cdots & A_k & 0\cdots \\ \vdots & \ddots & \ddots & & \ddots \end{bmatrix} .$$

To derive the transition matrix it is convenient to introduce the Λ matrix, which represents the evolution of all sources. By indicating with $S_i(C)$ the transition matrix (of size 2^C) which describes the i-th source, it is easy to observe that Λ is the Kronecker product of $S_i(C)$, i.e., $\Lambda = S_1(C) \otimes ... \otimes S_k(C)$, where, given matrices R and S, the Kronecker product is defined as

$$R \otimes S = \begin{bmatrix} r_{11}S & \cdots & r_{1n}S \\ \cdots & \cdots & \cdots \\ r_{\lfloor 1n \rfloor}S & \cdots & r_{nn}S \end{bmatrix} .$$

Finally, from Λ matrix, the elements A_j of the transition matrix P are obtained. To this end, let D_x $(0 \le x \le k)$, be diagonal matrices of size 2^{kC}, defined as

$$|D_x|_{ii} = \begin{cases} 1 & \text{if row } i \text{ in } \Lambda \text{ matrix represents a status for which } N_{Pack} = x \\ 0 & \text{otherwise} \end{cases}$$

then matrices A_x can be obtained by the following relationship $A_x = D_x\Lambda$.

3.4 APPLICATION OF NODE-IN-ISOLATION MODELS: WORST-CASE ANALYSIS

The modeling and performance analysis of MAC protocols with a *cyclic behavior* -- such as FDDI (see Chapter 5) CRMA (see Section 9.1.1) -- is known to be very difficult, as an exact model of the network is more complex than a polling system with an Exhaustive-limited (or Gated-limited) service discipline for which no exact solution is known ([147],[44]).

To overcome these difficulties, in the literature quite often a simplified model of the network (*worst-case model*), which can be analytically solved and yet still provides useful information, is defined [50]. Specifically, the worst-case model focuses on a specific node (*tagged node*), and assumes that the remaining nodes operate in asymptotic conditions (i.e., all the network nodes, except the tagged node, always have cells[1] ready for transmission). On the basis of these hypotheses the worst-case model is characterized by the following assumptions

- each station Q_i is fed with high- and low-priority traffic[2];
- in each station Q_i the queue of low-priority traffic is never empty. This means that in each cycle, each station Q_i, transmits l_i cells.

According to the previous hypotheses, the cycle length of the network is always equal to its maximum length (C_{max}).

The aim of the worst-case model analysis is to study the quality of service experienced by the high-priority traffic in the tagged station.

The worst-case model is described via an embedded Markov chain for which two solution techniques are normally used. The first is based on the z-transform technique and provides a closed formula for the PGF of the distribution of the number of cells in the system (see Section 3.2.1). The second is based on the theory developed for $M/G/1$-*type* Markov chains (see Section 3.3).

Below, first the accuracy of the worst-case model is discussed to understand under what conditions it provides results which are close to the real model. The worst-case model is then analyzed for two relevant service disci-

---

1. When packets are of constant length the word cell is often used to be consistent with ATM [58].

2. The high-priority traffic may represent real-time traffic (e.g., audio and video), while the low-priority traffic commonly represents classical data traffic (file-transfer, e-mail, remote login, etc.).

plines: E-limited and G-limited.

The concepts and results reported in the rest of this chapter are taken from [50].

3.4.1 Accuracy of the Worst-case Model Approach

The accuracy of the worst-case model is evaluated via simulation [50]. The performance indices obtained by solving the stochastic model (*real model*) of a network, which behaves like a polling system with an E-limited service discipline (e.g., FASNET [113]) are compared with the estimates obtained by solving the worst-case model of the network. Simulative results are obtained by assuming a network configuration with $K = 10$ nodes equally spaced. The network capacity is 150 Mbps, the cell size is 53 octets, and $l = 30$. The network nodes generate cells with the same rate and with exponential interarrival times.

Table 3.1 Worst-case performance figures for a generic node

	Real $E[N_q]$	Worst-case $E[N_q]$	Real $E[W]$	Worst-case $E[W]$
$OL = 0.90$	12.8	22.5	401	707
$OL = 0.95$	34.5	41.5	996	1225
$OL = 0.99$	58.2	62.7	1647	1782

Table 3.1 reports the estimates of the average waiting time expressed in milliseconds ($E[W]$) and the average queue length ($E[N_q]$) obtained from the real and worst-case models for a generic node in the network. As expected, the statistics obtained from the worst-case model always overestimate the performance figures of the network. However, as the offered load increases, the distance between the two models decreases. In fact, for $OL = 0.90$, the worst-case model overestimates the performance figures of the real model by about 40%, while for $OL = 0.99$ the distance between the two models is about 10%. Furthermore, the queue-length distribution in Figure 3.12 and in Figure 3.13 highlights that the worst-case model provides upper bounds for this distribution; and for $OL \geq 0.95$, these bounds closely approximate the tail of the queue length distribution. This implies that, as the QoS of the real time applications mainly depends on the tail of the distribu-

tion of the network performance figures, the worst-case model is suitable for studying the QoS provided by the network to the high-priority traffic.

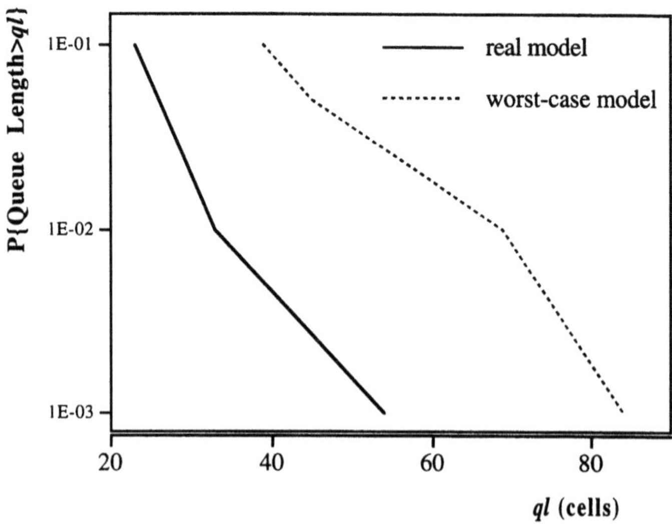

Figure 3.12: Worst-case model vs. real model (*OL*=0.90)

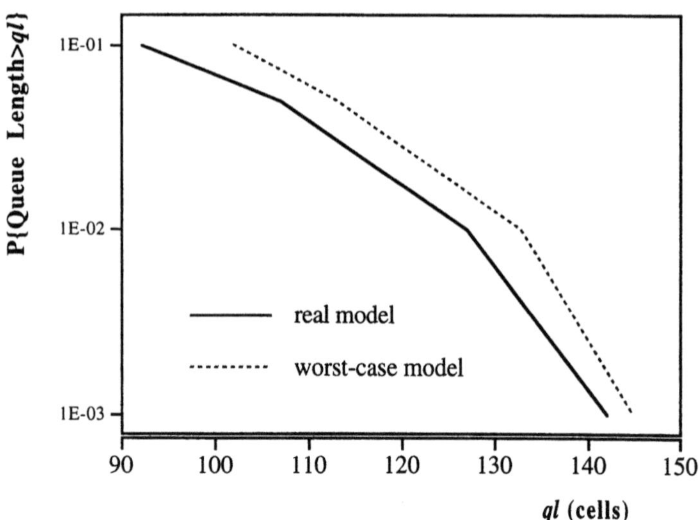

Figure 3.13: Worst-case model vs. real model (*OL*=0.95)

3.4.2 *E*-limited Service Discipline

According to the worst-case assumptions, the model which describes the number of high priority cells in the tagged node is a single server with vacation and an *E*-limited service discipline with parameter l. The distribution of the service time is deterministic and equal to a cell transmission time Δ. The vacation time is made up of two components: one is deterministic and is equal to $(C_{max} - l\Delta)$, while the other is a random variable and is equal to $(l - Sp_n)\,\Delta$, where Sp_n is the number of high-priority cells transmitted in the n-th cycle by the tagged station. $(C_{max} - l\Delta)$ represents the server vacation with respect to the tagged station, while $(l - Sp_n)$ represents the number of low priority cells transmitted by the tagged station. Hence, the cycle length is constant and equal to C_{max}. The input traffic to the queue is generated by the superposition of Bernoulli processes, and $A^{(i)}$ denotes the number of arrival at the i-th slot.

Following the approach presented in Section 3.2.1, the state of the system is analyzed with the embedding points technique. Two sets of embedding points are used: the vacation termination instants and the high-priority traffic service completion instants. A vacation begins either if the high priority queue becomes empty after the k-th transmission or after the l-th service completion instant, whichever occurs first. In the first case (see part (a) in Figure 3.14) the number of (high-priority cell) arrivals during a vacation will be denoted by $F^{(i+1)} + V$, where V is the number of arrivals in $(C_{max} - l\Delta)$ and $F^{(i+1)} = A^{(i+1)} + A^{(i+2)} + \dots + A^{(l)}$ is the number of arrivals while the low priority cells are being transmitted by the tagged station. In the second case (see part (b) in Figure 3.14) no low priority traffic is transmitted by the tagged station and thus the number of arrivals during a vacation is V.

The state of the system at the embedding points is described by a couple $\{\xi, N\}_n$. $\xi = j$ indicates the embedding point just after the j-th $(j = 1, \dots, l)$ transmission during the n-th cycle, and $\xi = 0$ indicates the vacation termination instant. The r.v. N is the number of cells in the tagged station at an embedding point.

To ensure system stability the average number of arrivals in a cycle should be less than l. Hereafter the stability condition is assumed to hold, and thus the following steady-state joint probabilities exist

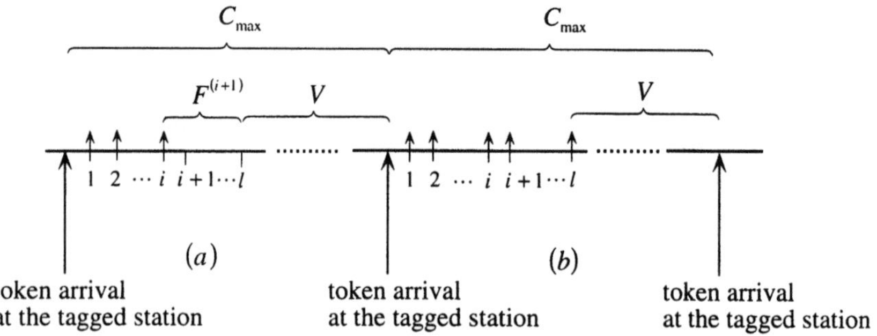

Figure 3.14: Cell arrivals

$$\pi_k^{(m)} \; = \; \lim_{n \to \infty} P\{\xi = m, N = k\}_n, \quad m = 0, 1, \ldots, l, \quad k = 0, 1, \ldots \; . \quad (3.89)$$

The steady state probability that an embedded point is the vacation termination instant and that there are k cells in the queue is

$$\pi_k^{(0)} \; = \; \sum_{h=0}^{l-1} \pi_0^{(h)} \langle F^{(h+1)} + V \rangle_k + \sum_{m=0}^{k} \pi_m^{(l)} \langle V \rangle_{k-m} \; , \qquad (3.90)$$

where the notation $\langle X_1 + X_2 + \ldots + X_n \rangle_j$ indicates the probability that the sum of the random variables in brackets is equal to j.

The steady-state probability that an embedded point is the h-th service completion instant and that there are k cells in the system is

$$\pi_k^{(h)} \; = \; \sum_{m=1}^{k+1} \pi_m^{(h-1)} \langle A^{(h)} \rangle_{k-m+1} \; . \qquad (3.91)$$

To compute the steady-state probabilities of the embedded Markov chain two techniques will be applied: z-transform and matrix analytic.

Z-TRANSFORM. This section focuses on the z-transform solution technique. First, a closed formula for the PGF of the number of (high-priority) cells queued in the tagged station is derived. Some performance figures are then estimated from the PGF.

By defining

$$\Pi_m(z) = \sum_{k=0}^{\infty} \pi_k^{(m)} z^k, \quad m = 0, 1, ..., l \quad ,$$ (3.92)

after routine algebraic manipulations the following relations are obtained

$$\Pi_0(z) = \frac{V(z)}{z^l - C(z)} \cdot \sum_{k=0}^{l-1} [z^l - z^k] \, \pi_0^{(k)} \, F^{(k+1)}(z) \quad ,$$ (3.93)

$$\Pi_m(z) = \frac{\Pi_0(z)}{z^m} \cdot \prod_{j=1}^{m} A^{(j)}(z) - \sum_{k=0}^{m-1} \pi_0^{(k)} \prod_{j=1}^{m} \left[\frac{A^{(j)}(z)}{z} \right], \quad m = 1, 2, ..., l \quad ,$$ (3.94)

where $A^{(j)}(z)$, $F^{(k+1)}(z)$, and $V(z)$ are the PGFs of the r.v. $A^{(j)}$, $F^{(k+1)}$, and V, respectively. Furthermore, $C(z) = F^{(1)}(z)V(z)$ is the PGF of the number of cell arrivals in a cycle.

Relations (3.93) and (3.94) completely define $\Pi_m(z)$ $(m = 0, 1, ..., l)$ in terms of l boundary probabilities $\{\pi_0^{(k)}, k = 0, 1, ..., l\}$. To determine these boundary probabilities the approach outlined in Section 3.2.1 is used. For further details see [50].

Table 3.2 Buffer size statistics

Time	Average	Standard Deviation
0	25.6	5.34
5	20.6	5.34
10	15.6	5.34
15	10.74	5.2
20	6.74	4.6
25	4.27	3.76
30	3.21	3.21

The PGF of the distribution of the number of cells in the tagged station at an embedding point is $\Pi_m(z)/\Pi_m(1)$,[1] from which all the moments can be

---

1. $\Pi_m(1)$ is the probability that the embedded Markov chain is at the m-th embedding point. Hence, $\Pi_m(z)/\Pi_m(1)$ is the PGF of the number of packets in the queue given that $\{\xi = m\}$.

obtained. For example, Table 3.2 reports the average and standard deviation of the queue length at the embedding points for a network with $C_{max} = 300$, $\Delta = 2.82\mu sec$, $l = 30$ and the number of arrivals in the i-th slot sampled from a binomial distribution

$$A^{(i)}(z) = \begin{cases} [1 - p + zp]^{10} & 25 < i \le 275 \\ 1 & otherwise \end{cases}$$

where $p = 0.01$.

The distribution of the buffer occupancy is now estimated using the so called *two moments approximation* [103], [161]. This approximation is obtained by choosing a distribution with a known form whose first and second moments match those computed analytically. Since Table 3.2 clearly shows that the squared coefficient of variation is less than one and that the distribution is a discrete time distribution, the negative binomial is selected. To measure the effectiveness of the proposed approximations, the worst-case model was also solved via simulation. Figure 3.15 shows that the binomial approximation gives estimates which are conservative but very close to real figures.

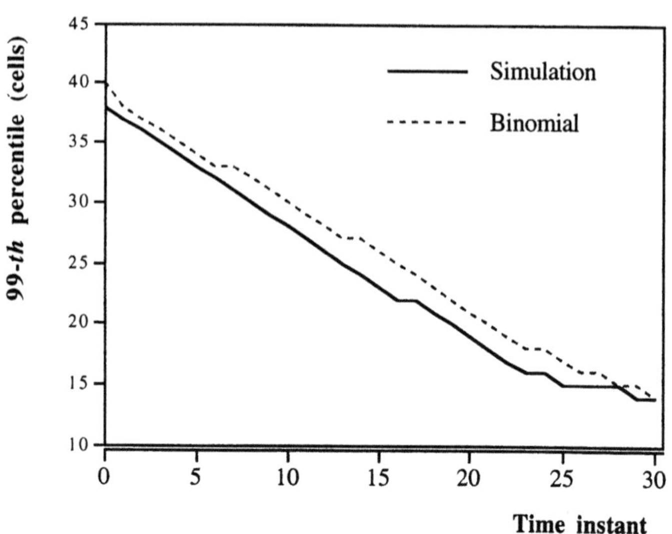

Figure 3.15: 99-th percentile of the queue length estimated via simulation and with the negative binomial approximation

MATRIX ANALYTIC. As was shown in the previous section, the queue length distribution cannot be accurately derived with the PGF approach. In this section it is shown how to apply the matrix-analytic method to compute this distribution. In fact, the embedded Markov chain of the worst case model is of $M/G/1$-*type* since

- in a transition the r.v. N decreased at most by one;
- with the exception of the boundary level (i.e., states with $\{\xi = \cdot, N = 0\}$) the transition mechanism is spatially homogeneous, i.e., the transition probabilities from $N = i$ to $N = i + k$ are the same for all $i \geq 1$ and $k \geq -1$.

The transition matrix has an $M/G/1$-*type* structure

$$
P = \begin{array}{c} N = 0 \\ N = 1 \\ N = 2 \\ \cdots \\ \cdots \end{array}
\begin{bmatrix}
B_0 & B_1 & B_2 & B_3 & \cdots \\
A_0 & A_1 & A_2 & A_3 & \cdots \\
0 & A_0 & A_1 & A_2 & \cdots \\
\cdots & \cdots & & A_0 & A_1 & \cdots \\
\cdots & \cdots & \cdots & \cdots & \cdots
\end{bmatrix} , \tag{3.95}
$$

where, B_k and A_k $(k \geq 0)$ are $(l+1) \times (l+1)$ matrices, with the following structure

$$
B_k = \begin{array}{c} \xi = 0 \\ \xi = 1 \\ \cdots \\ \xi = l-1 \\ \xi = l \end{array}
\begin{bmatrix}
\langle F^{(1)} + V \rangle_k & 0 & \cdots & 0 & 0 \\
\langle F^{(2)} + V \rangle_k & 0 & \cdots & 0 & 0 \\
\cdots & \cdots & \cdots & \cdots & \cdots \\
\langle F^{(l)} + V \rangle_k & 0 & \cdots & 0 & 0 \\
\langle V \rangle_k & 0 & \cdots & 0 & 0
\end{bmatrix} ,
$$

$$
A_k = \begin{array}{c} \xi = 0 \\ \xi = 1 \\ \cdots \\ \xi = l-1 \\ \xi = l \end{array}
\begin{bmatrix}
0 & a_k^{(1)} & 0 & \cdots & 0 \\
0 & 0 & a_k^{(2)} & \cdots & 0 \\
\cdots & \cdots & \cdots & \cdots & 0 \\
0 & 0 & 0 & \cdots & a_k^{(l)} \\
v_k & 0 & 0 & \cdots & 0
\end{bmatrix} ,
$$

Figure 3.16 reports the tail of buffer-occupancy distribution for the network configuration previously analyzed with the z-transform technique. As expected, values related to the 99-th percentile (i.e., the curve tagged with

$1 - 10^{-2}$) match simulative results (see Figure 3.15). Furthermore, the matrix analytic approach can be used to estimate very low tail probabilities (up to $1 - 10^{-8}$ in this example).

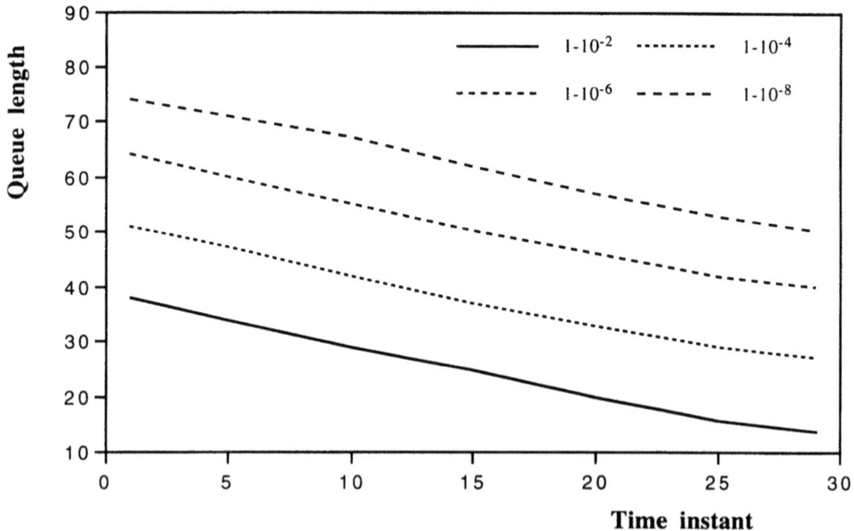

Figure 3.16: Percentiles of the tagged station buffer size estimated via matrix analytic techniques

3.4.3 *G*-limited Service Discipline

Protocols for metropolitan area networks may require that a station reserves in advance the bandwidth it needs for packet transmission (e.g., Cyclic Reservation Multiple Access [122]). In this case the node-in-isolation model has a *G*-limited service discipline and the resulting queueing model is shown in Figure 3.17. When the server arrives at the tagged station, up to l packets queued in the station are authorized for transmission[1]. After these packets are transmitted the server goes on vacation again.

In the worst-case model for this queueing system the cycle length is deterministic and is equal to its maximum value (C_{max}) . Furthermore, since most of the metropolitan area networks are slotted, packets are assumed to be of constant length.

---

1. In the literature, this operation is referred to by saying that "gate S1 opens".

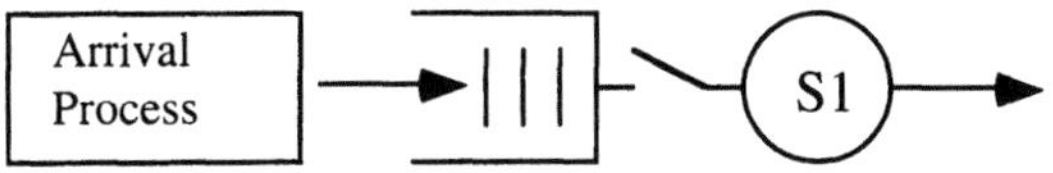

Figure 3.17: *G*-limited queueing model

The arrival process is assumed to be modulated by a Markov chain, in which transitions occur when the server arrives at the queue. As a result of the *G*-limited service policy, without any loss of generality, it is assumed that all the packets generated during a cycle are queued just before the gate opens. Depending on the state of the Markov chain, the number of cells generated in a cycle is sampled from a general distribution associated with that state.

In the following examples, this type of arrival process is used to characterize a variable bit rate (*VBR*) video traffic as proposed by Sengupta and Ramamurthy [130]. According to their proposal, the arrival process is assumed to be modulated by the three-state Markov chain $\{Ph\}_n$ ($Ph = 0, 1, 2$), whose transition graph is shown in Figure 3.18. The number of packets which arrive during a cycle in which the Markov chain is in state $\{Ph = x\}_n$ will be denoted by $A(Ph = x)$.

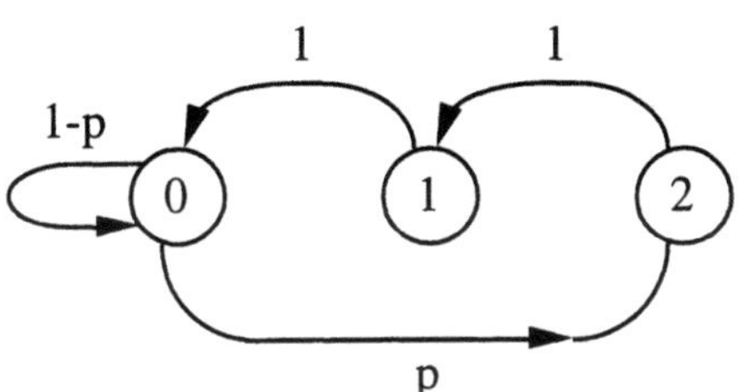

Figure 3.18: Modulating process

Most of the time the Markov chain is in state 0, which represents the codec output between two scene changes. A transition (with probability *p*) to state 2 models a scene change. Finally, state 1 is needed to model the effects of the scene change in the next picture.

This type of worst-case model was used in [7] to characterize the quality of service of a VBR video application in a CRMA network. The resulting stochastic model of the system (Figure 3.17) is a single server with vacation and *G*-limited service discipline with parameter *l* [149]. The distribution of the cycle length is deterministic and equal to C_{max}. This system can be analyzed by studying a Markov chain embedded immediately after the arrival of the server at the tagged station. The state of the system at the embedding points is defined by the number of packets in queue (K), and the state Ph of the Markov chain that modulates the arrivals.

$$\{K = x, Ph = \varphi\}_n \rightarrow \{K = (x - l)^+ + A(\varphi), Ph = \psi\}_{n+1} \qquad (3.96)$$

Equation (3.96) shows the relationship between the state of the system at the n-th and at the $(n + 1)$th embedding points. As stated before, the state of the Markov chain that modulates the arrival process determines the number of packets queued just before the next embedding point.

The embedded Markov chain $\{K, Ph\}_n$ is clearly irreducible and aperiodic. In addition, to guarantee the positive recurrence of the embedded Markov chain, the average number of cells arrival in a cycle (α) must be strictly less than l, i.e., $\alpha < l$, where

$$\alpha = E[A(Ph = 0)] \frac{1}{1 + 2p} + (E[A(Ph = 1)] + E[A(Ph = 2)]) \frac{p}{1 + 2p} .$$

Below, it is assumed that this condition holds. The following steady-state joint probabilities thus exist

$$\pi_k^{(i)} = P\{K = k, Ph = i\}, \qquad i = 0, 1, 2; \ k \geq 0 ,$$

and satisfy the following equations

$$\pi_k^{(0)} = \left[\sum_{j=0}^{l-1} \pi_j^{(0)} a_k^{(0)} + \sum_{j=l}^{k+l} \pi_j^{(0)} a_{k-j+l}^{(0)}\right] \cdot (1 - p) +$$
$$\left[\sum_{j=0}^{l-1} \pi_j^{(1)} a_k^{(1)} + \sum_{j=l}^{k+l} \pi_j^{(1)} a_{k-j+l}^{(1)}\right] ,$$

$$\pi_k^{(1)} = \left[\sum_{j=0}^{l-1} \pi_j^{(2)} a_k^{(2)} + \sum_{j=l}^{k+l} \pi_j^{(2)} a_{k-j+l}^{(2)}\right] , \qquad (3.97)$$

and

$$\pi_k^{(2)} = \left[\sum_{j=0}^{l-1} \pi_j^{(0)} a_k^{(0)} + \sum_{j=l}^{k+l} \pi_j^{(0)} a_{k-j+l}^{(0)}\right] \cdot p ,$$

where $a_k^{(i)}$ $(i = 0, 1, 2; \ k \geq 0)$ is the probability that k packets are generated in a cycle when the process that modulates the arrivals is in state $Ph = i$.

Below it is shown how performance figures of interest can be derived from equations (3.97) by exploiting the z-transform and matrix-analytic techniques.

Z-TRANSFORM. By defining $\Pi_i(z) = \sum_{k=0}^{\infty} \pi_k^{(i)} z^k$ $(i = 0, 1, 2)$, equations (3.97), after routine algebraic manipulations, provide a linear system of three equations among the unknowns $\Pi_i(z)$ $(i = 0, 1, 2)$. By solving this system, expressions for $\Pi_i(z)$ $(i = 0, 1, 2)$ are derived, which contain as unknowns the boundary probabilities $\pi_k^{(0)}$, $\pi_k^{(1)}$ and $\pi_k^{(2)}$ with $0 \le k < l$. These unknowns are derived by exploiting the analyticity of $\Pi_i(z)$ $(i = 0, 1, 2)$. For example, to derive the $3l$ boundary probabilities, the zeros in the denominator of

$$\Pi_0(z) = \frac{\psi(\pi_0^{(0)}, \pi_1^{(0)}, \ldots \pi_{l-1}^{(0)}, \pi_0^{(1)}, \pi_1^{(1)}, \ldots \pi_{l-1}^{(1)}, \pi_0^{(2)}, \pi_1^{(2)}, \ldots \pi_{l-1}^{(2)}, z)}{z^{3l} - [\,(1-p)\, z^{2l} A^{(0)}(z) + pA^{(0)}(z)A^{(1)}(z)A^{(2)}(z)\,]} \tag{3.98}$$

are used following to the approach outlined in Section 3.2.1. The explicit formula of $\psi(\cdot)$, in (3.98) and details of its derivation can be found in [50].

As was shown in the analysis of E-limited systems, all the moments of the buffer occupancy at the embedding points can be obtained from the PGFs. In addition, approximations of the distribution can also be derived (see Section 4.1.3 in [26]). Below it is shown how to derive the distribution by using the matrix-analytic technique.

MATRIX ANALYTIC. The Markov chain $\{K, Ph\}_n$ is clearly not of $M/G/1$-type with K as a level and Ph as a phase, since in a transition the number of cells K may decrease of more than one.

$$P = \begin{matrix} L = 0 \\ L = 1 \\ L = 2 \\ \cdots \\ \cdots \end{matrix} \begin{bmatrix} B_0 & B_1 & B_2 & B_3 & \cdots \\ A_0 & A_1 & A_2 & A_3 & \cdots \\ 0 & A_0 & A_1 & A_2 & \cdots \\ & \cdots & \cdots & A_0 & A_1 & \cdots \\ \cdots & \cdots & \cdots & \cdots & \cdots \end{bmatrix}$$

Figure 3.19: Transition probability matrix

To overcome this problem, each state $\{K = k, Ph = j\}_n$ of the embedded Markov chain is coded with the triple

$$\{L = \lfloor k/l \rfloor, J_1 = (k - \lfloor k/l \rfloor l), J_2 = j\} \quad ,$$

where $L \times l + J_1$ stands for the number of packets in the system, and J_2 is the state of the Markov chain which modulates the arrivals, it can easy be seen that the chain is of *M/G/1-type*. The transition probability matrix P of the Markov chain has the form shown in Figure 3.19, where L stands for the number of blocks of l packets in the system.

The square matrices A_k and B_k have a size of $3l$. Specifically, B_k is partitioned into 3×3 square blocks of size l

$$
B_k = \begin{array}{c} J_2 = 0 \\ J_2 = 1 \\ J_2 = 2 \end{array}
\begin{bmatrix}
(1-p) \cdot \tilde{B}_k^{(0)} & 0 & p \cdot \tilde{B}_k^{(0)} \\
\tilde{B}_k^{(1)} & 0 & 0 \\
0 & \tilde{B}_k^{(2)} & 0
\end{bmatrix} ,
$$

where each block has the following special structure

$$
\tilde{B}_k^{(x)} = \begin{array}{c} (J_1 = 0) \\ (J_1 = 1) \\ (J_1 = 2) \\ \cdots \\ (J_1 = l-1) \end{array}
\begin{bmatrix}
a_{kl}^{(x)} & a_{kl+1}^{(x)} & a_{kl+2}^{(x)} & \cdots & a_{kl+(l-1)}^{(x)} \\
a_{kl}^{(x)} & a_{kl+1}^{(x)} & a_{kl+2}^{(x)} & \cdots & a_{kl+(l-1)}^{(x)} \\
a_{kl}^{(x)} & a_{kl+1}^{(x)} & a_{kl+2}^{(x)} & \cdots & a_{kl+(l-1)}^{(x)} \\
\cdots & \cdots & \cdots & \cdots & \cdots \\
a_{kl}^{(x)} & a_{kl+1}^{(x)} & a_{kl+2}^{(x)} & \cdots & a_{kl+(l-1)}^{(x)}
\end{bmatrix} .
$$

In fact, when the embedded Markov chain is in a boundary state (i.e., $L = 0$) the packets in the queue are all transmitted when the gate opens. The number of packets in the queue at the next embedding points is thus equal to the number of arrivals in a cycle, which only depends on the state $(J_2 = x)$ of the chain which modulates the arrivals.

Similarly to B_k, also A_k is partitioned into 3×3 l size square blocks

$$
A_k = \begin{array}{c} J_2 = 0 \\ J_2 = 1 \\ J_2 = 2 \end{array}
\begin{bmatrix}
(1-p) \cdot \tilde{A}_k^{(0)} & 0 & p \cdot \tilde{A}_k^{(0)} \\
\tilde{A}_k^{(1)} & 0 & 0 \\
0 & \tilde{A}_k^{(2)} & 0
\end{bmatrix} ,
$$

where each block has the following structure

$$\tilde{A}_k^{(x)} = \begin{matrix} (J_1 = 0) \\ (J_1 = 1) \\ (J_1 = 2) \\ \cdots \\ (J_1 = l-1) \end{matrix} \begin{bmatrix} a_{kl}^{(x)} & a_{kl+1}^{(x)} & a_{kl+2}^{(x)} & \cdots & a_{kl+(l-1)}^{(x)} \\ a_{kl-1}^{(x)} & a_{kl}^{(x)} & a_{kl+1}^{(x)} & \cdots & a_{kl+(l-2)}^{(x)} \\ a_{kl-2}^{(x)} & a_{kl-1}^{(x)} & a_{kl}^{(x)} & \cdots & a_{kl+(l-3)}^{(x)} \\ \cdots & \cdots & \cdots & \cdots & \cdots \\ a_{kl-(l-1)}^{(x)} & a_{kl-(l-2)}^{(x)} & a_{kl-(l-3)}^{(x)} & \cdots & a_{kl}^{(x)} \end{bmatrix} , \quad (3.99)$$

in which $a_j^{(x)} = 0$, if $j < 0$.

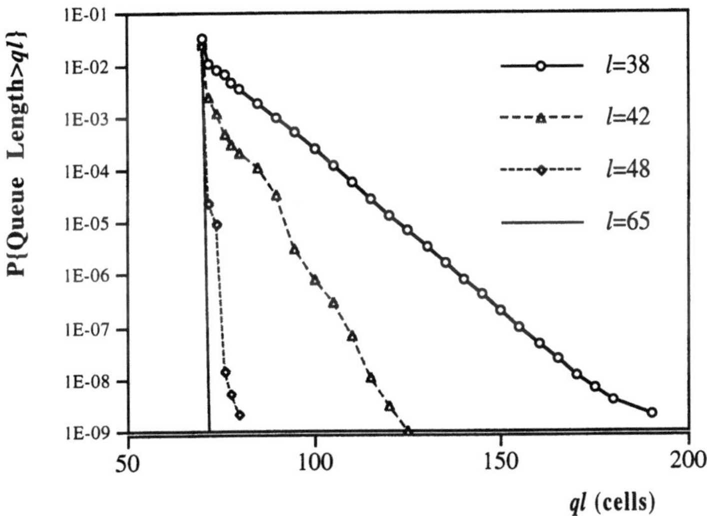

Figure 3.20: Queue-length tail distribution for several values of l

The structure (3.99) is derived by taking into consideration that, whenever the embedded Markov chain is in a state with $L > 0$, l packets are removed from the queue when the gate opens. The number of packets in the queue at the next embedding points is therefore equal to $(L \cdot l + J_1) - l$ plus the number of arrivals in a cycle, which depends on the state $(J_2 = x)$.

By exploiting the *M/G/1-type* structure of the Markov chains its steady-state probabilities can be calculated by applying the solution methodology presented in Section 3.3, from which the tail of the queue length distribution can be computed. Specifically, as shown in Figure 3.20, the model can be used to investigate the relationship between the number of cells (l)

that the tagged station can transmit in a cycle and the distribution of the buffer occupancy. The results show that the worst-case model is appropriate for estimating some bounds on the quality of service experienced by real-time applications such as VBR video. These applications can tolerate only a very low packet loss rate (e.g., in the order of 10^{-8}) which cannot be estimated via a simulative approach due to the computational cost.

4 Methods for the Analysis of Network-wide Models: Polling Models

Polling models are used throughout this book to represent the behavior of MAC protocols with a cyclic behavior, such as Token Ring (see Chapter 1) and FDDI (see Chapter 5). To guarantee fair access to all the network stations and to limit access delay, MAC protocols with a cyclic behavior are based on limited service disciplines. For this reason hereafter only polling systems with an l-limited (gated or exhaustive) service discipline are taken into consideration. For a complete overview of the analysis of polling systems with gated or exhaustive service disciplines see Takagi ([145], [146], [147], [151]).

Analyzing a polling system with an l-limited service discipline is very difficult, if not impossible. Exact results have been obtained only for polling systems with 1-limited service discipline[1] operating under special configurations, e.g. two active nodes, symmetric systems, single buffer queue. Specifically, for symmetric non-exhaustive polling systems an exact analysis of the average waiting time was performed independently by Furhmann [66] and Takagi [144]. An exact analysis for a polling system with two active nodes is presented in Boxma [22] for a symmetric system, and is extended in Boxma and Groenendijk [25] to non-symmetric systems. In the general case no exact analytical solution exists, however, by exploiting pseudo-conservation laws some approximate results have been derived. Nevertheless, even in the special cases where exact results do exist, simple approximations based on pseudo-conservation laws are very much preferred, since calculating these exact results often requires considerable computing effort. These waiting

---

1. In the literature, this discipline is also referred to as *non-exhaustive*.

time approximations are based on several lemmas which are quite lengthy and tedious to prove. For this reason, in Section 4.1 lemmas are presented without proof. The interested reader can find the omitted proofs in Section 4.3.

In addition to the approximations based on the pseudo-conservation law, numerical algorithms have been proposed to perform the waiting time analysis of polling systems operating with a limited service discipline. In [153] Tran-Gia and Raith present an iterative methodology for an approximate analysis of 1-limited polling systems in which each queue has a finite buffer (see Section 4.2.1). On the other hand, the Power Series Analysis algorithm ([19], [20]), developed by Blanc, provides an exact numerical solution of those polling models which can be described by means of a stochastic process with a quasi-birth-death (QBD) structure (see Section 4.2.2).

4.1 FROM PSEUDO-CONSERVATION LAWS TO WAITING TIME ANALYSIS

In this section pseudo-conservation laws are applied to derive, for each station Q_i, the average waiting time experienced by a tagged packet which arrives at Q_i at a random point in time. First the methods for the analysis of 1-limited polling systems are presented, and then their extension to the analysis of l-limited polling systems is described.

The analysis of polling systems with limited service discipline is very complex mainly due to the interdependence among the number of packets stored in the queues. To make the study analytically tractable, approximate analyses presented in the literature generally focus on a tagged queue (or station) and represent the dependence between the tagged station and the other stations via a simplified cycle length process. Once this process has been derived, the tagged queue is analyzed in isolation (see Chapter 3).

For ease of reading the following notations are introduced

- $C_{i,n}$ is the random variable associated with the duration of the n-th cycle observed by Q_i, i.e., the time interval between the $(n-1)$th and n-th arrival of the server at Q_i;
- $Sp_{i,n}$ is the length of service period at Q_i during the n-th cycle;
- $I_{i,n}$ is the length of intervisit period at Q_i during the n-th cycle.

The sequences $\{C_{i,n}\}$, $\{Sp_{i,n}\}$, $\{I_{i,n}\}$ are assumed to be stationary; and C_i, Sp_i, I_i denote the generic random variable for $\{C_{i,n}\}$, $\{Sp_{i,n}\}$, and $\{I_{i,n}\}$, respectively. In addition, let X be one of the above random variables, X_+, (X_-) will denote the residual (elapsed) time of X (see Section 2.3).

By exploiting the notation so far introduced, the tagged packet arriving at Q_i during the n_0-th cycle experiences the following waiting time (see Figure 4.1)

$$W_i = C_{i, n_0+} + \sum_j C_{i,\,(n_0+j)} + \sum_m B_i \;. \tag{4.1}$$

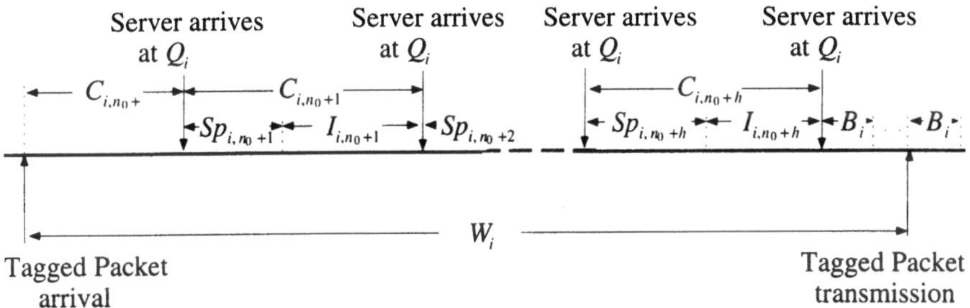

Figure 4.1: Waiting time at Q_i with a l-limited service discipline

Specifically, W_i includes

- the *residual cycle* from the arrival of the tagged packet until the server returns to Q_i (C_{i,n_0+}) ;
- h cycles each containing l_i packet transmissions at Q_i; and
- m $(m < l_i)$ packets transmission at Q_i before the tagged packet is transmitted.

To simplify the notation, the cycle in which the tagged packet arrives will hereafter be denoted as 0-th cycle.

The cycle length analysis is first carried out for the 1-limited polling system, and then is extended to the l-limited polling system.

4.1.1 Cycle Length Analysis: 1-limited Polling System

The distribution function of the r.v. C that models the cycle length is unknown except for its mean $E[C]$ (see Section 2.1.3). Furthermore, it can be shown that while the average cycle length does not depend on the node index (i.e., $E[C_j] = E[C_i]$, $\forall i, j$) the higher moments of the cycle length depend on the node index [23].

A simple approximation of the distribution of C can be obtained by assuming that the stations in a polling system behave independently. In this case, by denoting with $C^*(s)$, $S^*(s)$, and $Sp_i^*(s)$ the LSTs of the cycle length, total switch-over time, and service period observed by Q_i, respectively, the following relations can be derived

$$Sp_i^*(s) = [\alpha_i B_i^*(s) + 1 - \alpha_i] \quad , \tag{4.2}$$

and

$$C^*(s) = S^*(s) \cdot \prod_{i=1}^{N} Sp_i^*(s) \quad , \tag{4.3}$$

where $\alpha_i = \lambda_i E[C_i]$ is the probability that Q_i transmits a packet during a cycle (see (2.96)).

It can be observed that the independence assumption which leads to (4.3) provides an exact expression for $E[C]$ but, as shown via simulative results [102], it underestimates the variance of the cycle length. The underestimation of the cycle length variance is expected since a positive correlation exists between the number of transmissions the stations perform in a cycle. For example, if Q_i transmits a packet in a cycle, the probability that Q_{i+1} does not have a packet ready for transmission in that cycle decreases with respect to the case in which Q_i does not transmit.[1] To capture this type of dependency, in [102], the concept of *conditional cycle* was introduced. Given a tagged station Q_i, the cycle length is conditioned on the number of transmissions performed by Q_i in that cycle. By denoting with $C_{i|0}$ and $C_{i|1}$ the random variables which represent the cycle length conditioned on the occurrence of 0 and 1 Q_i transmission (in the cycle), respectively,

$$E[C] = E[C_{i|0}]\left(1 - \frac{\lambda_i s}{1-\rho}\right) + E[C_{i|1}]\frac{\lambda_i s}{1-\rho} \quad , \tag{4.4}$$

---

1. During the Q_i transmission, packets may arrive at Q_{i+1} and the occurrence of this event decreases the probability of an empty queue at Q_{i+1} during this cycle.

where, as shown in Section 2.6.3, $\lambda_i s/(1-\rho)$ is the steady-state probability that Q_i transmits in a cycle.

To study the statistics of $C_{i|0}$ and $C_{i|1}$ two auxiliary polling systems are introduced.

$C_{i|0}$ ANALYSIS. To perform this computation the polling system is modified by assuming that traffic never arrives at Q_i. Then the average cycle length of the first auxiliary polling system satisfies the following equation

$$E[C_{i|0}] = s + \sum_{j \neq i} \alpha_{ji|0} b_j \quad , \tag{4.5}$$

where $\alpha_{ji|0} = min\{\lambda_j E[C_{i|0}], 1\}$ is the probability that Q_j $(j \neq i)$ transmits in a cycle given that Q_i does not transmit in the cycle.

The stability condition $\lambda_j E[C] < 1$ (see Section 2.1.3) guarantees that $\forall j \; \alpha_{ji|0} < 1$, and hence (4.5) can be rewritten as

$$E[C_{i|0}] = \frac{s}{1-\rho+\rho_i} \quad . \tag{4.6}$$

Finally, by using an independence assumption, the $C_{i|0}$ LST can be derived

$$C_{i|0}^*(s) = S^*(s) \prod_{j \neq i} [\alpha_{ji|0} \cdot B_j^*(s) + 1 - \alpha_{ji|0}] \quad . \tag{4.7}$$

$C_{i|1}$ ANALYSIS. To compute the $C_{i|1}$ statistics it is assumed that Q_i is never empty. In this case the average cycle length satisfies the following equation

$$E[C_{i|1}] = s + b_i + \sum_{j \neq i} min\{\lambda_j E[C_{i|1}], 1\} b_j \quad , \tag{4.8}$$

where the probability that Q_j $(j \neq i)$ transmits in that cycle is approximated by

$$\alpha_{ji|1} = min\{\lambda_j E[C_{i|1}], 1\} \quad . \tag{4.9}$$

If Q_i is never empty, it may happen that some queue is not stable in the modified polling system (i.e., $\exists j: \lambda_j E[C_{i|1}] \geq 1$), even if the original polling system satisfies the stability condition (i.e., $\lambda_j E[C] < 1$). Thus, two different situations may occur

(i) $\forall j, \lambda_j E[C_{i|1}] < 1$, then

$$E[C_{i|1}] = \frac{s + b_i}{1 - \rho + \rho_i} \quad ; \tag{4.10}$$

(ii) $\exists j, \lambda_j E[C_{i|1}] \geq 1$, then $E[C_{i|1}]$ is computed according to an iterative scheme based on (4.8). Specifically, by denoting with $E[C_{i|1}(n)]$ the average length of the cycle computed at the end of the n-th iteration, the following iterative scheme is used to compute the next value

$$E[C_{i|1}(n+1)] = s + b_i + \sum_{j \neq i} \min\{\lambda_j E[C_{i|1}(n)], 1\} b_j \quad . \tag{4.11}$$

Finally, by using an independence assumption, the $C_{i|1}$ LST is derived

$$C_{i|1}^*(s) = S^*(s) B_i^*(s) \prod_{j \neq i} [\alpha_{ji|1} \cdot B_j^*(s) + 1 - \alpha_{ji|1}] \quad . \tag{4.12}$$

The proof of the convergence of the above iterative scheme (4.11) is based on the following lemma which is proved in Section 4.3.

LEMMA 4.1   *For all starting values $E[C_{i|1}(0)] > 0$ it follows that*

$$\forall n, E[C_{i|1}(n)] > 0 \quad , \tag{4.13}$$

and

$$\left| E[C_{i|1}(n+1)] - E[C_{i|1}(n)] \right| < \left| E[C_{i|1}(n)] - E[C_{i|1}(n-1)] \right| \quad . \tag{4.14}$$

$\diamond$

Lemma 4.1 guarantees that the sequence $\{d_n\}$ defined as $\{d_n\} = \{E[C_{i|1}(n+1)] - E[C_{i|1}(n)]\}$ is monotone decreasing. Since $\{d_n\}$ is lower bounded by zero it follows that the iterative procedure is convergent, and there exists a value $E[\tilde{C}_{i|1}]$ such that

$$E[\tilde{C}_{i|1}] = \lim_{n \to \infty} E[C_{i|1}(n)] \quad .$$

To speed up the convergence of the iterative scheme (4.11), in ([75], [76]) it is suggested to set

$$E[C_{i|1}(0)] = (s + b_i) / (1 - \rho + \rho_i) \quad .$$

Throughout, to simplify the notation, the expression (4.15) will be used to approximate $E[C_{i|1}]$

$$E[C_{i|1}] \approx \frac{s + b_i}{1 - \rho + \rho_i} \ . \tag{4.15}$$

The symbol $\approx$ indicates that the iterative scheme (4.11) must be applied to obtain an approximation of $E[C_{i|1}]$ whenever $\exists j:$ $\lambda_j (s + b_i) / (1 - \rho + \rho_i) > 1$.

GENERALIZATION OF THE CONDITIONAL CYCLE CONCEPT. The conditional cycle concept, introduced by Kuhen, was extended to characterize the length of cycles that contain either a *special service time* or *a special switchover time*[1] ([76], [139]).

In a system in which a given station Q_i experiences a special service time with an average length $\overline{b}_i$, the average length of a cycle is approximated by

$$E[C_{i|(1, \overline{b}_i)}] = \frac{s + \overline{b}_i}{1 - \rho + \rho_i} \tag{4.16}$$

if, $\forall j \neq i, \lambda_j E[C_{i|(1, \overline{b}_i)}] < 1$, otherwise the iterative scheme (4.11) (in which b_i is set equal to $\overline{b}_i$) is applied.

On the other hand, the average length of a cycle in which a special switch-over time occurs between Q_i and Q_{i+1} (whose average length will be denoted by $\overline{s}_i$) is approximated by

$$E[C_{|\overline{s}_i}] = \frac{\overline{s}_i + \sum_{j \neq i} s_j}{1 - \rho} \ , \tag{4.17}$$

if, $\forall j, \lambda_j E[C_{|\overline{s}_i}] < 1$, otherwise (i.e., $\exists j, \lambda_j E[C_{|\overline{s}_i}] < 1$) the computation should be based on the following iterative scheme

$$E[C_{|\overline{s}_i}(n)] = \overline{s}_i + \sum_{j \neq i} s_j + \sum_j min \ \{\lambda_j E[C_{|\overline{s}_i}(n - 1)], 1\} \ b_j \ . \tag{4.18}$$

4.1.2 Cycle Length Analysis: *l*-limited Polling System

Chang and Sandhu [32] extended the conditional cycle concept to character-ize the cycle length in *l*-limited polling systems in which a given station

---

1. "Special" means that in this cycle a service time (switchover time between two consecutive stations) is sampled from a distribution which differs from the station service time (switchover time) distribution.

transmits a fixed number of packets. Specifically, in this type of system each station Q_i can transmit up to l_i packets, and the concept of conditional cycle is used to compute

(i)   the average length of a cycle in which a given station Q_i transmits $\bar{l}_i$ ($\bar{l}_i \leq l_i$) packets;

(ii)   the average length of a cycle in which a given station Q_i transmits $\bar{l}_i$ ($\bar{l}_i \leq l_i$) packets, and one of these packets requires a special service time.

In case (i), the average length of a cycle given that Q_i transmits $\bar{l}_i$ packets is

$$E[C_{\mid \bar{l}_i}] = \frac{s + \bar{l}_i b_i}{1 - \rho + \rho_i} \tag{4.19}$$

if $\forall j \neq i, \lambda_j E[C_{\mid l_i}] < l_j$, otherwise the computation is performed by applying the following iterative scheme

$$E[C_{\mid \bar{l}_i}(n)] = s + \bar{l}_i b_i + \sum_{j \neq i} \dot{m}in\,\{\lambda_j E[C_{\mid \bar{l}_i}(n-1)], l_j\}\, b_j \quad . \tag{4.20}$$

In case (ii), a given station Q_i transmits $\bar{l}_i$ packets and one of these packets requires a (special) service time whose average length is $\bar{b}_i$, then

$$E[C_{\mid (l_i, \bar{b}_i)}] = \frac{s + (\bar{l}_i - 1)\, b_i + \bar{b}_i}{1 - \rho + \rho_i} \tag{4.21}$$

provided that, $\forall j \neq i, \lambda_j E[C_{\mid (l_i, \bar{b}_i)}] < l_j$. Otherwise $E[C_{\mid (l_i, \bar{b}_i)}]$ is derived through the following iterative scheme

$$E[C_{\mid (l_i, \bar{b}_i)}(n)] = s + (\bar{l}_i - 1)\, b_i + \bar{b}_i + \tag{4.22}$$

$$\sum_{j \neq i} min\,\{\lambda_j E[C_{\mid (l_i, \bar{b}_i)}(n-1)], l_j\}\, b_j \quad .$$

4.1.3 Waiting Time Analysis for 1-limited Polling Systems

Most of the approximations proposed for the analysis of the average waiting time of a Q_i tagged packet are obtained by extending and refining a technique developed by Boxma and Meister [23]. This technique exploits both the conditional cycle concept and the pseudo-conservation law. As shown in Figure 4.2, the average waiting time of a tagged packet is

$$E[W_i] = E[C_{i,0+}] + \sum_{n=1}^{Nq_i} E[C_{i,n}] \quad . \tag{4.23}$$

REMARK. The average number of packets, observed by the tagged packet at its arrival at Q_i, is given by $E[N_i] = E[Nq_i] + \rho_i$. In any case, when the server return to Q_i only $E[Nq_i]$ still have to be served before the tagged packet is served. The service of the packet which may be in service when the tagged packet arrives contributes to $E[C_{i,0+}]$.

$\Diamond$

Given that each cycle contains a Q_i packet transmission, in Boxma and

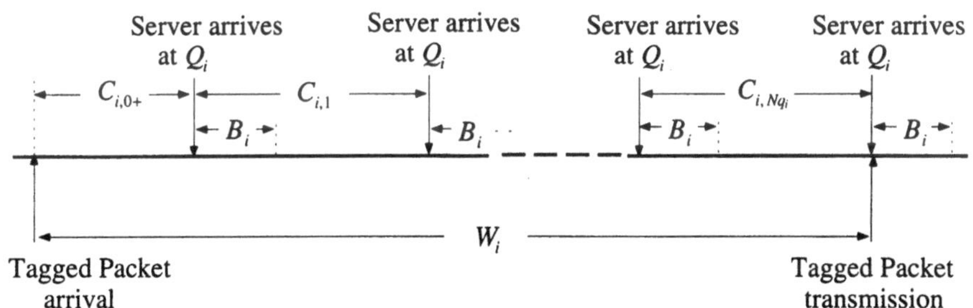

Figure 4.2: Waiting time in a non-exhaustive polling system

Meister [32] it is assumed that cycle lengths $C_{i,1}, C_{i,2}, ..., C_{i,Nq_i}$ form a sequence of i.i.d. random variables distributed as $C_{i|1}$ (see (4.12)). Hence from (4.23), the average waiting time of the tagged packet is

$$E[W_i] = E[C_{i,0+}] + E[Nq_i] \cdot E[C_{i|1}] \quad . \tag{4.24}$$

Due to the PASTA property (see Theorem 2.3), the average number of packets queued at Q_i at an arrival instant is equal to the mean queue length at an arbitrary time, and then, by applying Little's theorem $(E[Nq_i] = \lambda_i E[W_i])$, (4.24) can be re-written as

$$E[W_i] = \frac{E[C_{i,0+}]}{1 - \lambda_i E[C_{i|1}]} \quad . \tag{4.25}$$

Assuming that $E[C_{i,0+}]$ does not depend on the index i (i.e., $E[C_{i,0+}] = E[C_{j,0+}]$, $\forall i, j$), and denoting the residual cycle length with $E[C_+]$, (4.25) becomes

$$E[W_i] = \frac{E[C_+]}{1 - \lambda_i E[C_{i|1}]} \quad . \tag{4.26}$$

Finally, using (4.26) and the pseudo-conservation law for 1-limited polling systems (see (2.100)), after some algebraic manipulations

$$E[W_i] = \frac{1 - \rho + \rho_i}{1 - \rho - \lambda_i s} \cdot \frac{1 - \rho}{(1 - \rho)\rho + \sum_{j=1}^{K} \rho_j^2} \cdot \tag{4.27}$$
$$\left[\frac{\rho}{2(1 - \rho)} \sum_{j=1}^{K} \lambda_j b_j^{(2)} + \frac{\rho}{2s} \sum_{j=1}^{K} [s_j^{(2)} - s_j^2] + \frac{s}{2(1 - \rho)} \left[\rho + \sum_{j=1}^{K} \rho_j^2 \right] \right] \quad .$$

REMARK. For symmetric non-exhaustive polling systems an exact analysis of the average waiting time was performed independently by Furhmann [66] and Takagi [144]. Furhmann's analysis exploits decomposition properties (see Section 2.5.3), while Takagi derives the solution by applying the z-transform technique. Formula (4.27), obtained from the pseudo-conservation law, includes the result derived by Furhmann and Takagi as a special case. In fact, for symmetric polling systems, λ_i, b_i, s_i, $s_i^{(2)}$, and ρ_i do not depend on the index i ($i = 1, 2, \ldots, K$), and $s = Ks_i$, $\rho = K\rho_i$, hence (4.27) can be rewritten as

$$E[W_i] = \frac{1 - \rho}{1 - \rho - \lambda_i s} \cdot \left[\frac{K\lambda_i b_i^{(2)}}{2(1 - \rho)} + \frac{s_i^{(2)}}{2s_i} - \frac{s_i}{2} \left[\frac{1 - 2K\rho_i - K}{1 - \rho} \right] \right] \quad . \tag{4.28}$$

$\Diamond$

REMARK. Formula (4.27) is exact when $K = 1$. For a single queue system $s = s_i$, $s^{(2)} = s_i^{(2)}$, and $\rho = \rho_i$, hence (4.27) becomes

$$E[W_i] = \frac{1 - \rho}{1 - \rho - \lambda_i s} \cdot \left[\frac{\lambda_i b_i^{(2)}}{2(1 - \rho)} + \frac{s^{(2)}}{2s} + \frac{s\rho}{(1 - \rho)} \right] \quad , \tag{4.29}$$

and the average system response time $(E[R_i])$ is

$$E[R_i] = E[W_i] + E[b_i] =$$

$$\frac{1-\rho}{1-\rho-\lambda_i s} \cdot \left[E[R_{M/G/1}] + \frac{s^{(2)}}{2s} \right] = E[R_{M/G/1_{1-limited}}] \quad ,$$

where $E[R_{M/G/1}]$ and $E[R_{M/G/1_{1-limited}}]$ are the average response times in the corresponding $M/G/1$ and $M/G/1_{1-limited}$ systems, respectively (see Section 2.5.2).

◊

Equation (4.27) provides an explicit and simple formula to perform a quantitative analysis of polling systems with 1-limited service discipline over a wide range of parameters. In fact, as shown above, (4.27) is exact for symmetric polling systems and for single queue systems. Furthermore, (4.27) provides good approximations for low and medium offered loads [23]. However, under asymmetric high offered loads the approximation errors become greater than 20% [23]. The errors introduced by the technique proposed by Boxma and Meister come from the approximations used to derive formula (4.27)

(i)   the characterization of the residual cycle time $(C_{i,0+})$; and
(ii)   the assumption that the cycles $C_{i,1}$, $C_{i,2}$, ..., C_{i,Nq_i} are independent and distributed like $C_{i|1}$.

Srinivasan [139] and Groenendijk ([75], [76]) propose alternative methods, based on a more accurate analysis of points (i) and (ii) above, to derive an expression of the average waiting time that provides a better approximation under high-load conditions. Specifically, they perform a detailed analysis of the average length of the residual cycle $(C_{i,0+})$ in which the tagged packet arrives, and take into consideration the dependency between the arrival time of the tagged packet and the length of the next intervisit period at Q_i. The lengths of the remaining $(Nq_i - 1)$ cycles are assumed (as in Boxma and Meister [23]) to be independent and distributed like $C_{i|1}$ (see (4.12)).

Before presenting these analyses it is useful to introduce some notations. The state of the system as observed by Q_i, during the m-th cycle, is described by a couple $\{\xi_i, N_i\}_m$. ξ_i denotes the position of the server with respect to Q_i, and N_i is the number of packets at Q_i. $\xi_i = 0$ and $\xi_i = 1$

indicate the intervisit period and the service period at Q_i, respectively.

Under the assumption that the stability criteria are satisfied (see (2.25)) the following steady-state joint probabilities can be defined

$$q_i(j,k) \;=\; \lim_{m \to \infty} P\{\xi_i = j, N_i = k\}_m \;=\; P\{\xi_i = j, N_i = k\} \quad , \qquad (4.30)$$

obviously, $j = 1$ implies $k > 0$. Furthermore,

$$p_i(k) \;=\; P\{N_i = k\} \;=\; \sum_j P\{\xi_i = j, N_i = k\} \quad . \qquad (4.31)$$

The steady-state probability that the server is at Q_i ($P\{\xi_i = 1\}$) can be formally derived by exploiting renewal/regenerative results (see Section 2.3). Specifically, regenerative points of a polling system are the time instants at which the server polls a given station Q_j and all queues are empty[1]. By denoting with M the number of polling cycles in a regenerative cycle it can be easily verified that (see Theorem 2.4)

$$P\{\xi_i = 1\} \;=\; \frac{\sum_{m=1}^{M} E[Sp_{i,m}]}{\sum_{m=1}^{M} E[C_m]} \;=\; \frac{E[Sp_i]}{E[C]} \;=\; \rho_i \quad , \qquad (4.32)$$

where C_m and $Sp_{i,m}$ are the lengths of the m-th polling cycle and of the m-th Q_i service period in a regenerative cycle, respectively.

Finally, it is useful to partition the event $\{\xi_i = 0\}$ into the following set of disjoint events

$$\{\xi_i = 0\} \equiv \{\bigcup_{j \neq i} \{\xi_j = 1\}\} \cup \{\bigcup_{j=1}^{K} \{\zeta_{j,j+1} = 1\}\} \quad , \qquad (4.33)$$

where $\zeta_{j,j+1}$ is equal to 1 if the server is switching between Q_j and Q_{j+1}, otherwise $\zeta_{j,j+1} = 0$.

$P\{\zeta_{j,j+1} = 1\}$ can be derived by applying the same line of reasoning used in the derivation of (4.32)

$$P\{\zeta_{j,j+1} = 1\} \;=\; \frac{E[S_j]}{E[C]} \;=\; \frac{s_j}{E[C]} \quad , \qquad (4.34)$$

and hence, as expected,

---

1. Without any loss of generality, $j = 1$ can be assumed.

$$P\{\bigcup_{j \neq i}\{\xi_j = 1\}\} \cup \{\bigcup_{j=1}^{K}\{\zeta_{j,j+1} = 1\}\} = \tag{4.35}$$

$$\sum_{j \neq i}\rho_j + \frac{s}{E[C]} = 1 - \rho_i = P\{\xi_i = 0\} \quad .$$

Due to the PASTA theorem (see Theorem 2.3), probabilities (4.32) and (4.34) are equal to the steady-state probabilities that, when the tagged packet arrives, the server is at Q_i or is switching between Q_j and Q_{j+1}, respectively. Furthermore,

$$P\{\xi_i = 1 | \xi_j = 0\} = \frac{\rho_i}{1 - \rho_j} \quad , \tag{4.36}$$

and

$$P\{\zeta_{j,j+1} = 1 | \xi_j = 0\} = \frac{s_j}{(1 - \rho_j)\,E[C]} \quad . \tag{4.37}$$

Exploiting the above results, the waiting time analysis is performed by conditioning the waiting time on the number of packets at Q_i when the tagged packet arrives

$$E[W_i] = E[W_i | N_i = 0]\,P\{N_i = 0\} + \tag{4.38}$$
$$E[W_i | N_i > 0]\,P\{N_i > 0\} \quad .$$

Due to the PASTA property, the steady-state probability of the event $\{N_i = 0\}$ is equal to the steady-state probability, $p_i(0)$, that Q_i is empty at a random point in time, hence,

$$E[W_i] = E[W_i | N_i = 0] \cdot p_i(0) + \tag{4.39}$$
$$E[W_i | N_i > 0] \cdot (1 - p_i(0)) \quad .$$

To compute $E[W_i | N_i > 0]$ a conditioning on the state of the server (serving Q_i, or in vacation with respect to Q_i) when the tagged packet arrives is introduced.

$$E[W_i | N_i > 0] = E[W_i | \xi_i = 0, N_i > 0]\,P\{\xi_i = 0 | N_i > 0\} + \tag{4.40}$$
$$E[W_i | \xi_i = 1, N_i > 0]\,P\{\xi_i = 1 | N_i > 0\} \quad .$$

Noting that $P\{\xi_i = 0 | N_i = 0\} = 1$, it follows that

$$P\{\xi_i = 0\} = P\{\xi_i = 0 | N_i > 0\}\,(1 - p_i(0)) + 1 \cdot p_i(0) \quad . \tag{4.41}$$

Since $P\{\xi_i = 0\} = 1 - \rho_i$, it can be verified that

$$P\{\xi_i = 0 | N_i > 0\} = \frac{1 - \rho_i - p_i(0)}{1 - p_i(0)} \quad , \tag{4.42}$$

and

$$P\{\xi_i = 1 | N_i > 0\} = \rho_i / (1 - p_i(0)) \quad . \tag{4.43}$$

Substituting (4.40), (4.42) and (4.43) into (4.39), the following expression for the average waiting time is finally obtained

$$\begin{aligned}
E[W_i] = {} & E[W_i | N_i = 0] \cdot p_i(0) + \\
& E[W_i | \xi_i = 0, N_i > 0] (1 - \rho_i - p_i(0)) + \\
& E[W_i | \xi_i = 1, N_i > 0] \rho_i \quad .
\end{aligned} \tag{4.44}$$

Therefore the waiting time analysis is reduced to the analysis of $p_i(0)$ and of the average waiting time for the following three scenarios

(i)   empty queue, i.e., $\{N_i = 0\}$ (i.e., $E[W_i | N_i = 0]$);
(ii)   not-empty queue and server on vacation (i.e., $E[W_i | \xi_i = 0, N_i > 0]$);
(iii) not-empty queue and server at Q_i (i.e., $E[W_i | \xi_i = 1, N_i > 0]$).

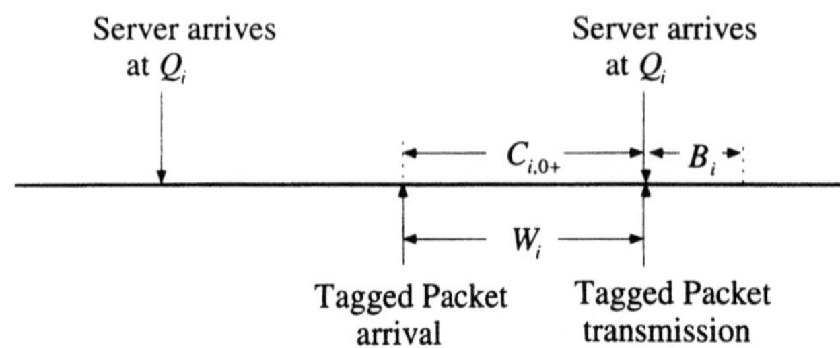

Figure 4.3: Tagged-packet waiting time when Q_i is empty

The average waiting time in case (i) (see Figure 4.3) is equal to the average residual length of a cycle in which Q_i is empty when the tagged packet arrives

$$E[W_i | N_i = 0] = E[C_{i,0+} | N_i = 0] \quad . \tag{4.45}$$

In case (ii) the average waiting time experienced by the tagged packet is (see Figure 4.4)

$$E[W_i|\xi_i = 0, N_i > 0] \;=\; E[C_{i,0+}|\xi_i = 0, N_i > 0] + b_i + \tag{4.46}$$
$$E[I_{i,1}^{(v)}] + E[N_i - 1|\xi_i = 0, N_i > 0]\,E[C_{i|1}] \quad,$$

where $I_{i,1}^{(v)}$ denotes the length of the first intervisit period following the tagged-packet arrival (i.e., $E[I_{i,1}^{(v)}] = E[I_{i,1}|\xi_i = 0, N_i > 0]$). In fact, according to the hypothesis that the tagged-packet arrival only affects the length of the next intervisit period, the cycle lengths $C_{i,2}, \ldots, C_{i,N_i}$ are assumed to be distributed like $C_{i|1}$.

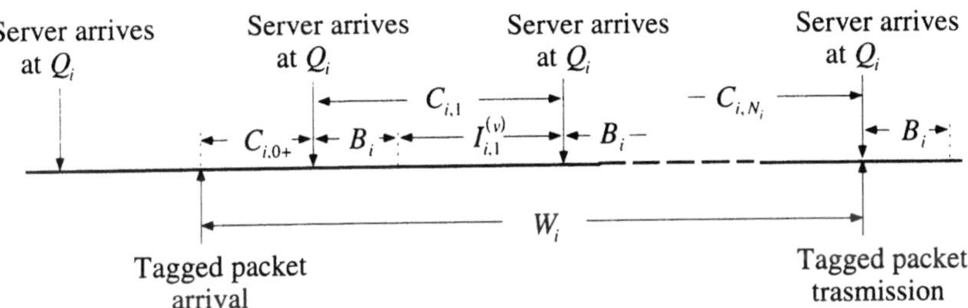

Figure 4.4: Tagged-packet arrives when Q_i is not empty and the server is on vacation

The waiting time of a tagged packet which arrives during a service period (case (iii)) is shown in Figure 4.5. The tagged-packet waiting time includes the residual service time at Q_i, the successive intervisit period ($I_{i,0}^{(s)}$), and $(N_i - 1)$ cycles (i.e., $C_{i,1}, \ldots, C_{i,N_i-1}$) distributed like $C_{i|1}$

$$E[W_i|\xi_i = 1, N_i > 0] \;=\; \frac{b_i^{(2)}}{2b_i} + E[I_{i,0}^{(s)}] + \tag{4.47}$$
$$E[N_i - 1|\xi_i = 1, N_i > 0]\,E[C_{i|1}] \quad,$$

where $E[I_{i,0}^{(s)}] = E[I_{i,0}|\xi_i = 1, N_i > 0] = E[I_{i,0}|\xi_i = 1]$.

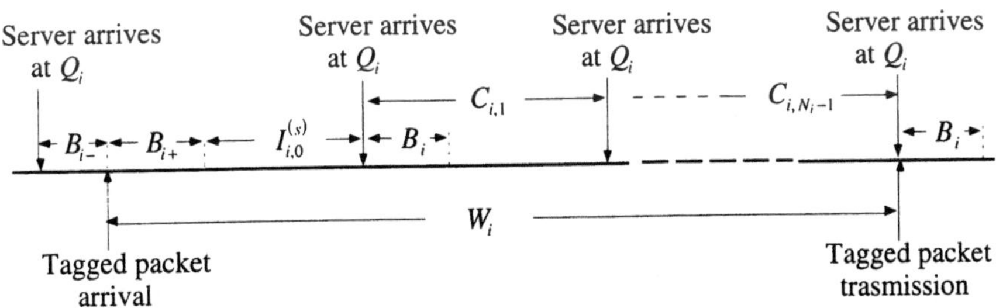

Figure 4.5: The tagged packet arrives during a service at Q_i

Since

$$E[C_{i,0+}] = E[C_{i,0+}|N_i = 0] \, p_i(0) + \tag{4.48}$$
$$E[C_{i,0+}|\xi_i = 1, N_i > 0] \, \rho_i$$
$$E[C_{i,0+}|\xi_i = 0, N_i > 0] \, (1 - \rho_i - p_i(0)) \quad ,$$

and

$$E[C_{i,0+}|\xi_i = 1, N_i > 0] = b_i^{(2)}/2b_i + E[I_{i,0}^{(s)}] \quad , \tag{4.49}$$

substituting (4.45)-(4.47) into (4.44), it follows that

$$E[W_i] = E[C_{i,0+}] + (b_i + E[I_{i,1}^{(v)}]) \, (1 - \rho_i - p_i(0)) + \tag{4.50}$$
$$E[C_{i|1}] \, E[N_i - 1|N_i > 0] \, (1 - p_i(0)) \quad .$$

Finally, by exploiting the following relationship

$$E[N_i - 1|N_i > 0] \, (1 - p_i(0)) = E[Nq_i] - (1 - \rho_i - p_i(0)) = \tag{4.51}$$
$$\lambda_i E[W_i] - (1 - \rho_i - p_i(0)) \quad ,$$

after some algebraic manipulations (4.50) can be rewritten as

$$E[W_i] \, (1 - \lambda_i E[C_{i|1}]) = E[C_{i,0+}] + \tag{4.52}$$
$$(E[I_{i,1}^{(v)}] - E[I_{i|1}]) \, (1 - \rho_i - p_i(0)) \quad ,$$

where $E[I_{i|1}] = E[C_{i|1}] - b_i$.

The average waiting time is therefore

$$E[W_i] = Y_i E[C_{i,0+}] + Z_i \quad , \tag{4.53}$$

where

1. $Y_i = 1/(1 - \lambda_i E[C_{i|1}])$;

2. $Z_i = (E[I_{i,1}^{(v)}] - E[I_{i|1}]) \, (1 - \rho_i - p_i(0)) / (1 - \lambda_i E[C_{i|1}])$; and

3. $E[C_{i,0+}] = E[C_{i,0+}|\xi_i = 0] \, (1 - \rho_i) + (b_i^{(2)}/2b_i + E[I_{i,0}^{(s)}]) \, \rho_i$.

Note that, if $Z_i = 0$, expression (4.53) is equal to (4.25) which was derived by Boxma and Meister.

Formula (4.53) reduces the computation of the average waiting time $E[W_i]$ to the computation of $p_i(0)$, $E[C_{i,0+}|\xi_i = 0]$, $E[I_{i,0}^{(s)}]$ and $E[I_{i,1}^{(v)}]$. Approximate expressions for these unknowns are derived in Section 4.3.2. Expression for $p_i(0)$, $E[I_{i,0}^{(s)}]$ and $E[I_{i,1}^{(v)}]$ depend only on system parameters and on $E[C_{i,0+}|\xi_i = 0]$. In the following, by exploiting the pseudo-conservation law, an iterative procedure is outlined to produce an approximation for $E[C_{i,0+}|\xi_i = 0]$ which is better than the one provided in Section 4.3.2. This leads to an improvement in the average waiting time approximation.

Specifically, by assuming that $E[C_{i,0+}]$ does not depend on the index i, and by substituting (4.53) in (2.100) (i.e., the pseudo-conservation law for 1-limited polling systems) after some algebraic manipulation it results that

$$E[C_{i,0+}] = \frac{X - \sum_{n=1}^{K} Z_i \cdot \theta_i}{\sum_{n=1}^{K} \theta_i} , \tag{4.54}$$

where

- X is the right-hand side of the equation (2.100); and
- $\theta_i = [\rho_i(1 - \lambda_i E[C_i])] / (1 - \lambda_i E[C_{i|1}])$.

From (4.54) and equation 3 in (4.53) it is possible to obtain an improved approximation for $E[C_{i,0+}|\xi_i = 0]$. The new improved estimate of $E[C_{i,0+}|\xi_i = 0]$ is then applied to compute a new estimate of $E[W_i]$. This procedure is repeated over and over until the estimate of $E[C_{i,0+}|\xi_i = 0]$ in two consecutive steps differs less than a fixed quantity.

As pointed out before the Srinivasan approximation was introduced to better approximate the average delay in 1-limited polling systems under asymmetric high-load conditions. The results presented in Figure 4.6 quantify this improvement [139]. Specifically, the figure reports the station average delay vs node index for a 16 station polling system operating under heavy load conditions, $\rho = 0.8$. Service times are exponentially distributed, and the average service time of stations 1 and 7 is three times greater than the average service time of the other stations.

REMARK. Following the same line of reasoning as Srinivasan, Groenendijk ([75], [76]) proposed the following approximation of the average waiting

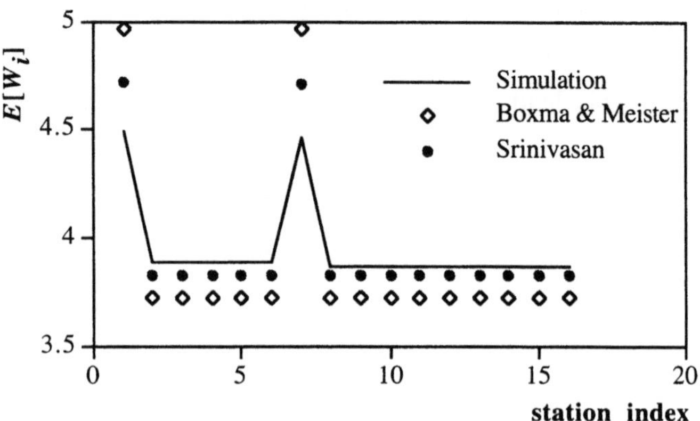

$$\lambda_1 = \lambda_2 = \ldots = \lambda_{16}$$
$$b_1 = b_7, b_i = b_1/3, i \neq 1, 7$$

Figure 4.6: Numerical results for an asymmetric polling system

time at Q_i

$$E[W_i] = \frac{E[C_{i,0+}]}{(1 - \lambda_i E[C_{i|1}])} + H_i \quad , \tag{4.55}$$

where

$$H_i = Z_i + \frac{(1 - p_i(1)/(1 - p_i(0)))}{1 - \lambda_i E[C_{i|1}]} p_i (E[C_{2,n}|\xi_i = 1, N_i > 0] - E[C_{i|1}]) \quad .$$

Hence the main difference between (4.55) and (4.53) is the second adden-
dum in the expression of H_i. An additional difference in the analyses performed by Srinivasan and Groenendijk is in the approximation of $p_i(0)$. In
fact, Groenendijk represents the system with an $M/G/1$ queueing system
with exceptional first service but he provides a different characterization of
the exceptional first service in order to capture the dependency of $C_{i,1}$ from
$C_{i,0+}$.

$\Diamond$

A comparison between the different methods, presented in this section to
approximate the average waiting time in a 1-limited polling system, is

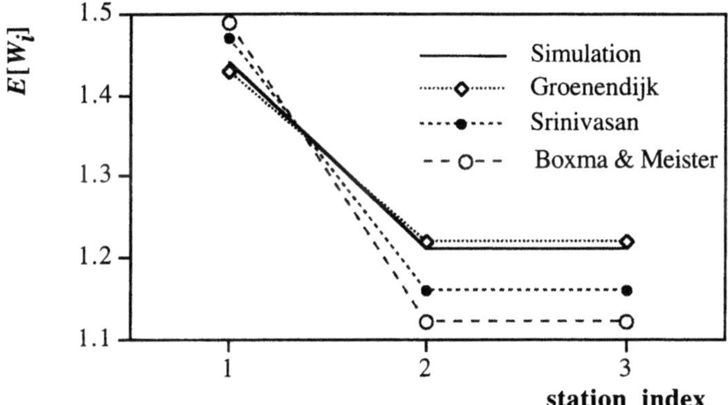

Figure 4.7: Numerical results for an asymmetric polling system, $\rho = 0.5$

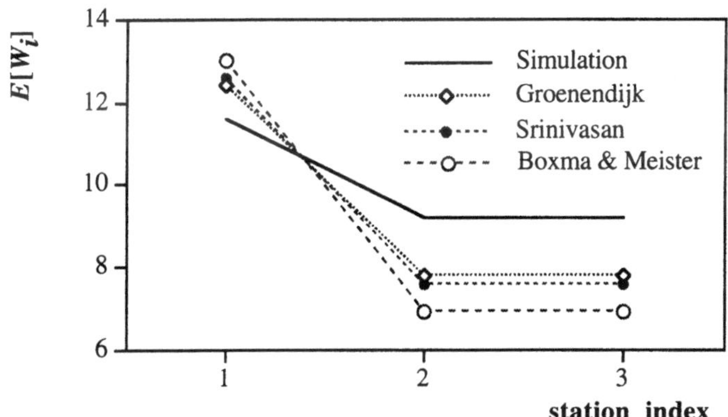

Figure 4.8: Numerical results for an asymmetric polling system, $\rho = 0.8$

reported in Figures 4.7 and 4.8. The results shown in these figures are taken from [75] and refer to a three station polling system with $\lambda_1 = \lambda_2 = \lambda_3$, $b_2 = b_3 = b_1/3$ and $s_1 = s_2 = s_3 = 0.1$. Service times and switchover times are exponentially distributed.

The results reported in Figures 4.7 and 4.8 clearly show that Groenendijk's approach improves the accuracy of Srinivasan's approximation under medium load conditions ($\rho = 0.5$), whereas under heavy load conditions ($\rho = 0.8$) the results obtained with the two approximations are very close.

4.1.4 Waiting Time Analysis for Polling Systems with Mixed Polling Strategies

A polling model with mixed service strategies can be used to represent a network in which stations have different roles. For example, one station is the gateway interconnecting two or more networks, while the remaining stations connect the user's equipment to the network. In this case the gateway can transmit according to an exhaustive (or gated) service discipline, while the other stations transmit according to a 1-limited service policy. For the analysis of these systems, Groenendijk ([74], [75], [76]) proposes a methodology based on the following steps

(i)   the average waiting time at each station Q_i is expressed as a function of the residual cycle length observed by a tagged packet which arrives at this station (i.e., $E[C_{i, 0+}]$ according to the notation introduced in the previous section);

(ii) an approximate expression of $E[C_{i, 0+}]$ is then obtained by exploiting formula (2.135), i.e., the pseudo-conservation law for a mixed service discipline.

This method is briefly outlined below, for further details see ([74], [75], [76]).

As far as step (i) is concerned, the following expressions define the Q_i average waiting time for various service disciplines.

1-LIMITED SERVICE DISCIPLINE. As shown in the previous section

$$E[W_i] \approx \frac{E[C_{i, 0+}]}{(1 - \lambda_i E[C_{i|1}])} + R_i \quad , \tag{4.56}$$

where the value of R_i depends on the approximation used to characterize a 1-limited polling system

- $R_i = 0$ in Boxma and Meister's approximation (see (4.25)).
- $R_i = Z_i$, where Z_i is defined in 2 of (4.53) in Srinivasan's approximation;
- $R_i = H_i$, where H_i is given by (4.55) in Groenendijk's approximation.

GATED SERVICE DISCIPLINE. According to the gated policy, a tagged packet is transmitted in the cycle which follows the one in which the tagged packet arrives, and its transmission occurs after all the packets which have already arrived in the same cycle. According to Theorem 2.5 the average residual cycle length is equal to the average elapsed cycle length when the tagged packet arrives (i.e., $E[C_{i,0+}] = E[C_{i,0_-}]$). Hence the average waiting time (see (4.57)) is equal to $E[C_{i,0+}]$, plus the average time required to serve the packets that arrived during the elapsed cycle, i.e., $\lambda_i E[C_{i,0_-}] b_i$.

$$E[W_i] = (1 + \rho_i) E[C_{i,0+}] \quad . \tag{4.57}$$

EXHAUSTIVE SERVICE DISCIPLINE. Equation (4.58) was proposed by Groenendijk ([75], [76]) after proving that $E[W_i] = (1 - \rho_i) E[\hat{C}_{i,0+}]$, where $E[\hat{C}_{i,0+}]$ is the average residual time of a cycle which starts when the server departs from Q_i. Equation (4.58) is obtained by ignoring the differences between $E[\hat{C}_{i,0+}]$ and $E[C_{i,0+}]$.

$$E[W_i] \approx (1 - \rho_i) E[C_{i,0+}] \quad . \tag{4.58}$$

The second step of the approximation (see (ii) on page 154) is an iterative procedure based on the pseudo-conservation law.

According to Theorem 2.5 it is possible to verify that (if the system is stable)

$$E[C_{i,0+}] = \frac{E[C_i^2]}{2E[C_i]} \quad , \tag{4.59}$$

where $E[C_i^2]$ is the second moment of the cycle length distribution.

As shown in Section 2.1.3, the average cycle length does not depend on the node index. On the other hand, the second moments are dependent on it though the differences are generally small [23]. To obtain an expression for $E[C_{i,0+}]$ from the pseudo-conservation law, in ([75], [76]), it is assumed that second moments are independent of the node index. Hence, by substituting the expressions (4.56)-(4.58) of the average waiting time into the l.h.s. of the pseudo-conservation law (2.135), and assuming that $E[C_{i,0+}]$ does not depend on the node index i, an improved approximation of the residual cycle length, $E[C_{i,0+}^{(1)}]$, is obtained.[1] $E[C_{i,0+}^{(1)}]$, together with (4.56)-(4.58), is then used to obtain improved approximations for the average waiting time which are again substituted into the pseudo-conservation law to obtain a new

---

1. Throughout, the superscript "(n)" after a quantity denotes its value at the end of the n-th step of the iterative algorithm.

approximation of the residual cycle length $E[C_{i,0+}^{(2)}]$. The iterative procedure ends when the estimates of the average waiting time do not significantly differ from the estimates computed in the previous step.

4.1.5 Waiting Time Analysis for l-limited Polling Systems

The methodology for the waiting time analysis developed for 1-limited polling systems (see Section 4.1.3) was extended by Chang and Sandhu [32] to study the average waiting time in l-limited polling systems. Specifically, Chang and Sandhu first derive an approximate expression for the average waiting time at Q_i as a function of the residual cycle length, $E[C_{i+}]$, i.e.,

$$E[W_i] \approx Y_i \cdot E[C_{i+}] + Z_i \quad , \tag{4.60}$$

where Y_i and Z_i are functions of the system parameters. They then refine the average waiting time approximation by exploiting the pseudo-conservation law for l-limited polling systems (see Section 2.6.4).

The main effort in the work of Chang and Sandhu [32] is the derivation of expressions for Y_i and Z_i, which is performed by analyzing the average delay experienced by a tagged packet arriving at Q_i at a random point in time. In performing this computation they assume (as proposed by Srinivasan [139]) that the arrival of the tagged packet generates a length-biasing phenomenon (see [55]) of the length of the cycle in which the arrival occurs ($C_{i,0}$), which propagates to the next cycle as well (i.e., $C_{i,1}$). On the other hand, the length-biasing effect is assumed to be almost negligible on the length of successive cycles (i.e., $C_{i,2}$, $C_{i,3}$, ...). In addition, to compute the average length of the different types of cycles (i.e., the cycle in which the arrival occurs, the next cycle and successive cycles) the authors make extensive use of the conditional cycle concept introduced by Kuhen (see Section 4.1.1).

To perform the analysis of the waiting time experienced by the tagged-packet, similarly to the 1-limited service discipline (see Section 4.1.3), the state of the polling system in the n-th polling cycle observed by Q_i is described by a couple of random variables $\{\xi_i, N_i\}_n$. $\xi_i = 0$ indicates that, with respect to Q_i, the server is on vacation, while $\xi_i = j$ ($j = 1, 2, ..., l_i$)

indicates that Q_i is transmitting the j-th packets during the n-th service period. Finally, the r.v. N_i denotes the number of packets queued at Q_i.

Assuming that the polling system satisfies the stability criteria (see Section 2.1.3), the following steady-state joint probabilities exist

$$q_i(j, k) = \lim_{n \to \infty} P\{\xi_i = j, N_i = k\}_n, \quad j = 0, \dots, l_i, \quad k = 0, 1, \dots \quad . \quad (4.61)$$

Obviously, in (4.61) if $j \neq 0 \Rightarrow k > 0$. In addition, following the same line of reasoning used in Section 4.1.3 to derive (4.32) and (4.35), it can be shown that

$$\sum_{k=0}^{\infty} q_i(0, k) = 1 - \rho_i \quad \text{and} \quad \sum_{j=1}^{l_i}\sum_{k=1}^{\infty} q_i(j, k) = \rho_i \quad . \quad (4.62)$$

$q_i(j, k)$ are the steady-state probabilities of the system at the tagged packet arrival instant since, and due to PASTA,

$$P\{\xi_i = j, N_i \geq 0 | tagged\ packet\ arrival\} = P\{\xi_i = j, N_i \geq 0\} \quad .$$

Hence, by noting that when the tagged packet arrives at Q_i the server can either be on vacation or serving the j-th packet at Q_i during this cycle, it follows that

$$E[W_i] = \sum_{k=0}^{\infty} E[W_{i|\xi_i = 0, N_i = k}]\, q_i(0, k) + \quad\quad (4.63)$$

$$\sum_{j=1}^{l_i}\sum_{k=1}^{\infty} E[W_{i|\xi_i = j, N_i = k}]\, q_i(j, k) \quad .$$

The unknown quantities of (4.63) are derived in Lemma 4.2 and Lemma 4.3. The proofs of these lemmas can be found in Section 4.3.2.

LEMMA 4.2   *Let $\phi_{i|\xi_i = 0}$ be the joint probability that the tagged packet arrives during an intervisit period and that it will not be served in the next cycle; the following relationship thus holds:*

$$\sum_{k=0}^{\infty} E\left[W_{i|\xi_i=0, N_i=k}\right] q_i(0,k) = E\left[I_{i+}\right](1-\rho_i) + \tag{4.64}$$

$$b_i E\left[N_{i|\xi_i=0}\right] P\{\xi_i=0\} + \left(E\left[C_{i,1|\xi_i=0, N_i=k}\right] - \right.$$

$$E\left[C_{i|l_i}\right]) \; \phi_{i|\xi_i=0} + \left(E\left[C_{i,j|l_i}\right] - l_i b_i\right) \sum_{k=l_i}^{\infty} \lfloor k/l_i \rfloor q_i(0,k) \quad .$$

$$\Diamond$$

LEMMA 4.3   *Let $\phi_{i|\xi_i=j}$ be the joint probability that the tagged packet arrives during a service period and that it will not be served in that cycle; the following relationship thus holds:*

$$\sum_{j=1}^{l_i} \sum_{k=1}^{\infty} E\left[W_{i|\xi_i=j, N_i=k}\right] q_i(j,k) = \frac{b_i^{(2)}}{2b_i}\rho_i + \tag{4.65}$$

$$b\left(E\left[N_{i|\xi_i=0}\right] P\{\xi_i=0\} - \rho_i\right) + E\left[I_{i,1|\xi_i=j, N_i>l_i-j}\right] \phi_{i|\xi_i=j} +$$

$$\left(E\left[C_{i|l_i}\right] - l_i b_i\right) \sum_{j=1}^{l_i} \sum_{k>l_i-j} \left(\left\lfloor \frac{(k+j-1)}{l_i} \right\rfloor - 1\right) q_i(j,k) \quad .$$

$$\Diamond$$

By substituting (4.64) and (4.65) into (4.63), and since $b_i\left(E\left[N_i\right] - \rho_i\right) = \rho_i E\left[W_i\right]$, after some routine algebraic manipulations, it can be shown that

$$E\left[W_i\right](1-\rho_i) = E\left[I_{i+}\right](1-\rho_i) + \frac{b_i^{(2)}}{2b_i}\rho_i + \tag{4.66}$$

$$\left(E\left[C_{i,1|\xi_i=0, N_i\geq l_i}\right] - E\left[C_{i|l_i}\right]\right) \phi_{i|\xi_i=0} +$$

$$\left(E\left[C_{i,1|\xi_i=j, N_i>l_i-j}\right] - E\left[C_{i|l_i}\right]\right) \phi_{i|\xi_i=j} +$$

$$\left(E\left[C_{i|l_i}\right] - l_i b_i\right) E\left[\Theta_i\right] \quad ,$$

where $E\left[\Theta_i\right]$ is the average number of server departures from Q_i observed by the tagged packet

$$E[\Theta_i] \;=\; \sum_{k=l_i}^{\infty} \left\lfloor \frac{k}{l_i} \right\rfloor \cdot q_i(0,k) + \sum_{j=1}^{l_i} \sum_{k=l_i+j+1}^{\infty} \left(\left\lfloor \frac{(k+j-1)}{l_i} \right\rfloor \right) q_i(j,k) \quad .$$

To derive the unknown quantities contained in (4.66) (e.g., $E[I_{i+}]$, $\phi_{i|\xi_i=j}$, and $\phi_{i|\xi_i=0}$) Chang and Sandhu (see [32]) introduce an auxiliary Markov chain embedded at the end of each packet transmission at Q_i.[1] The state of the system at the embedding points is described by the couple of random variables $\{\Psi_i, K_i\}_n$. $\Psi_i = j$ $(j = 1, 2, ..., l_i)$ denotes the end of the j-th packet transmission during the n-th service period at Q_i, and K_i denotes the number of packets queued at Q_i.

Assuming that the Markov chain is irreducible and positive recurrent, the following steady-state probabilities exist

$$p_{i,k,m} \;=\; \lim_{n \to \infty} P\{\Psi_i = m, K_i = k\}_n, \quad m = 1, ..., l_i, \quad k = 0, 1, ... \quad . \quad (4.67)$$

To derive an approximate estimate of the above steady-state probabilities, Chang and Sandhu, in [31] and [32], analyzed station Q_i in isolation (see Chapter 3). Specifically, the behavior of the polling system, as observed by Q_i, is represented via a *corresponding M/G/1* queueing system with vacation and E-limited service discipline (see Section 3.2.1), where the vacations represent the intervisit periods at Q_i. In [31] the vacations (in the corresponding *M/G/1* system) are assumed to be independent and sampled from an exponential distribution with average $E[I_i]$. This characterization was refined in [32] where vacations are assumed to be distributed according to the gamma distribution which includes both hypoexponential and hyperexponential distributions [85]. This refinement was proposed by observing, via simulation, that the distribution of the length of the intervisit periods and the cycle lengths (depending on the load and system parameters) can either be hypoexponential or hyperexponential (i.e, coefficient of variation less or greater than 1).

It is worth recalling that a gamma density function with parameters α and β is defined as

---

1. This Markov chain has already been used by Chang and Sandhu to derive the pseudo-conservation law for *l*-limited polling systems, see Section 2.6.4.

$$\Gamma_{\alpha, \beta}(x) = \begin{cases} \dfrac{\beta^{\alpha}}{\Gamma(\alpha)} x^{\alpha-1} e^{-\beta x} & x > 0 \\ 0 & x \leq 0 \end{cases} \qquad (4.68)$$

where $\Gamma(\alpha)$ is the so-called gamma function i.e., $\Gamma(\alpha) = \int_{0}^{\infty} y^{\alpha-1} e^{-y} dy$.

For each station Q_i, the α_i and β_i parameters are obtained by matching the moments of the cycle and intervisit time. Specifically, as shown in [32], the following relationships hold

$$\beta_i = \frac{(E[C])^2}{E[C_i^2] - (E[C])^2} , \qquad (4.69)$$

and

$$\alpha_i = \frac{\beta_i}{E[I_i]} . \qquad (4.70)$$

The *corresponding M/G/1* queueing system with vacation and E-limited service discipline is analyzed by applying the techniques presented in the previous chapter for the analysis of single server queueing systems (see Section 3.2.1). Specifically, the solution technique (z-transform) explicitly provides the numerical values for the boundary probabilities $\{p_{i, 0, j} , j = 1, 2, ..., l_i - 1\}$. By exploiting these boundary probabilities Chang and Sandhu (see [32]) propose the following approximations for $\phi_{i|\xi_i = 0}$, $\phi_{i|\xi_i = j}$ and $E[\Theta_i]$.

$$\phi_{i|\xi_i = 0} \approx \left(1 - \sum_{k = 0}^{l_i - 1} k p_{i, 0, k} \right) [1 - \rho_i - q_i(0, 0)] , \qquad (4.71)$$

where $q_i(0, 0) = \sum_{k = 0}^{l_i} p_{i, 0, k}$;

$$\phi_{i|\xi_i = j} \approx \left(1 - \sum_{k = 0}^{l_i - 1} k p_{i, 0, k} \right) \rho_i ; \qquad (4.72)$$

and

$$E[\Theta_i] \approx \frac{\lambda_i E[W_i]}{l_i} + \Gamma_i , \qquad (4.73)$$

where

$$\Gamma_i = \frac{1}{2l_i}\left[(1-\rho_i)\sum_{j=1}^{l_i-1} j\,(l_i-j)\,p_{i,0,j} - [l_i-1-(l_i+1)\rho_i]\right] \quad.$$

Finally, by noting that

$$E[C_{i+}] = E[C_{i+|\xi_i=0}]\,(1-\rho_i) + E[C_{i+|\xi_i>0}]\,\rho_i = \tag{4.74}$$
$$E[I_{i+}]\,(1-\rho_i) + E[C_{i+|\xi_i>0}]\,\rho_i \quad,$$

and by exploiting (4.73), (4.66) can be rewritten as

$$E[W_i]\cdot(1-\lambda_i E[C_{i|l_i}]/l_i) = E[C_{i+}] - E[C_{i+|\xi_i>0}]\,\rho_i + \frac{b_i^{(2)}}{2b_i}\rho_i + \tag{4.75}$$

$$(E[C_{i,1|\xi_i=0,N_i\geq l_i}] - E[C_{i|l_i}])\,\phi_{i|\xi_i=0} +$$

$$(E[C_{i,1|\xi_i=j,N_i>l_i-j}] - E[C_{i|l_i}])\,\phi_{i|\xi_i=j} +$$

$$(E[C_{i|l_i}] - l_i b_i)\,\Gamma_i \quad.$$

From (4.75) it results that $E[W_i]$ can be expressed as a function of the residual cycle time:

$$E[W_i] \approx Y_i\cdot E[C_{i+}] + Z_i \quad, \tag{4.76}$$

where $Y_i = 1/(1-\lambda_i E[C_{i|l_i}]/l_i)$, and the only unknown of Z_i, $E[C_{i+|\xi_i>0}]$, is derived in the following lemma which is proved in Section 4.3.3.

LEMMA 4.4   *Let θ_i be an r.v. which denotes the number of packets served at Q_i during a service period, and let $\{A_i(\theta_i)=1\}$ be an r.v. which denotes the event "the tagged packet arrived at Q_i during the service period θ_i" then*

$$E[C_{i+|\xi_i>0}] = \frac{b_i^{(2)}}{2b_i} + \sum_{j=1}^{l_i-1} E[C_i|\theta_i=j,A_i(\theta_i)=1]\,j p_{i,0,j} + \tag{4.77}$$

$$E[C_i|\theta_i=l_i,A_i(\theta_i)=1]\left(1-\sum_{j=1}^{l_i-1} j p_{i,0,j}\right) -$$

$$\frac{b_i}{2}\left[(l_i+1) + \sum_{j=1}^{l_i-1}(l_i-j)\,j p_{i,0,j}\right] \quad,$$

where

$$E[C_i | \theta_i = j, \xi_i = h] \approx \frac{s + (j-1) b_i + b_i^{(2)} / b_i}{1 - \rho + \rho_i} \quad .$$

$\Diamond$

Formulas (4.71)-(4.73) and (4.77) clearly show that the computation of the average waiting time depends on the solution of the corresponding $M/G/1$ system which in turn depends on the assumption on the vacation time distribution. The iterative algorithm presented in Table 4.1 starts by assuming an exponential distribution for the vacation time (i.e., $\alpha_i = 1/E[I_i], \beta_i = 1$), and in each step refines the α_i and β_i values until the algorithm converges.

Table 4.1 Iterative scheme for the computation of the average waiting time in an l-limited polling system

---

/* *initialization section*

$n = 0$

for each Q_i do $\alpha_i^{(0)} = 1/E[I_i], \beta_i^{(0)} = 1$;

/* *iterative section*

$n = n + 1$

for each Q_i do

begin

a)   compute $p_{i,0,k}^{(n)}, k = 1, ..., l_i - 1$ from the corresponding $M/G/1$;

b)   compute $E[W_i^{(n)}]$ from (4.75);

end

for each Q_i do

  begin

c)   compute $E[C_{i+}^{(n)}]$ from the pseudo-conservation law;

d)   compute $\alpha_i^{(n)}$ from (4.70);

e)   compute $\beta_i^{(n)}$ from (4.69);

  end

if $n = 1$ then go to /* iterative section

convergence=*true*;

f)   for each Q_i do if $\left| E[W_i^{(n)}] - E[W_i^{(n-1)}] \right| > \varepsilon$ then convergence=*false*

g)   if convergence=*false* then repeat /* iterative section

end

Specifically, in each iterative step (i.e., the "*/ iterative section" in Table 4.1) the following computations are performed[1]

(i) analysis of the corresponding $M/G/1$ system to derive its boundary probabilities (point a in Table 4.1);

(ii) analysis of the average waiting time at Q_i by applying (4.75) and the boundary probabilities computed in the previous step (point b);

(iii) application of the pseudo-conservation law to estimate $E[C_{i+}]$ (point c). Specifically, this is done by neglecting the dependency of $E[C_{i+}]$ on the station index i (i.e., $E[C_{i+}] = E[C_+]$), and by substituting (4.76) into the pseudo-conservation law for exhaustive-limited polling systems (see (2.117)). From $E[C_{i+}] = E[C_i^2]/2E[C_i]$ an estimate of $E[C_i^2]$ it is also immediately obtained;

(iv) refinement of the α_i, β_i estimates by exploiting the $E[C_i^2]$ value computed in the previous step (points d and e);

(v) convergence test (point f); if the desired accuracy is obtained the algorithm terminates, otherwise the next iterative step is performed (point g).

4.2 NUMERICAL METHODS

4.2.1 Analysis of Finite Capacity Systems

In real networks buffers are of finite capacity. In the investigation of finite capacity polling systems the analysis methods based on pseudo-conservation laws cannot be applied since, for finite capacity systems, pseudo-conservation laws do not hold. For non-exhaustive polling systems with finite capacity Tran-Gia and Raith [153] have developed an approximate analysis method based on a numerical algorithm to compute, for each queue Q_j, the steady-state probabilities of the number of packets queued at Q_j just before the server arrives at this queue (throughout, *scanning epoch*). By denoting with m_j the Q_j buffer size and with $P_j^{(n)}(k)$, $k = 0, 1, ..., m_j$ the probability mass function of the number of packets queued at Q_j at the n-th scanning epoch, the objective of the algorithm is to compute

$$P_j(k) = \lim_{n \to \infty} P_j^{(n)}(k), \ \ k = 0, 1, ..., m_j, \ j = 1, 2, ..., K \ \ . \qquad (4.78)$$

---

1. In Table 4.1 the superscript "(n)" to a quantity denotes its value at the end of the n-th step of the iterative algorithm.

To derive the limiting probabilities the algorithm builds up, for each Q_j, the n-sequence of vectors $\{P_j^{(n)}(k),\ k = 0, 1, \ldots, m_j\}$, and the limiting probability mass function is approximated by the vector $P_j^{(\tilde{n})}(k)$ such that $\forall k\ |P_j^{(\tilde{n})}(k) - P_j^{(\tilde{n}+1)}(k)| \leq \varepsilon$, where ε is the approximation level.

To construct the sequences $\{P_j^{(n)}(k),\ k = 0, 1, \ldots, m_j\}$, the following recursion is established between the probability mass functions of the number of packets queued at Q_j at two consecutive scanning epochs:

$$P_j^{(n+1)}(k) = P_j^{(n)}(0) \cdot a_{j,\,k|0}^{(n)} + \sum_{i=1}^{k+1} P_j^{(n)}(i) \cdot a_{j,\,k-i+1|1}^{(n)}, \quad k = 0, \ldots, m_j - 1 \quad (4.79)$$

$$P_j^{(n+1)}(m_j) = P_j^{(n)}(0) \sum_{i=m_j}^{\infty} a_{j,\,i|0}^{(n)} + \sum_{i=1}^{m_j} P_j^{(n)}(i) \sum_{k=m_j-i+1}^{\infty} a_{j,\,k|1}^{(n)}, \quad (4.80)$$

where $a_{j,\,k|1}^{(n)}$ $(a_{j,\,k|0}^{(n)})$ is the probability that k packets arrive at Q_j in the n-th cycle given that Q_j transmits 1 (0) packet in that cycle. It can be verified that

$$a_{j,\,k|h}^{(n)} = \int_0^{\infty} \left[\frac{(\lambda_j t)^k e^{-\lambda_j t}}{k!} \right] dC_{j,\,n|h}(t), \quad h = 0, 1 \quad, \quad (4.81)$$

where $C_{j,\,n|h}(t)$ is the distribution function of the r.v. $C_{j,\,n|h}$ which indicates the length of the cycle measured by Q_j at the $(n+1)$th scanning epoch conditioning on an empty (i.e., $h = 0$), or not-empty (i.e., $h = 1$), queue at the n-th scanning epoch.

Due to dependencies between the service periods at $\{Q_1, Q_2, \ldots, Q_K\}$, the exact computation of $C_{j,\,n|h}(t)$ is very difficult, if not impossible. To overcome this problem, Tran-Gia and Raith adopt the approach developed by Kuhen (see Section 4.1.1), i.e., they assume independence between service periods at the various stations. Under this assumption, the LSTs of $C_{j,\,n|0}(t)$ and $C_{j,\,n|1}(t)$ can be derived

$$C_{j,\,n|0}^*(s) = S^*(s) \prod_{i=1}^{j-1} Sp_{i,\,n+1}^*(s) \prod_{i=j+1}^{K} Sp_{i,\,n}^*(s) \quad, \quad (4.82)$$

$$C_{j,\,n|1}^*(s) = S^*(s) B_j^*(s) \prod_{i=1}^{j-1} Sp_{i,\,n+1}^*(s) \prod_{i=j+1}^{K} Sp_{i,\,n}^*(s) \quad, \quad (4.83)$$

where $Sp^*_{i,h}(s) = [B^*_i(s)(1 - P^{(h)}_i(0)) + P^{(h)}_i(0)]$.

In principle, from (4.82) and (4.83), by using an LST inversion procedure $C_{j,n|0}(t)$ and $C_{j,n|1}(t)$ can be derived. However, this approach is computational demanding since K inversion procedures (i.e., one for each station) must be performed for each step of the iterative scheme. This problem is overcome by deriving moments from the LSTs (4.82) and (4.83), and then fitting a convenient distribution to these moments ([161], [162]). Specifically, from (4.82) and (4.83) the average and the coefficient of variation of $C_{j,n|0}$ and $C_{j,n|1}$ can be computed and then an approximation for their pdfs is obtained with the *two-moment approximation* shown in Section 4.3.4. Thus, the recursive scheme specified by (4.79) and (4.80) is completely defined. From an operational standpoint, the recursive scheme is described in Table 4.2.

Table 4.2 Iterative scheme for the computation of $P_j(k)$

Procedure

/ initialization section*

For each Q_j do
initialize the probability vector $\{P^{(0)}_j(0), P^{(0)}_j(1), ..., P^{(0)}_j(m_j)\}$;
$n=0$;
   / iterative section*

For $j=1$ to K do
Begin
compute $C_{j,n|0}(t)$ and $C_{j,n|1}(t)$ by using (4.82), (4.83), and the two-moment approximation;
compute $\{P^{(n+1)}_j(0), P^{(n+1)}_j(1), ..., P^{(n+1)}_j(m_j)\}$ by using (4.79), (4.80);
End
compute $\Delta = \sum_{j=1}^{K} \left| \sum_{k=0}^{m_j} kP^{(n+1)}_j(k) - \sum_{k=0}^{m_j} kP^{(n)}_j(k) \right|$;
$n=n+1$
If $\Delta > \varepsilon$ repeat *iterative section*

End.

Table 4.2 clearly shows that the procedure is stopped after $n+1$ iterations when the queue length at any station is stationary (i.e., it does not depend on the iteration index). Specifically, Δ is the sum through all the stations of the differences between the average number of packets in a station at the n and $n+1$ iterations. The procedure ends when Δ is lower than the value of the convergence factor ε.

The outputs of the above procedure are

(i)   the steady-state probabilities of the number of packets at Q_j at the scanning epochs, i.e., $\{P_j(0), P_j(1), ..., P_j(m_j)\}$;

(ii)   the LST of the steady-state distribution functions of the length of the cycle measured by Q_j conditioning on an empty and not-empty queue at a scanning epoch, respectively, i.e., $C_{j|0}(t)$ and $C_{j|1}(t)$. In the following, the cycles with distribution function $C_{j|0}(t)$ and $C_{j|1}(t)$ will be denoted as *type* 1 and *type* 2 cycles, respectively.

To derive the relevant system performance measures (e.g., packet-loss probabilities), for each Q_j, the steady-state probabilities at an arbitrary time t, denoted by $\{p_j(0), p_j(1), ..., p_j(m_j)\}$, need to be derived. This computation can be performed by conditioning on the cycle type the observation instant falls:

$$p_j(k) \ = \ p_j(k|type\ 1\ cycle)p_j^{(1)} + p_j(k|type\ 2\ cycle)p_j^{(2)} \ , \qquad (4.84)$$

where $p_j^{(1)}$, and $p_j^{(2)}$ are the probabilities that a random point in time falls in a *type* 1 and *type* 2 cycle, respectively.

In the next subsections the unknown quantities in (4.84) will be derived.

$p_j^{(1)}$ AND $p_j^{(2)}$ COMPUTATION. To compute $p_j^{(1)}$ and $p_j^{(2)}$ it is useful to introduce the following processes

- $\{C_{j.n}, n \geq 0\}$ i.e., the sequence of cycle length observed by Q_j, and
- $\{C_{j.n|0}, n \geq 0\}$ i.e., the sequence of cycle length observed by Q_j conditioned on Q_j empty,
- $\{C_{j.n|1}, n \geq 0\}$ i.e., the sequence of cycle length observed by Q_j conditioned on Q_j not empty.

Note that $\{C_{j.n}, n \geq 0\}$, $\{C_{j.n|0}, n \geq 0\}$ and $\{C_{j.n|1}, n \geq 0\}$ are regenerative processes with respect to the time instants at which the overall system is empty when the server arrives at Q_j. Hence, by applying the main theorem for this class of processes (see Theorem 2.4) it results that

$$p_j^{(1)} = \frac{E\left[\sum_{i=1}^{H_1} C_{j,i|0}\right]}{E\left[\sum_{i=1}^{H} C_{j,i}\right]} \quad , \tag{4.85}$$

where H (H_1) is a random variable which counts the number of cycles (conditioned on Q_j empty) between two regenerative points.

From Theorem 2.4, it follows that

$$E[C_j] = E[C] = E\left[\sum_{i=1}^{H} C_{j,i}\right]/E[H] \quad , \tag{4.86}$$

$$E[C_{j|0}] = E\left[\sum_{i=1}^{H_1} C_{j,i|0}\right]/E[H_1] \quad , \tag{4.87}$$

$$p_{j,0} = E[H_1]/E[H] \quad , \tag{4.88}$$

and hence (4.85) can be rewritten as follows

$$p_j^{(1)} = (p_{j,0}E[C_{j|0}])/E[C] \quad . \tag{4.89}$$

Finally,

$$E[C] = p_{j,0}E[C_{j|0}] + (1-p_{j,0})E[C_{j|1}] \quad , \tag{4.90}$$

and

$$p_j^{(2)} = 1 - p_j^{(1)}. \tag{4.91}$$

COMPUTATION OF $p_j(k|type\ 1\ cycle)$ AND $p_j(k|type\ 2\ cycle)$. By indicating with $a_{j,n|0}$ and $a_{j,n|1}$ the probability of n arrivals at Q_j in the elapsed time from the last scanning epoch up to a random point in time t, conditioned on t belonging to a *type* 1 and *type* 2 cycle, respectively, it can be observed that

$$p_j(k|type\ 1\ cycle) = a_{j,k|0} \quad , \tag{4.92}$$

$$p_j(k|type\ 2\ cycle) = \sum_{n=1}^{k+1} \frac{P_j(n)}{1-P_j(0)} \cdot a_{j,k-n+1|1} \quad . \tag{4.93}$$

Clearly, to compute $a_{j,n|0}$ and $a_{j,n|1}$, the elapsed time distribution must be derived for both type of cycles, throughout denoted by $C_{j|0_-}(t)$ and $C_{j|1_-}(t)$. From Theorem 2.5, it results

$$C_{j|0\_}(t) = \frac{\int_0^t [1 - C_{j|0}(u)]\, du}{E\,[C_{j|0}]} \quad ,$$

(4.94)

$$C_{j|1\_}(t) = \frac{\int_0^t [1 - C_{j|1}(u)]\, du}{E\,[C_{j|1}]} \quad ,$$

(4.95)

hence

$$a_{j,\,n|0} = \int_0^\infty \left[\frac{(\lambda_j t)^n e^{-\lambda_j t}}{n!} \right] \cdot dC_{j|0\_}(t) \quad ,$$

(4.96)

$$a_{j,\,n|1} = \int_0^\infty \left[\frac{(\lambda_j t)^n e^{-\lambda_j t}}{n!} \right] \cdot dC_{j|1\_}(t) \quad .$$

(4.97)

PERFORMANCE FIGURES. From the steady-state probabilities the Q_j packet-loss probability $P_L(j)$ can be derived by noting that due to the PASTA property (see Theorem 2.3)

$$P_L(j) = p_j(m_j) \quad .$$

(4.98)

Furthermore, from Little's theorem (see Theorem 2.1) the Q_j response time ($E\,[R_j]$) is

$$E\,[R_j] = \frac{\sum_{k=1}^{m_j} k \cdot p_j(k)}{\lambda_j \cdot (1 - P_L(j))} \quad .$$

(4.99)

4.2.2 Power Series Algorithm

The *Power Series Algorithm* (*PSA*) [20] is an iterative algorithm that can be used to evaluate steady state probabilities of multiqueue systems which can be modeled as multidimensional QBD processes [125]. PSA requires a Markovian representation of multiqueue systems, and this is often achieved by using the supplementary variable technique. The state of the multiqueue system is described by means of a vector whose components represent the number of packets in each queue and, whenever necessary, one or more supplementary components to make the process Markovian. The latter may be needed, for example, to model the Coxian service time distribution or to model the state of a server (i.e., switching or serving in a polling system).

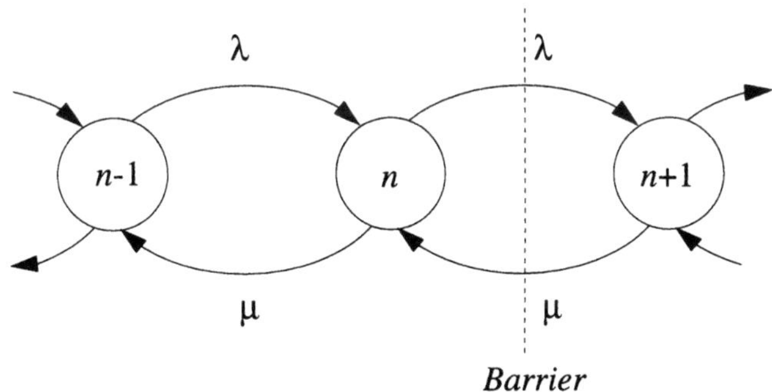

Figure 4.9: Rate transition diagram for *M/M/*1 queue

An important class of systems which can be modeled via multidimensional QBDs are the polling systems. Examples of PSA applied to polling systems are developed throughout the book to show the potential of this algorithm. Other examples are reported in [20].

Before showing the key ideas behind PSA, solution steps for one-dimensional birth-dead (*BD*) processes are reviewed. Specifically, for the *M/M/*1 queue with arrival process rate λ and service rate μ, a "short-cut" method for the computation of the stationary probabilities $p(n)$ of n packets in the *M/M/*1 system goes through the following steps:

(i)   for any given state (Figure 4.9) the "probability flow" [99] must be conserved, i.e., "flow into" a state must equate "flow out" of that state

$$(\lambda + \mu I_{\{n \geq 1\}})\, p(n) \;=\; \lambda I_{\{n \geq 1\}}\, p(n-1) + \mu p(n+1) \quad , \qquad (4.100)$$

where $I_{\{n \geq 1\}}$ is the indicator function of the event $\{n \geq 1\}$;

(ii)   the stationary probabilities $p(n)$ obtained by solving the linear equations system (4.100) must add up to one

$$\sum_{n=0}^{\infty} p(n) \;=\; 1 \quad ; \qquad (4.101)$$

(iii)   the (unique) solution to linear equations system (4.100) satisfying (4.101) has the following simple structure

$$p(n) = (1-\rho)\rho^n \ , \tag{4.102}$$

where $\rho = \lambda/\mu$ is the traffic intensity which must be less than one to have a stable $M/M/1$ system.

From (4.102) it can be readily verified that

$$p(n) = o(\rho^n), \rho \to 0 \ . \tag{4.103}$$

Equations (4.100), obtained by equating total flow into a state to total flow out of the state, are also referred to as *global balance equations*. For the $M/M/1$ queue, or more generally for the class of BDs permitting only near-est-neighbor transitions, the so-called *local* or *detailed* balance equations hold. As shown in Figure 4.9 these are obtained by looking at the probability flow through a barrier between state n and state $n + 1$. In steady-state conditions, the flow to the right of this barrier must be equal to the flow to the left, and this condition leads to the following equations $\lambda p(n) = \mu p(n + 1)$. Therefore for BDs permitting only nearest-neighbor transitions (e.g., the $M/M/1$ queue), global balance equations (4.100) are equivalent to

$$\lambda p(n) = \mu p(n + 1) \ . \tag{4.104}$$

REMARK. From the elementary queueing theory it is known that the problem of computing $p(n)$ is greatly simplified when the *local balance equations* (4.104) hold. The recursive structure of (4.104) allows the computation of $p(n)$ in an iterative way which leads to a closed formula for $p(n)$.

$$\Diamond$$

For QBD processes the global balance equations (4.100) still hold. Therefore, the steady-state probabilities, which satisfy (4.101) and a property similar to (4.103)

$$p(n_1, ..., n_s) = o(\rho^{n_1 + ... + n_s}), \quad \rho \to 0 \ , \tag{4.105}$$

can be computed from (4.100).

REMARK. It can be recognized that (4.105) holds for the stationary probabilities of a product form queueing network which can be modeled by a multidimensional BD process [54].

$$\Diamond$$

In multi-dimensional QBDs, local balance equations of type (4.104) usually do not hold due to the multiplicity of paths which may exist between pairs of neighbouring states. Hence, the recursive structure (4.104) cannot be exploited in the computation of the stationary probabilities.

The key idea behind PSA is the transformation of the non-recursively (infinite) set of global balance equations into a different set of recursive equations by adding one or more dimensions to the state space. This transformation is realized by means of power-series expansions of the steady-state probabilities as functions of the offered load of the system. The following concepts and notations are introduced before beginning the discussion of PSA.

The state of a multidimensional QBD process can be described by the couple $(\mathbf{N}, F)$ where:

- $\mathbf{N} = (N_1, N_2, ..., N_s)$ is a vector where the N_i component describes the number of packets in the i-th queue $(i = 1, ..., s)$ and

- F is a supplementary variable which takes a finite number of values in the set Ω.

The supplementary variable, which can be a scalar or a vector, may be used, for example, to model Coxian service time distributions [77] or to indicate in a polling system whether the server is switching or serving.

The steady-state probability of a multidimensional QBD process in state $(\mathbf{n}, \varphi) \in \mathbb{N}^s \times \Omega$ will be denoted by

$$p(\mathbf{n}, \varphi) = P\{(\mathbf{N}, F) = (\mathbf{n}, \varphi)\} \ .$$

For $(\mathbf{n}, \varphi) \in \mathbb{N}^s \times \Omega$, $j = 1, ..., s$, $\psi \in \Omega$, the one step transition rates of the multidimensional QBD are defined as follows

- $\chi \cdot a_j(\mathbf{n}, \varphi, \psi)$: arrival rate at queue j, which leads the system from state $(\mathbf{n}, \varphi)$ to state $(\mathbf{n} + \mathbf{e}_j, \psi)$;

- $d_j(\mathbf{n}, \varphi, \psi)$: departure rate from queue j, which leads the system from state $(\mathbf{n}, \varphi)$ to state $(\mathbf{n} - \mathbf{e}_j, \psi)$; $d_j(\mathbf{n}, \varphi, \psi) = 0$ if $n_j = 0$;

- $u(\mathbf{n}, \varphi, \psi)$: the phase transition rate from state $(\mathbf{n}, \varphi)$ to state $(\mathbf{n}, \psi)$.

In the previous definitions $\mathbf{e}_j$ indicates the vector with zero entries except for the j-th entry $(j = 1, 2, ..., s)$ which is equal to one. Parameter χ is used as a variable in the power series expansion of $p(\mathbf{n}, \varphi)$. Furthermore, $a_j(\mathbf{n}, \varphi, \psi)$ are the relative arrival rates which are assumed to be normalized in such a way that the system is stable for $\chi < 1$.

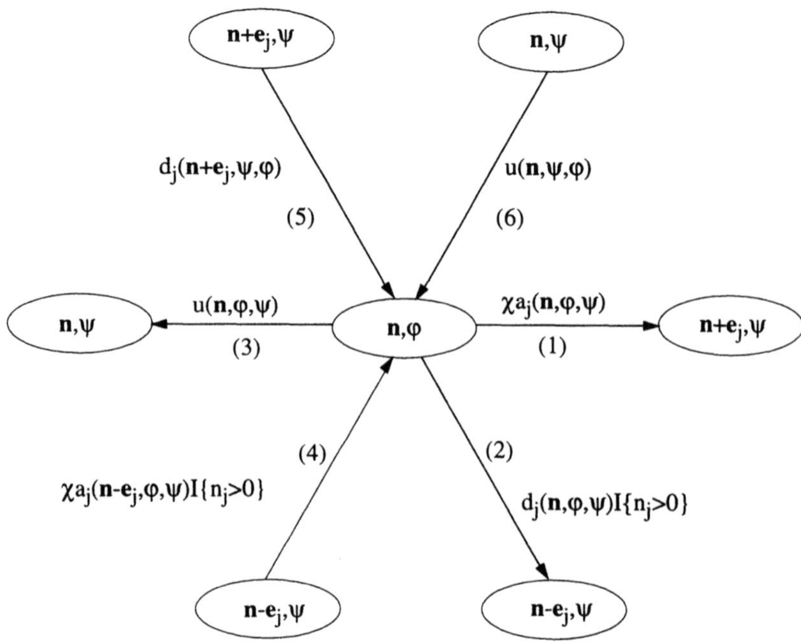

Figure 4.10: Rate transition diagram

Figure 4.10 depicts the transition rates from and onto a generic state $(\mathbf{n}, \varphi)$. Specifically, state $(\mathbf{n}, \varphi)$ is left if

1. an arrival occurs at one of the queues (transition 1);

2. a service at one of the queues is completed (transition 2);

3. there is a change in the phase (transition 3).

On the other hand, state $(\mathbf{n}, \varphi)$ is entered if

4. an arrival occurs at Q_j $(j = 1, 2, ..., s)$ and the process was in state $(\mathbf{n} - \mathbf{e}_j, \psi)$ (transition 4). This transition is a possible state only if $n_j > 0$);

5. a service is completed at queue j and the process was in state $(\mathbf{n} + \mathbf{e}_j, \psi)$ (transition 5);

6. a phase change from ψ to φ occurs and the process was in state $(\mathbf{n}, \psi)$ (transition 6).

Following the classical approach for birth-death processes, the global balance equations for the flows out of and into state $(\mathbf{n}, \varphi)$ can be derived

$$\left\{ \sum_{j=1}^{s} \sum_{\psi \in \Omega} [\chi a_j(\mathbf{n}, \varphi, \psi) + d_j(\mathbf{n}, \varphi, \psi)] + \sum_{\psi \in \Omega} u(\mathbf{n}, \varphi, \psi) \right\} \cdot p(\mathbf{n}, \varphi) \qquad (4.106)$$

$$= \chi \sum_{j=1}^{s} \sum_{\psi \in \Omega} a_j(\mathbf{n} - \mathbf{e}_j, \psi, \varphi) \cdot p(\mathbf{n} - \mathbf{e}_j, \psi) \cdot I_{\{n_j > 0\}}$$

$$+ \sum_{j=1}^{s} \sum_{\psi \in \Omega} d_j(\mathbf{n} + \mathbf{e}_j, \psi, \varphi) \cdot p(\mathbf{n} + \mathbf{e}_j, \psi) + \sum_{\psi \in \Omega} u(\mathbf{n}, \psi, \varphi) \cdot p(\mathbf{n}, \psi) \ .$$

In (4.106) $I_{\{n_j > 0\}}$ is the indicator function of the event $\{n_j > 0\}$.

To transform (4.106) into a set of recursively solvable equations, PSA expands $p(\mathbf{n}, \varphi)$ into the following power-series

$$p(\mathbf{n}, \varphi) = \chi^{\{n_1 + n_2 + \ldots + n_s\}} \sum_{k=0}^{\infty} \chi^k b(k; \mathbf{n}, \varphi), \ \forall \, (\mathbf{n}, \varphi) \in \mathbb{N}^s \times \Omega \ . \qquad (4.107)$$

Substituting (4.107) into (4.106) and equating the coefficients of corresponding powers of χ, the following sets of equations in terms of the coefficients of the power series (4.107) are obtained for $(\mathbf{n}, \varphi) \in \mathbb{N}^s \times \Omega$, and for $k = 0, 1, 2, \ldots,$

$$\left\{ \sum_{\psi \in \Omega} \left[u(\mathbf{n}, \varphi, \psi) + \sum_{j=i}^{s} d_j(\mathbf{n}, \varphi, \psi) \right] \right\} \cdot b(k; \mathbf{n}, \varphi) \qquad (4.108)$$

$$= \sum_{\psi \in \Omega} u(\mathbf{n}, \psi, \varphi) \cdot b(k; \mathbf{n}, \psi)$$

$$+ \sum_{j=1}^{s} \sum_{\psi \in \Omega} [\, a_j(\mathbf{n} - \mathbf{e}_j, \psi, \varphi) \cdot b(k; \mathbf{n} - \mathbf{e}_j, \psi) \cdot I_{\{n_j > 0\}}$$

$$- a_j(\mathbf{n}, \varphi, \psi) \cdot b(k - 1; \mathbf{n}, \varphi) \cdot I_{\{k > 0\}} \,]$$

$$+ \sum_{j=1}^{s} \sum_{\psi \in \Omega} d_j(\mathbf{n} + \mathbf{e}_j, \psi, \varphi) \cdot b(k - 1; \mathbf{n} + \mathbf{e}_j, \psi) \cdot I_{\{k > 0\}} \ .$$

As shown in Figure 4.11 the set of equations (4.108) can be solved iteratively as there is a partial ordering with respect to the components $(k;\mathbf{n})$.

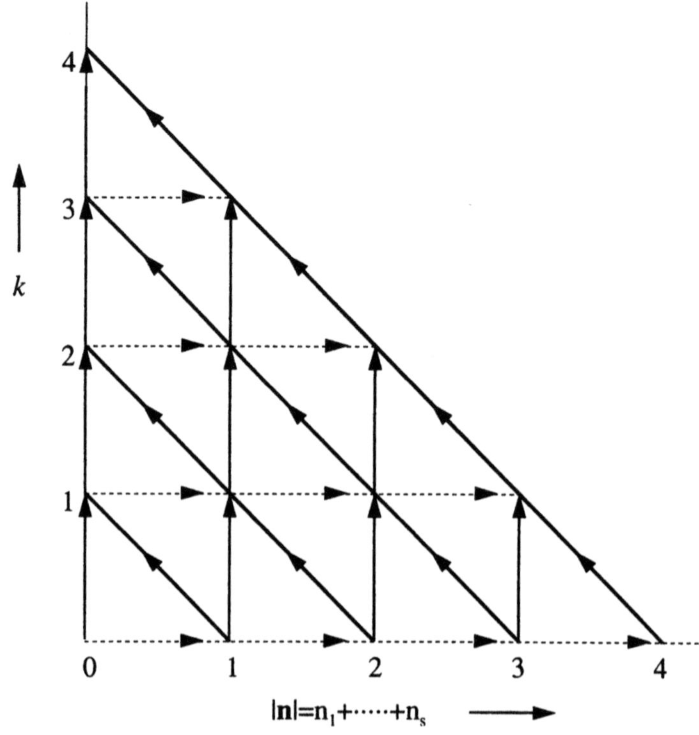

Figure 4.11: Dependencies between the power series coefficients

From (4.108) can be seen that $b(k;\mathbf{n}, \varphi)$ depends on lower order coefficients, i.e., $b(k-1;\mathbf{n}, \varphi)$, $b(k;\mathbf{n}-\mathbf{e}_j, \varphi)$, $b(k-1;\mathbf{n}+\mathbf{e}_j, \psi)$ and on coefficients of the same order $b(k;\mathbf{n}, \psi)$. Hence, for any $(k;\mathbf{n})$, (4.108) forms a system of at most $|\Omega|$ linear equations from which the $b(k;\mathbf{n}, \varphi)$ coefficients can be derived from lower order coefficients. It can be shown [20] that if the system is stable the solution of the above system is unique.

States where $\mathbf{n} = \mathbf{0}$ must be handled with special care. In this case as departure rates vanish for $\varphi \in \Omega$ and for $k = 0, 1, 2, ...,$ (4.108) is reduced to:

$$\sum_{\psi \in \Omega} u(\mathbf{0}, \varphi, \psi) \cdot b(k;\mathbf{0}, \varphi) \tag{4.109}$$

$$= \sum_{\psi \in \Omega} u(\mathbf{0}, \psi, \varphi) \cdot b(k;\mathbf{0}, \psi) - \sum_{j=1}^{s} \sum_{\psi \in \Omega} a_j(\mathbf{0}, \varphi, \psi) \cdot b(k-1;\mathbf{0}, \varphi) \cdot I_{\{k>0\}}$$

$$- \sum_{j=1}^{s} \sum_{\psi \in \Omega} d_j(\mathbf{e}_j, \psi, \varphi) \cdot b(k-1;\mathbf{e}_j, \psi) \cdot I_{\{k>0\}} \quad .$$

It can be shown that, for each k, equation (4.109) forms a set of dependent equations for the unknown coefficients $b(k;\mathbf{0}, \varphi)$. Hence, to find a unique solution of the system (4.109) an additional equation is needed. This is provided by the normalization condition, i.e.,

$$\sum_{\mathbf{n} \in \mathbb{N}^s} \sum_{\varphi \in \Omega} p(\mathbf{n}, \varphi) = 1 \quad . \tag{4.110}$$

Substituting (4.107) into (4.110) and equating coefficients of corresponding powers of χ the following equations, one for each k $(k = 0, 1, 2, ...,)$, are obtained

$$\sum_{\varphi \in \Omega} b(0;\mathbf{0}, \varphi) = 1$$

$$\sum_{\varphi \in \Omega} b(k;\mathbf{0}, \varphi) = - \sum_{0 < n_1 + n_2 + ... + n_s \leq k} \left(\sum_{\varphi \in \Omega} b(k - n_1 - n_2 - ... - n_s;\mathbf{n}, \psi) \right) \quad . \tag{4.111}$$

As can be seen from the r.h.s of (4.111), $b(k;\mathbf{0}, \varphi)$ is defined in terms of lower order coefficients. Hence, (4.111) allows the iterative solution of (4.109). To conclude, for $\mathbf{n} = \mathbf{0}$ the set of equations (one set for each k) to be solved is defined by (4.111) and by all but one equation (4.109). For each $\mathbf{n} \neq \mathbf{0}$ the set of equations to be solved, (one set for each k), is defined by equation (4.108).

REMARK. Figure 4.11 summarizes the dependencies between the power series coefficients that must be taken into account in the definition of the iterative schema. The figure shows that there appears to be some freedom in choosing the order in which the coefficients $b(k;\mathbf{n}, \varphi)$ can be computed. Disregarding phase φ, one possible approach could be the following: compute $b(0;\mathbf{n})$ recursively for increasing values of $|\mathbf{n}|$ up to $|\mathbf{n}| = M$ (for some values of M) then compute $b(1;\mathbf{n})$ recursively for increasing values of $|\mathbf{n}|$ up to $|\mathbf{n}| = M - 1$, and so on, until $b(M;\mathbf{0})$ is reached. Another approach could be to compute $b(k;\mathbf{n})$ according to increasing values $m = k + |\mathbf{n}|$ for $m = 0, 1, 2, ..., M$, where at each level m the coefficients have to be computed in increasing order of k, for $k = 0, 1, 2, ..., m$.

$\Diamond$

POLLING MODEL ANALYSIS VIA PSA. To show its potential, PSA is applied to the analysis of a cyclic polling model with limited service discipline [20]. For example, an FDDI network where stations transmit only synchronous frames, fits this model.

The polling model assumes that packets arrive at Q_j $(j = 1, ..., s)$ according to a Poisson process with arrival rate $\lambda_j = \chi a_j$. The aggregate arrival process is Poisson with rate $\lambda = \chi \sum_{j=1}^{s} a_j = \chi A$, where $A = \sum_{j=1}^{s} a_j$. Service times of packets arriving at Q_j are assumed to be exponentially distributed with rate μ_j. The local offered load at Q_j is $\rho_j = \lambda_j / \mu_j$, and the total offered load to the system is $\rho = \sum_{j=1}^{s} \rho_j$. The service discipline is exhaustive l-limited, see Section 2.1. During a visit of the server to Q_j at most l_j packets will be served. The server will switch to the next queue if this number has been reached or Q_j has been emptied, whichever occurs first. The time needed by the server to switch from Q_{j-1} to Q_j is assumed to be exponentially distributed with rate v_j $(j = 1, ..., s)$. Blanc gives the following condition for stability for this system [20]

$$\chi = \rho + E[S] \cdot Max_{\{j=1, ..., s\}} \{\lambda_j / l_j\} < 1 \ , \tag{4.112}$$

where $E[S]$ is the mean total switchover time during a cycle of the server along the queues.

Two supplementary variables, hereafter H and Z, are used to make the queue length process Markovian. The resulting state will be denoted by the triple $(\mathbf{N}, H, Z)$ where:

- $\mathbf{N} = (N_1, N_2, ..., N_s)$ is a vector in which the N_i component describes the number of packets in the i-th queue $(i = 1, ..., s)$;
- H indicates the server position, i.e., the queue to which the server is switching or the queue at which the server is serving; and

- Z represents the state of the server. More precisely, $Z = 0$ means that the server is switching to queue H, while $Z = \kappa$, $\kappa = 1, 2, ..., l_H$, indicates that the server is performing the κ-th service during the current visit to queue H.

The state probabilities of the resulting multidimensional QBD process will be denoted by $p(\mathbf{n}, h, \kappa)$.

The set of global balance equations for the QBD process under study can be partitioned into two subsets, depending upon whether, with respect to station j, the server is switching to the station $(Z = 0)$ or serving at the station $(Z = \kappa, \quad \kappa = 1, 2, ..., l_h)$.

In the former case, for $\mathbf{n} \in \mathbb{N}^s$, $h = 0, 1, ..., s - 1$ Figure 4.12 depicts the transition rates from and onto a generic state $(\mathbf{n}, h + 1, 0)$.

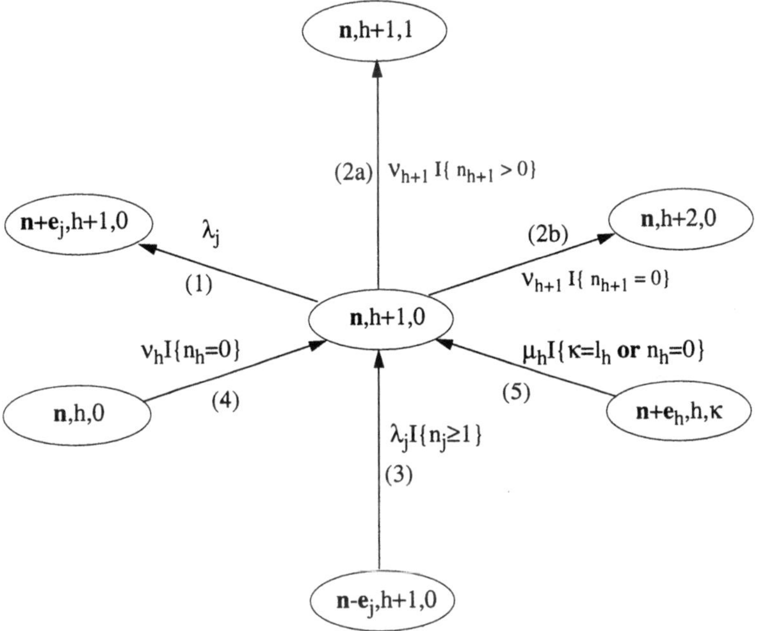

Figure 4.12: Rate transition diagram when the server is switching

State $(\mathbf{n}, h + 1, 0)$ is left if either an arrival occurs at one of the queues (transition 1) or the switching from queue h to queue $h + 1$ is complete. In the latter case two possibilities may happen: queue $h + 1$ is not empty (transition 2a), or queue $h + 1$ is empty and thus the switching to queue $h + 2$ begins (transition 2b).

State $(\mathbf{n}, h + 1, 0)$ is entered if one of the following conditions occurs:

- the process was in state $(\mathbf{n} - \mathbf{e}_j, h + 1, 0)$ and an arrival occurs at Q_j (transition 3);
- the process was in state $(\mathbf{n}, h, 0)$ and station h was empty when the switching from station $h - 1$ to h is completed (transition 4);
- the process was in state $(\mathbf{n} + \mathbf{e}_h, h, \kappa)$ and a service is completed at queue h (transition 5). In this case, the switching to the station $h + 1$ can be due to the fact that either the maximum allowed number of packet transmissions (l_h) has been reached, or queue h has been emptied, whichever occurs first.

Hence the global balance equations for the flows out of and into state $(\mathbf{n}, h + 1, 0)$ can be derived

$$(\lambda + v_{h+1})\, p(\mathbf{n}, h + 1, 0) = \sum_{j=1}^{s} \lambda_j I_{\{n_j \geq 1\}}\, p(\mathbf{n} - \mathbf{e}_j, h + 1, 0) + \qquad (4.113)$$

$$v_h I_{\{n_h = 0\}}\, p(\mathbf{n}, h, 0) + \mu_h \sum_{\kappa=1}^{l_h} I_{\{\kappa = l_h \vee n_h = 0\}}\, p(\mathbf{n} + \mathbf{e}_h, h, \kappa) \quad .$$

On the other hand, when the server is serving at station h, for $\mathbf{n} \in \mathbb{N}^s$, $h = 1, ..., s$, $n_h \geq 1$, and $\kappa = 1, 2, ..., l_h$, Figure 4.13 depicts the transition rates from and onto a generic state $(\mathbf{n}, h, \kappa)$.

State $(\mathbf{n}, h, \kappa)$ is left if either an arrival occurs at one of the queues (transition 1) or a service is completed at queue h. (transitions 2a and 2b). In the latter case the server either starts serving the next packet to queue h. (transition 2a) or switches to station $h + 1$ (transition 2b). The switching period starts as soon as queue h. becomes empty or l_h transmissions have been completed at queue h., whichever occurs first.

State $(\mathbf{n}, h, \kappa)$ is entered if one of the following conditions occurs:

- the process was in state $(\mathbf{n} - \mathbf{e}_j, h, \kappa)$ and an arrival occurs at Q_j (transition 3);
- the process was in state $(\mathbf{n}, h, 0)$, the switching from station $h - 1$ to h is completed and the first packet transmission at Q_h starts (transition 4),
- the process was in state $(\mathbf{n} + \mathbf{e}_h, h, \kappa - 1)$ and the $(\kappa - 1)$-th service at Q_h is completed (transition 5).

In this case the global balance equations for the flows out of and into state $(\mathbf{n}, h, \kappa)$ are

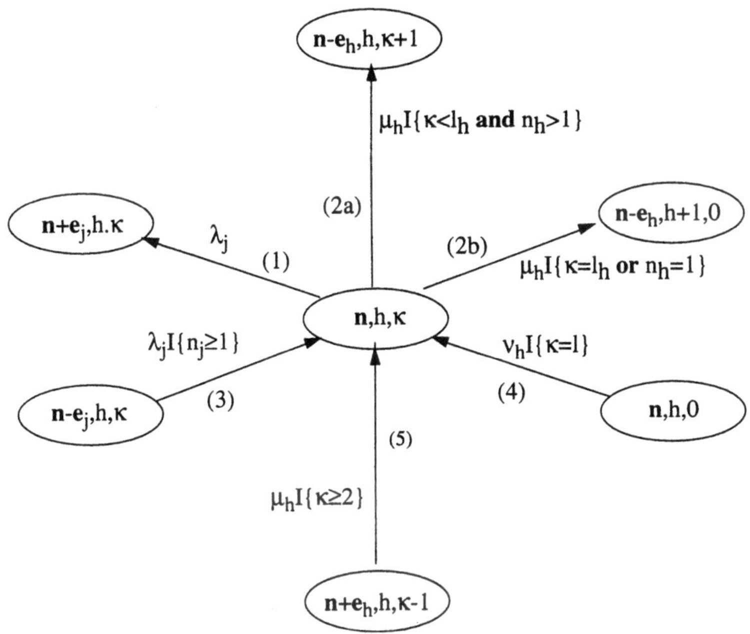

Figure 4.13: Rate transition diagram when the server is attending queue h

$$(\lambda + \mu_h)\, p(\mathbf{n}, h, \kappa) = \sum_{j=1}^{s} \lambda_j I_{\{n_j \geq 1\}}\, p(\mathbf{n} - \mathbf{e}_j, h, \kappa) + \tag{4.114}$$

$$v_h I_{\{\kappa=1\}}\, p(\mathbf{n}, h, 0) + \mu_h I_{\{\kappa \geq 2\}}\, p(\mathbf{n} + \mathbf{e}_h, h, \kappa-1) \quad .$$

Furthermore

$$\sum_{n_1=0}^{\infty} \cdots \sum_{n_s=0}^{\infty} \sum_{h=1}^{s} \sum_{\kappa=0}^{l_h} p(\mathbf{n}, h, \kappa) = 1 \quad . \tag{4.115}$$

Clearly, if $n_h = 0$, $p(\mathbf{n}, h, \kappa) = 0$ for all $\mathbf{n} \in \mathbb{N}^s$, $h = 1, \ldots, s$, and $\kappa = 1, 2, \ldots, l_h$. According to the general PSA theory developed above, the state probabilities $p(\mathbf{n}, h, \kappa)$ are expressed as a power series expansion of χ. Specifically, for $\mathbf{n} \in \mathbb{N}^s$, $h = 1, \ldots, s$, and $\kappa = 0, 1, 2, \ldots, l_h$,

$$p(\mathbf{n}, h, \kappa) = \chi^{|\mathbf{n}|} \sum_{k=0}^{\infty} \chi^k b(k; \mathbf{n}, h, \kappa) \quad . \tag{4.116}$$

It should be noted that the stability of the system requires that χ satisfies the following inequality: $0 \le \chi < 1$.

Substituting (4.116) into the global balance equations (4.113) and (4.114), and equating the coefficient of corresponding powers of χ in the resulting equations, leads to the following set of equations for the coefficients in (4.116).

1. for $k = 0, 1, 2, \ldots$, for $\mathbf{n} \in \mathbb{N}^s$, and $h = 0, 1, \ldots, s-1$,

$$v_{h+1} b(k;\mathbf{n}, h+1, 0) = \sum_{j=1}^{s} a_j I_{\{n_j \ge 1\}} b(k;\mathbf{n} - \mathbf{e}_j, h+1, 0) +$$

$$AI_{\{k \ge 1\}} b(k-1;\mathbf{n}, h+1, 0) + v_h I_{\{n_h = 0\}} b(k;\mathbf{n}, h, 0) +$$

$$\mu_h I_{\{k \ge 1\}} \sum_{\kappa=1}^{l_h} I_{\{\kappa = l_h \vee n_h = 0\}} b(k-1;\mathbf{n} + \mathbf{e}_h, h, \kappa) \quad ; \tag{4.117}$$

2. for $k = 0, 1, 2, \ldots$, $\mathbf{n} \in \mathbb{N}^s$, $h = 1, \ldots, s$, $n_h \ge 1$, and $\kappa = 1, 2, \ldots, l_h$,

$$\mu_h b(k;\mathbf{n}, h, \kappa) = \sum_{j=1}^{s} a_j I_{\{n_j \ge 1\}} b(k;\mathbf{n} - \mathbf{e}_j, h, \kappa) - AI_{\{k \ge 1\}} b(k-1;\mathbf{n}, h, \kappa) +$$

$$v_h I_{\{k \ge 1\}} b(k;\mathbf{n}, h, 0) + \mu_h I_{\{\kappa \ge 2, k \ge 1\}} b(k-1;\mathbf{n} + \mathbf{e}_h, h, \kappa - 1) \quad . \tag{4.118}$$

For $\mathbf{n} \ne \mathbf{0}$ equations (4.117) and (4.118) express the coefficients $b(k;\mathbf{n}, h, \kappa)$ in terms of the lower order with respect to the partial ordering $\prec$ of vectors $(k;\mathbf{n}, h, \kappa)$ defined as follows

(i) $(k;\mathbf{n}, h, \kappa) \prec (\hat{k};\hat{\mathbf{n}}, \hat{h}, \hat{\kappa})$ if

$$(k + n_1 + \ldots\ldots + n_s < \hat{k} + \hat{n}_1 + \ldots\ldots + \hat{n}_s) \quad \text{or}$$

$$(k + n_1 + \ldots\ldots + n_s = \hat{k} + \hat{n}_1 + \ldots\ldots + \hat{n}_s \text{ and } k < \hat{k}) \quad ;$$

(ii) $(k;\mathbf{n}, h, \kappa) \prec (k;\mathbf{n}, h, \hat{\kappa})$ if

$$\kappa < \hat{\kappa} \text{ for } \kappa = 0, 1, \ldots\ldots, l_h \quad .$$

The above partial ordering does not hold for the term $b(k;\mathbf{n}, h, 0)$ in (4.117), which only plays a role when $n_h = 0$ for some h, $h = 1, 2, \ldots, s$. However, if $\mathbf{n} \ne \mathbf{0}$, the set of coefficients $b(k;\mathbf{n}, h, 0)$, given k and $\mathbf{n}$, can still be recursively computed by starting at a value $h = j$ with $n_j > 0$ and by proceeding sequentially with the computations of the coefficients $b(k;\mathbf{n}, h, 0)$. Then the algorithm computes $b(k;\mathbf{n}, h, 0)$ for $h = j + 1, ..., s, 1, ..., j - 1$.

The only states which must be handled with care are those with $\mathbf{n} = \mathbf{0}$ and $\kappa = 0$. For these states equation (4.117) yields, for $k = 0, 1, 2, ..., \tilde{h} = 0, ..., s - 1$

$$v_{h+1} b(k;\mathbf{0}, h + 1, 0) = v_h b(k;\mathbf{0}, h, 0) + \tag{4.119}$$

$$I_{\{k \geq 1\}} \left[\mu_h \sum_{\kappa = 1}^{l_h} b(k - 1;\mathbf{e}_h, h, \kappa) - A \cdot b(k - 1;\mathbf{0}, h + 1, 0) \right] .$$

It can be verified that, for each k, $(k = 0, 1, 2, ...,)$ the equations defined by (4.119) are dependent. As mentioned before, a set of independent equations is obtained by exploiting the normalization condition. Specifically, by substituting (4.116) into (4.115) and equating the coefficients of corresponding powers of χ, in the resulting equation the following equations are obtained

$$\sum_{h=1}^{s} b(0;\mathbf{0}, h, 0) = 1 , \tag{4.120}$$

and, for $k = 1, 2, ...$

$$\sum_{h=1}^{s} b(k;\mathbf{0}, h, 0) = - \sum_{1 \leq |\mathbf{n}| \leq k} \sum_{h=1}^{s} \sum_{\kappa=0}^{l_h} b(k - |\mathbf{n}|;\mathbf{n}, h, \kappa) . \tag{4.121}$$

Equation (4.120) for $k = 0$ and equation (4.121) for $k = 1, 2, ...$ together with $s - 1$ equations defined by (4.119) form a set of s linear equations which allow one to uniquely determine the s coefficients $b(k;\mathbf{0}, h, 0)$.

BIBLIOGRAPHY ON THE POWER-SERIES ALGORITHM. The basic idea of PSA was introduced in Hooghiemstra, Keane, and van de Ree [86]. The algorithm has been further developed by Blanc which has led to more efficient implementations, faster convergence of the power series by means of the epsilon algorithm [167], and a broader scope of applications.

General conditions for applying the PSA to birth-and-death models are derived in [15]. The application of the PSA to exponential cyclic-polling systems with zero switching times and Bernoulli schedules as service disciplines is developed in [16]. The PSA has been extended to exponential cyclic polling systems with non-zero switching times in [17]. Computations with the PSA are compared with simulations in [17], [18]. It has turned out that (pseudo)conservation laws for mean waiting times are much better fulfilled by computations with the PSA than by estimations obtained by simulations of comparable duration as required by the PSA. [18] describes the PSA for polling systems without switchover times with general periodic polling order, with cyclic polling systems with limited service disciplines as special cases.

The review paper [19] discusses the PSA in its generality for quasi birth-and-death models, and in detail for periodic-polling systems with Bernoulli schedules and with Coxian distributed service and switching times. Also, it discusses the applicability and numerical complexity of the PSA for polling systems with other visit rules and service disciplines.

In all the above mentioned studies Poisson arrival processes are assumed. Generalization of the concept of the PSA to models with Batch Markovian Arrival Processes (BMAP) is the goal of Van den Hout & Blanc [158].

4.3 FOR FURTHER STUDY

For ease of reading, several lemmas in Section 4.1 were stated without proof. This section contains the omitted proofs. Specifically, Section 4.3.1 contains the proof related to the convergence of the iterative scheme used to approximate the cycle length in a polling system, Sections 4.3.2 and 4.3.3 contain the lemmas related to the waiting time analysis, based on the pseudo-conservation law, for the 1-limited and l-limited polling systems, respectively. Finally, this section includes the two-moment approximation method which is extensively used to approximate general distributions.

4.3.1 Cycle Length Analysis: Convergence of the Iterative Procedure

This section is devoted to proving Lemma 4.1 (see Section 4.1.1 on page 137) which defines the conditions under which the iterative scheme to compute the average cycle length converges.

LEMMA 4.1 *For all starting values $E[C_{i|1}(0)] > 0$ it follows that*

$$\forall n, E[C_{i|1}(n)] > 0 \quad , \tag{4.122}$$

and

$$\left| E[C_{i|1}(n+1)] - E[C_{i|1}(n)] \right| < \left| E[C_{i|1}(n)] - E[C_{i|1}(n-1)] \right| \quad . \tag{4.123}$$

PROOF

From (4.11)

$$E[C_{i|1}(n+1)] = s + b_i + \sum_{j \neq i} \min\{\lambda_j E[C_{i|1}(n)], 1\} b_j$$

it immediately follows that, if $\forall n \ E[C_{i|1}(n)] > 0$ then $E[C_{i|1}(n+1)] > 0$. Hence $E[C_{i|1}(0)] > 0 \Rightarrow E[C_{i|1}(n)] > 0, \forall n$.

Let $\left| E[C_{i|1}(n)] - E[C_{i|1}(n-1)] \right| = \varepsilon$; this can occur if and only if

(i) $E[C_{i|1}(n)] = E[C_{i|1}(n-1)] + \varepsilon$; or

(ii) $E[C_{i|1}(n)] = E[C_{i|1}(n-1)] - \varepsilon$.

In case (i), i.e., $E[C_{i|1}(n)] = E[C_{i|1}(n-1)] + \varepsilon$,

$$\left| E[C_{i|1}(n+1)] - E[C_{i|1}(n)] \right| = \tag{4.124}$$

$$\left| \sum_{j \neq i} [\min\{\lambda_j(E[C_{i|1}(n-1)] + \varepsilon), 1\} - \min\{\lambda_j E[C_{i|1}(n-1)], 1\}] b_j \right| \quad .$$

Let

$$\Theta^{(n)} = \{j \mid j \neq i, \lambda_j E[C_{i|1}(n-1)] \leq 1\} \quad ,$$

and

$$\Gamma^{(n)} = \{j \mid (j \neq i, \lambda_j E[C_{i|1}(n-1)] > 1)\} \quad .$$

According to hypothesis (i),

$$\Gamma^{(n)} \subseteq \Gamma^{(n+1)}, \text{ and } \Theta^{(n)} \supseteq \Theta^{(n+1)}.$$

Hence in the sum of the r.h.s. of (4.124) the only terms which differ from zero are those which correspond to the indices j, $j \in \Theta^{(n)} \cap \Gamma^{(n+1)}$, i.e.,

$$\left| E\left[C_{i|1}(n+1)\right] - E\left[C_{i|1}(n)\right] \right| = \tag{4.125}$$

$$\left| \sum_{j \in \Theta^{(n)} \cap \Gamma^{(n+1)}} \left[1 - \lambda_j E\left[C_{i|1}(n-1)\right]\right] b_j \right| \quad .$$

By observing that $\forall j$, $j \in \Theta^{(n)} \cap \Gamma^{(n+1)}$, $\lambda_j \left(E\left[C_{i|1}(n-1)\right] + \varepsilon\right) > 1$ it immediately follows that

$$\left[1 - \lambda_j E\left[C_{i|1}(n-1)\right]\right] b_j < \lambda_j \varepsilon b_j \leq \rho_j \quad . \tag{4.126}$$

Finally, by substituting (4.126) in (4.125)

$$\left| E\left[C_{i|1}(n+1)\right] - E\left[C_{i|1}(n)\right] \right| < \left| \sum_{j \in \Theta^{(n)} \cap \Gamma^{(n+1)}} \rho_j \varepsilon \right| \leq \rho \varepsilon \quad , \tag{4.127}$$

and hence if the system is stable (i.e., $\rho < 1$) the first part of the lemma is proved.

The second part of the lemma can be proved by following the same line of reasoning. Specifically, assuming that $E\left[C_{i|1}(n)\right] = E\left[C_{i|1}(n-1)\right] - \varepsilon$, it immediately follows that $\forall j$, $j \in \Theta^{(n+1)} \cap \Gamma^{(n)}$, $\lambda_j E\left[C_{i|1}(n-1)\right] > 1$ and $\lambda_j \left(E\left[C_{i|1}(n-1)\right] - \varepsilon\right) < 1$ from which

$$\left| E\left[C_{i|1}(n+1)\right] - E\left[C_{i|1}(n)\right] \right| = \tag{4.128}$$

$$\left| \sum_{j \in \Theta^{(n+1)} \cap \Gamma^{(n)}} \left[\lambda_j \left(E\left[C_{i|1}(n-1)\right] - \varepsilon\right) - 1\right] b_j \right| =$$

$$\sum_{j \in \Theta^{(n+1)} \cap \Gamma^{(n)}} \left[1 - \lambda_j \left(E\left[C_{i|1}(n-1)\right] - \varepsilon\right)\right] b_j <$$

$$\sum_{j \in \Theta^{(n+1)} \cap \Gamma^{(n)}} \lambda_j \varepsilon b_j < \rho \varepsilon \quad .$$

4.3.2 1-limited Polling System: Details of the Computation

In Section 4.1.3 (see (4.53)) it is shown that

$$E\left[W_i\right] = Y_i E\left[C_{i,0+}\right] + Z_i \quad , \tag{4.129}$$

where

(i) $Y_i = 1/\left(1 - \lambda_i E\left[C_{i|1}\right]\right)$, and

(ii) $Z_i = (E[I_{i,1}^{(v)}] - E[I_{i|1}])(1 - \rho_i - p_i(0))/(1 - \lambda_i E[C_{i|1}])$,

(iii) $E[C_{i,0+}] = E[C_{i,0+}|\xi_i = 0](1 - \rho_i) + (b_i^{(2)}/2b_i + E[I_{i,0}^{(s)}])\rho_i$.

Formula (4.129) reduces the computation of the average waiting time $E[W_i]$ to the computation of $p_i(0)$, $E[C_{i,0+}|\xi_i = 0]$, $E[I_{i,0}^{(s)}]$ and $E[I_{i,1}^{(v)}]$. Throughout this section, expressions are derived for these quantities.

LEMMA 4.5   *In a 1-limited polling system the probability $p_i(0)$ that the station Q_i is empty at a random point in time is*

$$p_i(0) = \frac{1 - \lambda_i E[C_{i|1}]}{1 - \lambda_i(E[B_i] + 2E[C_{i,0+}|\xi_i = 0])} . \tag{4.130}$$

PROOF

From a Q_i standpoint the polling model is viewed as an *M/G/1* system with an exceptional service for the packet which arrives at an empty system [100]. Specifically, a packet which arrives when the system is not empty experiences a service time which is distributed as $C_{i|1} = B_i + I_{i|1}$. As shown in Figure 4.14, when a tagged packet arrives at a not-empty system it takes $C_{i|1}$ units of time from the time instant at which it arrives at the top of Q_i until the end of its transmission.

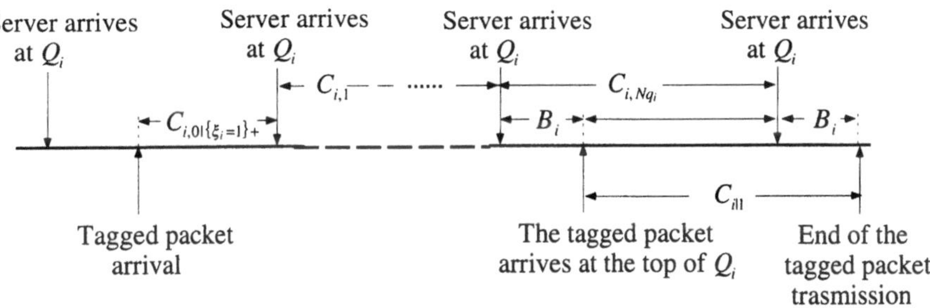

Figure 4.14: Service time for a packet which arrives at a not-empty queue

As soon as the system becomes empty the server goes on vacation. A random packet which arrives at Q_i when the server is on vacation has to wait on average, approximately, $E[C_{i,0+}|\xi_i = 0]$ before the server comes back to Q_i. In addition, the cycle in which the tagged packet arrives has an average length equal to $2 \cdot E[C_{i,0+}|\xi_i = 0]$ (see Theorem 2.5). Exploiting these observations, in [139], the average exceptional service time for the packet which arrives at an empty system is set to $2 \cdot E[C_{i,0+}|\xi_i = 0] + E[B_i]$.

In [160] it is shown that in an *M/G/1* system with an exceptional first service the probability that the system is empty at a random point in time is

$$P\{N = 0\} = \frac{1 - \lambda E[B]}{1 - \lambda E[B_0]} \, , \tag{4.131}$$

where

- λ is the arrival rate;
- B is the service time for a packet which arrives at a not-empty queue;
- B_0 is the service time for the first packet in a busy period.

Hence, applying (4.131) to Q_i it follows that

$$p_i(0) = \frac{1 - \lambda_i E[C_{i|1}]}{1 - \lambda_i(E[B_i] + 2E[C_{i,0+}|\xi_i = 0])} . \tag{4.132}$$

and hence the lemma is proved.

$$\Diamond$$

LEMMA 4.6 $E[C_{i,0+}|\xi_i = 0]$ *is given by*

$$E[C_{i,0+}|\xi_i = 0] = \sum_{j \neq i} E[C_{i,0+}|\xi_j = 1] P\{\xi_j = 1|\xi_i = 0\} +$$

$$\sum_{j=1}^{K} E[C_{i,0+}|\zeta_{j,j+1} = 1] P\{\zeta_{j,j+1} = 1|\xi_i = 0\} \, ,$$

where

- $P\{\xi_j = 1|\xi_i = 0\} = \rho_j/(1-\rho_i)$ (see (4.36));
- $P\{\zeta_{j,j+1} = 1|\xi_i = 0\} = s_j/((1-\rho_i)E[C])$ (see (4.37));
- $E[C_{i,0+}|\xi_j = 1] =$

$$= \begin{cases} \dfrac{b_j^{(2)}}{2b_j} + s_j + \sum_{h=j+1}^{i-1} (\alpha_{h|j} \cdot b_h + s_h) & j < i \\[4ex] \dfrac{b_j^{(2)}}{2b_j} + s_j + \sum_{h=j+1}^{K} (\alpha_{h|j} \cdot b_h + s_h) + \sum_{h=1}^{i-1} (\alpha_{h|j} \cdot b_h + s_h) & j > i \end{cases}$$

- $E[C_{i,0+}|\zeta_{j,j+1} = 1] =$

$$
\begin{cases}
\dfrac{s_j^{(2)}}{2s_j} + \displaystyle\sum_{h=j+1}^{i-1} (\alpha_{h|j \to j+1} \cdot b_h + s_h) & j < i \\[3ex]
\dfrac{s_j^{(2)}}{2s_j} + \displaystyle\sum_{h=j+1}^{K} (\alpha_{h|j \to j+1} \cdot b_h + s_h) + \displaystyle\sum_{h=1}^{i-1} (\alpha_{h|j \to j+1} \cdot b_h + s_h) & j > i
\end{cases}
$$

PROOF

The arrival of the tagged packet at Q_i qualifies the C_i polling cycle in which the tagged packet falls as *special* in the sense that it contains (at least) one packet arrival at Q_i.

$E[C_{i,0+}|\xi_i = 0]$ is derived by conditioning on the position of the server when the tagged packet arrives

$$
E[C_{i,0+}|\xi_i = 0] = \sum_{j \neq i} E[C_{i,0+}|\xi_j = 1] P\{\xi_j = 1|\xi_i = 0\} + \tag{4.133}
$$

$$
\sum_{j=1}^{K} E[C_{i,0+}|\zeta_{j,j+1} = 1] P\{\zeta_{j,j+1} = 1|\xi_i = 0\} \quad .
$$

The unknown components in the r.h.s. of (4.133) are separately derived below.

$E[C_{i,0+}|\xi_j = 1]$ ANALYSIS. The tagged-packet arrival occurs in a cycle which contains a service at Q_j. Furthermore, as the tagged-packet arrival occurs during a service at Q_j due to the length-biasing phenomenon [56] the average service time at Q_j is $b_j^{(2)}/b_j$, while the residual service at the tagged-packet arrival is $b_j^{(2)}/2b_j$ (see (2.36)).

As shown in (4.16) the average length of a cycle in which a given station Q_j experiences a special service whose average length is $\bar{b}_j = b_j^{(2)}/b_j$, i.e., $E[C_{j|(1,\bar{b}_j)}]$, is

$$
E[C_{j|(1,\bar{b}_j)}] \approx \frac{s + \bar{b}_j}{1 - \rho + \rho_j} \quad , \tag{4.134}
$$

and the probability that a station Q_h $(h \neq j)$ transmits in a cycle with average length $E[C_{j|(1,\bar{b}_j)}]$ is

$$\alpha_{h|j} \approx min\{\lambda_h E[C_{j|(1,\bar{b}_j)}], 1\} \quad . \tag{4.135}$$

Finally, by exploiting (4.134) and (4.135) the following approximation is obtained

$$E[C_{i,0+}|\xi_j = 1] = \tag{4.136}$$

$$\begin{cases} \dfrac{b_j^{(2)}}{2b_j} + s_j + \displaystyle\sum_{h=j+1}^{i-1} (\alpha_{h|j} \cdot b_h + s_h) & j < i \\[4mm] \dfrac{b_j^{(2)}}{2b_j} + s_j + \displaystyle\sum_{h=j+1}^{K} (\alpha_{h|j} \cdot b_h + s_h) + \displaystyle\sum_{h=1}^{i-1} (\alpha_{h|j} \cdot b_h + s_h) & j > i \end{cases}$$

$E[C_{i,0+}|\zeta_{j,j+1} = 1]$ ANALYSIS. The tagged-packet arrival occurs during a switching period between Q_j and Q_{j+1} whose average length (due to the length biasing phenomenon) is $s_j^{(2)}/s_j$, and hence the residual switching period is $s_j^{(2)}/2s_j$.

Exploiting (4.17)

$$E[C_{i,0}|\zeta_{j,j+1} = 1] \approx \frac{s + (s_j^{(2)}/s_j) - s_j}{1 - \rho} \quad , \tag{4.137}$$

and the probability that a station Q_h $(h \neq j)$ transmits in a cycle with average length $E[C_{i,0}|\zeta_{j,j+1} = 1]$ is

$$\alpha_{h|(j \to j+1)} \approx min\{\lambda_h E[C_{i,0}|\zeta_{j,j+1} = 1], 1\} \quad . \tag{4.138}$$

From (4.137) and (4.138) it follows that

$$E[C_{i,0+}|\zeta_{j,j+1} = 1] = \tag{4.139}$$

$$
\begin{cases}
\dfrac{s_j^{(2)}}{2s_j} + \displaystyle\sum_{h=j+1}^{i-1} (\alpha_{h|j\to j+1} \cdot b_h + s_h) & j < i \\[3ex]
\dfrac{s_j^{(2)}}{2s_j} + \displaystyle\sum_{h=j+1}^{K} (\alpha_{h|j\to j+1} \cdot b_h + s_h) + \sum_{h=1}^{i-1} (\alpha_{h|j\to j+1} \cdot b_h + s_h) & j > i
\end{cases}
$$

Finally, by substituting (4.136) and (4.139) into (4.133) the lemma is proved.

LEMMA 4.7 $E[I_{i,1}^{(v)}]$ *is given by*

$$E[I_{i,1}^{(v)}] = E[I_{i,1}|\xi_i = 0, N_i > 0] = \tag{4.140}$$

$$\sum_{j\neq i} E[I_{i,1}|\xi_j = 1, N_i > 0] \frac{\rho_j}{1-\rho_i} +$$

$$\sum_{j} E[I_{i,1}|\zeta_{j,j+1} = 1, N_i > 0] \frac{s_j}{(1-\rho_i) E[C]} \quad ,$$

where

- $E[I_{i,1}|\xi_j = 1, N_i > 0] = s + \sum_{h\neq i} b_h \cdot max\{\alpha_{h,i}, \alpha_{h|j}\}$;

- $E[I_{i,1}|\zeta_{j,j+1} = 1, N_i > 0] = s + \sum_{h\neq i} b_h \cdot max\{\alpha_{h,i}, \alpha_{h|j\to j+1}\}$;

- $\alpha_{h|(j\to j+1)} \approx min\{\lambda_h E[C_{i,0}|\zeta_{j,j+1} = 1], 1\}$;

- $\alpha_{h,i} \approx min\{\lambda_h E[C_{i,1}], 1\}$;

- $\alpha_{h|j} \approx min\{\lambda_h E[C_{j|(1,b_j)}], 1\}$.

PROOF

As shown in Figure 4.4, after a delay equal to $E[C_{i,0+}|\xi_i = 0]$ the server arrives at Q_i where it provides a (normal) service with average length b_i. The service at Q_i interrupts the effects (length biasing phenomenon) on the $C_{i,0}$ length generated by the tagged-packet arrival. However, in [139] it is assumed that the effect of the tagged arrival does not end with the service to

Q_i but it propagates into the next polling cycle as well (i.e., $C_{i,1}$). According to this approach the probability that Q_h transmits in cycle $C_{i,1}$ is determined either

(i)   by the tagged-packet arrival at Q_i, or
(ii)   by the transmission at Q_i,

whichever dominates. In the following the impacts of these two events are investigated.

EFFECTS OF THE TAGGED PACKET ARRIVAL AT Q_i. As shown in the analysis of $E[C_{i,0+}|\xi_i = 0]$ the tagged-packet arrival, depending on the position of the server at the arrival instants, gives rise to a cycle which contains either a special service time at Q_j $(j \neq i)$ or a special switching period between Q_j and Q_{j+1}. The probabilities that a station Q_h transmits in the cycle $C_{i,1}$ given the tagged packet arrival in the previous cycle are (see [139])

- $\alpha_{h|j}$, if the tagged packet arrives during a special service time at Q_j , and
- $\alpha_{h|(j \to j+1)}$ if the tagged packet arrives during a switchover between Q_j and Q_{j+1}.

EFFECTS OF THE Q_i TRANSMISSION. As $C_{i,1}$ contains a service at Q_i

$$E[C_{i,1}] \approx \frac{s + b_i}{1 - \rho + \rho_i} \quad , \tag{4.141}$$

thus the probability that a station Q_h transmits in that cycle is

$$\alpha_{h,i} = min\{\lambda_h E[C_{i,1}], 1\} \quad . \tag{4.142}$$

Taking into consideration the two effects (i.e., (i) and (ii)), and exploiting (4.36) and (4.37)

$$E[I_{i,1}^{(v)}] = E[I_{i,1}|\xi_i = 0, N_i > 0] = \tag{4.143}$$

$$\sum_{j \neq i} E[I_{i,1}|\xi_j = 1, N_i > 0] \frac{\rho_j}{1 - \rho_i} +$$

$$\sum_{j} E[I_{i,1}|\zeta_{j,j+1} = 1, N_i > 0] \frac{s_j}{(1 - \rho_i) E[C]} \quad ,$$

and

$$E[I_{i,1}|\xi_j = 1, N_i > 0] = s + \sum_{h \neq i} b_h \cdot max\{\alpha_{h,i}, \alpha_{h|j}\} \quad , \tag{4.144}$$

$$E[I_{i,1}|\zeta_{j,j+1} = 1, N_i > 0] = s + \sum_{h \neq i} b_h \cdot max\{\alpha_{h,i}, \alpha_{h|j \to j+1}\} \quad . \quad (4.145)$$

LEMMA 4.8 $E[I_{i,0}^{(s)}]$ *is approximated by the following expression*

$$E[I_{i,0}^{(s)}] \approx \frac{s + (b_i^{(2)}/b_i)}{1 - \rho + \rho_i} - (b_i^{(2)}/b_i) \quad . \quad (4.146)$$

PROOF

In this case the tagged packet arrival occurs when the server is at Q_i, and the effect of the tagged packet arrival propagates to the intervisit period following the residual service at Q_i, (see Figure 4.15). As in this case $C_{i,0}$ contains a special service at Q_i, according to (4.16)

$$E[C_{i,0}|\xi_i = 1, N_i > 0] \approx \frac{s + (b_i^{(2)}/b_i)}{1 - \rho + \rho_i} \quad . \quad (4.147)$$

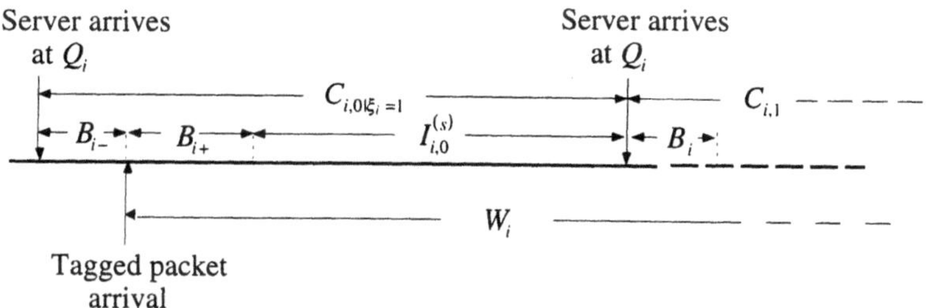

Figure 4.15: The tagged packet arrives during a service at Q_i

Hence, as shown in Figure 4.15

$$E[I_{i,0}^{(s)}] = E[C_{i,0}|\xi_i = 1, N_i > 0] - (E[B_{i+}] + E[B_{i.}]) = \quad (4.148)$$
$$E[C_{i,0}|\xi_i = 1, N_i > 0] - (b_i^{(2)}/b_i) \quad ,$$

and by substituting (4.147) into (4.148) the lemma is proved.

$$\lozenge$$

4.3.3 *l*-limited Polling System: Details of the Computation

This section focuses on the proofs of LEMMA 4.2, LEMMA 4.3 and LEMMA 4.4 used in Section 4.1.5 in the analysis of *l*-limited polling systems.

LEMMA 4.2 *Let* $\phi_{i|\xi_i=0}$ *be the joint probability that the tagged packet arrives during an intervisit period and that it will not be served in the next cycle, then the following relationship holds:*

$$\sum_{k=0}^{\infty} E\left[W_{i|\xi_i=0,\,N_i=k}\right] q_i\left(0, k\right) = E\left[I_{i+}\right](1-\rho_i) +$$

$$b_i E\left[N_{i|\xi_i=0}\right] P\{\xi_i = 0\} + \left(E\left[C_{i,\,1|\xi_i=0,\,N_i=k}\right] -\right.$$

$$E\left[C_{i|l_i}\right]) \; \phi_{i|\xi_i=0} + \left(E\left[C_{i,\,j|l_i}\right] - l_i b_i\right) \sum_{k=l_i}^{\infty} \lfloor k/l_i\rfloor q_i\left(0, k\right) \quad .$$

PROOF

As the server is on vacation when the tagged packet arrives, the waiting time of the tagged packet is equal to the residual length of a *special* intervisit period with respect to Q_i (I_{i+}), an integer number of polling cycles each containing l_i services at Q_i and finally the service of $m\,(0 \le m \le l_i - 1)$ packets at Q_i. Specifically, taking into consideration the length biasing phenomenon,

$$E\left[W_i|\xi_i = 0,\, N_i = k\right] = E\left[I_{i+}\right] + I_{(k<l_i)}kb_i + \tag{4.149}$$

$$I_{(k\ge l_i)}\left\{l_i b_i + E\left[I_{i,\,1}|\xi_i = 0,\, N_i = k\right] +\right.$$

$$\left.\sum_{j=2}^{\lfloor k/l_i\rfloor} E\left[C_{i,\,j}|l_i \text{ services at } Q_i\right] + \left(k - \left\lfloor\frac{k}{l_i}\right\rfloor l_i\right)b_i\right\} \quad .$$

The average length of a cycle which contains l_i services at Q_i is (see Section 4.1.1)

$$E\left[C_{i,\,j}|l_i \text{ services at } Q_i\right] \approx \frac{s + l_i b_i}{1 - \rho + \rho_i} \quad . \tag{4.150}$$

To derive $E[I_{i,1}|\xi_i = 0, N_i = k]$ it is assumed that the packets transmitted in $I_{i,1}$ by a station Q_j are those which arrive in the cycle which includes the special intervisit period at Q_i and l_i services at Q_i. Hence, $E[I_{i,1}|\xi_i = 0, N_i = k]$ is

$$E[I_{i,1}|\xi_i = 0, N_i = k] \approx s +$$

(4.151)

$$\sum_{j \neq i} min\{\lambda_j(2E[I_{i+}] + l_i b_i), l_j\} b_j \quad .$$

Finally, since the joint probability that the tagged packet arrives during an intervisit period and that it will not be served in the next cycle, i.e., $\phi_{i|\xi_i = 0}$, is given by

$$\phi_{i|\xi_i = 0} = \sum_{k = l_i}^{\infty} q_i(0, k) \quad ,$$

with routinary algebraic manipulations the lemma is proved.

$$\Diamond$$

LEMMA 4.3 *Let* $\phi_{i|\xi_i = j}$ *be the joint probability that the tagged packet arrives during a service period and that it will not be served in that cycle, then the following relationship holds:*

$$\sum_{j = 1}^{l_i} \sum_{k = 1}^{\infty} E[W_{i|\xi_i = j, N_i = k}] q_i(j, k) = \frac{b_i^{(2)}}{2b_i} \rho_i +$$

$$b_i(E[N_i|\xi_i > 0] - \rho_i) + E[I_{i,1}|\xi = j, N_i = k^+] \phi_{i|\xi_i = j} +$$

$$(E[C_{i|l_i}] - l_i b_i) \sum_{j = 1}^{l_i} \sum_{k > l_i - j} \left(\left\lfloor \frac{(k + j - 1)}{l_i} \right\rfloor - 1 \right) q_i(j, k) \quad .$$

PROOF

The arrival of the tagged packet occurs in a service period at Q_i. By taking into consideration the length biasing phenomenon the average waiting time is

$$E[W_i | \xi = j, N_i = k^+] = b_i^{(2)}/2b_i + I_{(k < l_i - j + 1)} \cdot (k - 1) b_i + \qquad (4.152)$$

$$I_{(k > l_i - j)} \left\{ E[I_{i,1} | \xi = j, N_i = k^+] + \right.$$

$$\left. \sum_{j=1}^{n-1} E[C_{i,j|l_i}] + [(k-1) - (n-1) l_i] b_i \right\},$$

where $n = \lfloor (k + j - 1)/l_i \rfloor$.

To compute $E[I_{i,1} | \xi_i = j, N_i > l_i - j]$ it is sufficient to note that

$$E[I_{i,1} | \xi_i = j, N_i > l_i - j] = \qquad (4.153)$$
$$E[C_{i,1} | \xi_i = j, N_i > l_i - j] - l_i b_i,$$

where, as shown in Section 4.1.2,

$$E[C_{i,1} | \xi_i = j, N_i > l_i - j] \approx \frac{s + b_i^{(2)}/b_i + (l_i - 1) b_i}{1 - \rho + \rho_i}.$$

Finally, since the joint probability that the tagged packet arrives during a service period and that it will not be served in the next cycle, i.e., $\phi_{i|\xi_i = j}$, is given by

$$\phi_{i|\xi_i = j} = \sum_{j=1}^{l_i} \sum_{k = l_i - j + 1}^{\infty} q_i(j, k),$$

after some algebraic manipulations, the lemma is proved.

$\Diamond$

LEMMA 4.4 *For a station Q_i in a polling system, let $\{ \theta_{i,n} | (\theta_{i,n} \in \{0, 1, ..., l_i\}) \}$ be a discrete time stochastic process, where the r.v. $\theta_{i,j}$ denotes the number of packets served at Q_i during the j-th service period. By denoting with θ_i the generic r.v. of the process $\{ \theta_{i,n} \}$ the following relationship holds*

$$E[C_{i+|\xi_i>0,\,N_i>0}] = \frac{b_i^{(2)}}{2b_i} + \sum_{j=1}^{l_i-1} E[C_i|\theta_i = j, A_i(\theta_i) = 1]\, j p_{i,\,0,\,j} + \qquad (4.154)$$

$$E[C_i|\theta_i = l_i, \xi_i = h]\left(1 - \sum_{j=1}^{l_i-1} j p_{i,\,0,\,j}\right) -$$

$$\frac{b_i}{2}\left[(l_i + 1) + \sum_{j=1}^{l_i-1}(l_i - j)\, j p_{i,\,0,\,j}\right] \quad,$$

where

$$E[C_i|\theta_i = j, \xi_i = h] \approx \frac{s + (j-1)\,b_i + b_i^{(2)}/b_i}{1 - \rho + \rho_i} \quad,$$

and $\{A_i(\theta_i) = 1\}$ *denotes the event "the tagged packet arrived at* Q_i *during the service period* θ_i*."*

PROOF

By denoting with $\{A_i(\theta_{i,\,n}) = 1\}$ the event that the tagged packet arrived at Q_i during the service period $\theta_{i,\,n}$, in Chang and Sandhu (see [32]) it is stated that

$$\lim_{n \to \infty} P\{\theta_{i,\,n} = j | \theta_{i,\,n} > 0, A_i(\theta_{i,\,n}) = 1\} = \qquad (4.155)$$

$$P\{\theta_i = j | \theta_i > 0, A_i(\theta_i) = 1\} = j p_{i,\,0,\,j} \quad, j = 1, \ldots, l_i - 1 \quad,$$

and hence

$$\lim_{n \to \infty} P\{\theta_{i,\,n} = l_i | \theta_{i,\,n} > 0, A_i(\theta_{i,\,n}) = 1\} = \left(1 - \sum_{j=1}^{l_i-1} j p_{i,\,0,\,j}\right) \quad. \qquad (4.156)$$

Equation (4.155) can" be formally derived by exploiting the theory of renewal and regenerative processes (see Section 2.3). The sequence $\{\theta_{i,\,n}\}$ is a regenerative process with respect to the sequence S of time instants at which the server arrives at Q_i when all the queues in the polling systems are empty. Hence, by denoting with

- M the number of service periods at Q_i in a renewal cycle;
- M_j the number of service periods at Q_i, in a renewal cycle, which contains exactly j services at Q_i;
- $\{Sp_n, n = 1, 2, \ldots, M\}$ the sequence of service periods in a renewal cycle; and

- $\{Sp_{n}^{(j)}, n = 1, 2, ..., M_j\}$ the sequence of service periods in a renewal cycle[1] which contains exactly j services

$$\lim_{n \to \infty} P\{\theta_{i,n} = j \mid \theta_{i,n} > 0, A_i(\theta_{i,n}) = 1\} = \tag{4.157}$$

$$\frac{1}{E\left[\sum_{h=1}^{M} Sp_i\right]} E\left[\sum_{h=1}^{M_j} Sp_h^{(j)}\right] \ .$$

By exploiting Theorem 2.4

$$E\left[\sum_{h=1}^{M_j} Sp_h^{(j)}\right] = E[M_j] E[Sp_h^{(j)}] = E[M_j] jb_i \quad , \tag{4.158}$$

$$E\left[\sum_{h=1}^{M} Sp_i\right] = E[M] E[Sp_i] = E[M] b_i E[\theta_i] \ , \text{ and} \tag{4.159}$$

$$E[M_j]/E[M] = P\{\theta_i = j\} \quad .$$

By noting that

$$E[M_j]/E[M] = P\{\theta_i = j\} =$$
$$P\{\theta_i = j \mid \theta_i > 0\} P\{\theta_i > 0\} \quad ,$$

$$E[\theta_i] = E[\theta_i \mid \theta_i > 0] P\{\theta_i > 0\} \quad ,$$

and by exploiting (2.121), (2.122) and (2.124),

$$P\{\theta_i = j \mid \theta_i > 0\} = p_{i,0,j} / \sum_{k=0}^{\infty} p_{i,k,1} \quad , \tag{4.160}$$

$$P\{\theta_i = l_i \mid \theta_i > 0\} = 1 - \sum_{j=1}^{l_i - 1} p_{i,0,j} / \sum_{k=0}^{\infty} p_{i,k,1} \quad , \tag{4.161}$$

$$E[\theta_i \mid \theta_i > 0] = 1 / \sum_{k=0}^{\infty} p_{i,k,1} \quad , \tag{4.162}$$

(4.155) is proved.

From (4.155), (4.160) and (4.161) with routinary algebraic manipulation $E[C_{i+\mid \xi_i > 0, N_i > 0}]$ can be derived. Specifically

$$E[C_{i+\mid \xi_i > 0, N_i > 0}] = E[C_{i+} \mid \theta_i > 0, A_i(\theta_i) = 1] \quad ,$$

and

---

1. It is easy to observe that $\{Sp_{n}^{(j)}, n = 1, 2, ..., M_j\}$ is a regenerative process with respect to the sequence S of renewal instants.

$$E[C_{i+}|\theta_i > 0, A_i(\theta_i) = 1] =$$

$$\sum_{j=1}^{l_i} E[C_{i+}|\theta_i = j, A_i(\theta_i) = 1] \, P\{\theta_i = j|\theta_i > 0, A_i(\theta_i) = 1\} \quad .$$

Furthermore

$$E[C_{i+}|\theta_i = j, A_i(\theta_i) = 1] =$$

$$\sum_{h=1}^{j} E[C_{i+}|\theta_i = j, \xi_i = h] \, P\{\xi_i = h|\theta_i = j, A_i(\theta_i) = 1\} \quad .$$

where

$$P\{\xi_i = h|\theta_i > 0, A_i(\theta_i) = 1\} = 1/j \quad .$$

To compute $E[C_{i+}|\theta_i = j, \xi_i = h]$ it is useful to note that

$$E[C_{i+}|\theta_i = j, \xi_i = h] = b_i^{(2)}/2b_i + E[C_i|\theta_i = j, \xi_i = h] - hb_i \quad ,$$

where by exploiting the conditional cycle concept (see (4.21)

$$E[C_i|\theta_i = j, \xi_i = h] \approx \frac{s + (j-1)b_i + b_i^{(2)}/b_i}{1 - \rho + \rho_i} \quad .$$

Hence,

$$E[C_{i+}|\theta_i > 0, A_i(\theta_i) = 1] = b_i^{(2)}/2b_i + \tag{4.163}$$

$$\sum_{j=1}^{l_i-1} \sum_{h=1}^{j} (E[C_i|\theta_i = j, \xi_i = h] - hb_i) \, p_{i,0,j} +$$

$$\sum_{h=1}^{l_i} (E[C_i|\theta_i = l_i, \xi_i = h] - hb_i) \frac{1}{l_i} \left(1 - \sum_{j=1}^{l_i-1} j p_{i,0,j} \right) \quad ,$$

which, after some routinary algebraic manipulations, is reduced to

$$E[C_{i+}|\theta_i > 0, A_i(\theta_i) = 1] = \tag{4.164}$$

$$\frac{b_i^{(2)}}{2b_i} + \sum_{j=1}^{l_i-1} E[C_i|\theta_i = j, A_i(\theta_i) = 1]\, j p_{i,0,j} +$$

$$E[C_i|\theta_i = l_i, A_i(\theta_i) = 1]\left(1 - \sum_{j=1}^{l_i-1} j p_{i,0,j}\right) -$$

$$\frac{b_i}{2}\left[(l_i + 1) + \sum_{j=1}^{l_i-1} (l_i - j)\, j p_{i,0,j}\right] \ .$$

4.3.4 Two-Moment Approximation of a Distribution

To reduce the computational complexity, the problem of approximating a distribution by m moments is often restricted to the case $m = 2$ [103]. For the *two-moment* approximation technique exponential building blocks are used. The way these building blocks are put together to construct the approximation depends upon the value of the coefficient of variation c: $c \leq 1$ and $c > 1$.

When $c \leq 1$, in order to achieve any coefficient of variation between 0 and 1, the sum of an exponential r.v., and a constant is commonly used. The resulting distribution is called *shifted exponential distribution* and has the following density

$$f(x) = \begin{cases} 0 & x < d \\ \lambda e^{-\lambda(x-d)} & x \geq d \end{cases} \tag{4.165}$$

where λ is the rate of the exponential distribution, and d is the constant. λ and d are related to the first, μ_1, and second, μ_2, moments by the following relationships

$$\mu_1 = \lambda^{-1} + d \ , \ \mu_2 - \mu_1^2 = \lambda^{-2} \ . \tag{4.166}$$

When $c > 1$, a mixture of two exponential distributions is used. The resulting (*hyperexponential*) density is

$$h(x) = p_1 \lambda_1 e^{-\lambda_1 x} + (1 - p_1) \lambda_2 e^{-\lambda_2 x}, \ x \geq 0 \ . \tag{4.167}$$

In (4.167) there are three parameters which can be fitted by using three moments. The two-moment approximation is greatly simplified by assuming balanced means, i.e., $p_1 \lambda_1^{-1} = (1 - p_1) \lambda_2^{-1}$ (see [103]). Under this assumption λ_1, λ_2 and p_1 are related to the first and second moments by the following relationships

$$p_1 = [1 + \sqrt{(c^2 - 1) / (c^2 + 1)}]/2 \quad , \tag{4.168}$$

$$\lambda_i = 2 p_i \cdot \mu_1^{-1}, \quad i = 1, 2 \quad , \tag{4.169}$$

where $c^2 = (\mu_2 - \mu_1^2) / \mu_1^2$, and $p_2 = 1 - p_1$.

In (4.107), there are three tables which contain some illustrative three moments. There was some more expression in this as great importance, as well as historical mean .... (4.107) .... (see also (4.101) ....)
.... $\phi_i(x,y)$, and given by all of the ... and vertical ... coordinates of the following type.

5 Fiber-Distributed Data Interface (FDDI)

The standard FDDI [4] describes a high-speed MAC protocol for LANs/ MANs which employs a timer-controlled, token-passing mechanism, called a *timed token protocol,* to control each station' s access to the shared medium.

The idea behind the timed token protocols is, firstly, to partition the services they can provide their users into two main classes: time-critical and non-time-critical; and secondly, to employ a token-passing MAC protocol with a cycle-dependent timing mechanism which limits the amount of data (organized into frames) transmitted by a station, per cycle, for each class of service. The non-time-critical class of service may be further divided into subclasses, according to a priority scheme which is usually optional.

5.1 INTRODUCTION

The FDDI standard was developed by X3T9.5, a task group of the technical committee X3T9 of the American National Standards Institute (*ANSI*). FDDI is based on a dual, fiber-optic ring with a maximum ring length of 100 Km and a transmission rate of 100 Mbit/sec (see Figure 5.1). A station connected to the ring may support up to three different classes of service. The first version of the standard, FDDI, provides for two classes of packet-switched traffic: synchronous traffic and asynchronous traffic with eight priorities. FDDI has been extended to FDDI-II, which additionally provides isochronous, i.e., circuit-switched, channels. FDDI-II is based on a 125 μs cycle, which is partitioned into 16 wide-band channels. The access to the medium in FDDI and that for the classes of packet-switched traffic in FDDI-II is controlled by a timed token protocol, which was first developed by Grow [78] and later

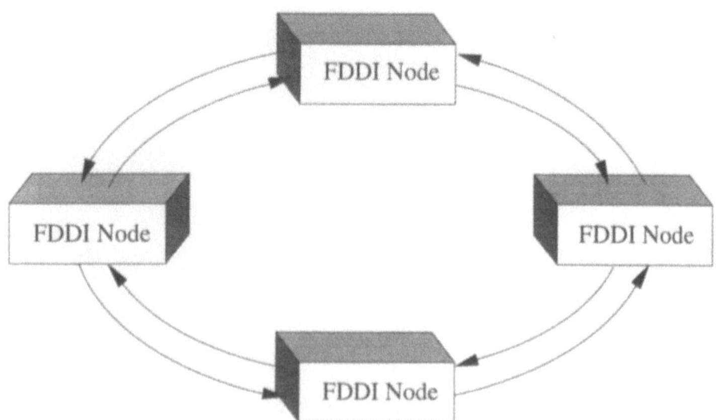

Figure 5.1: FDDI dual fiber-optic ring

standardized as the Medium Access Control (MAC) protocol for FDDI. It permits the integration of time-critical (synchronous) and non-time-critical (asynchronous) traffic.

Of FDDI and FDDI-II, this book will concentrate only on FDDI.

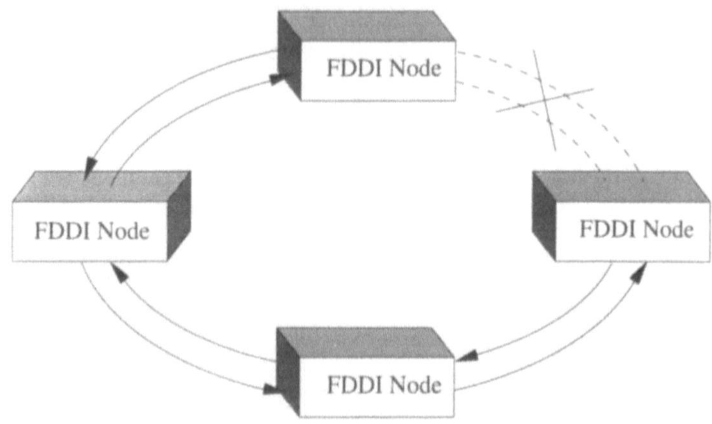

Figure 5.2: Wrapped FDDI network

In order to achieve high throughput in large FDDI configurations, the timed token protocol permits multiple packet transmissions per token arrival. However, stations are prevented from hogging the token by the introduction of timer-controlled limits on the service time.

FDDI can connect just a few nodes located in a computer room, or be a backbone of a complex network that interconnects stations located, for

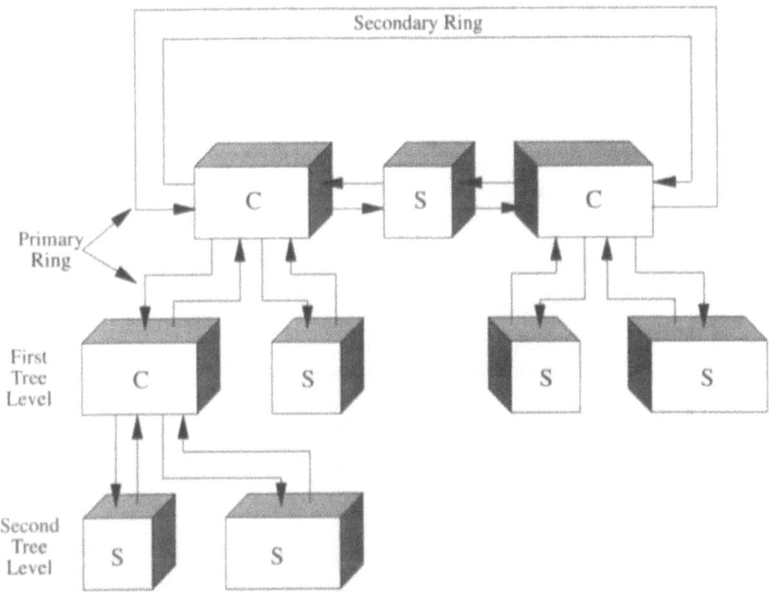

Figure 5.3: A possible FDDI network structure

example, throughout a metropolitan area. The number of stations supported by an FDDI network can vary from 1 up to 1,000.

In FDDI there are two rings, called the *primary* and *secondary* rings. In the most common configurations, FDDI would use the primary ring for normal operations, while the secondary ring would be used for redundancy.

The maximum length of the FDDI physical medium is 100 Km for each of the primary and secondary rings, which results in a total network length of up to 200 Km for a wrapped FDDI configuration (see Figure 5.2). This configuration results, for example, when the primary and the secondary links between two adjacent stations fail.

FDDI can also support a "dual ring of trees" topology like the one shown in Figure 5.3. Stations (denoted by S) are the end user equipment, while concentrators (denoted by C) are used to implement the tree network configurations. The tree structure unfolds below the primary ring only, with a maximum number of levels imposed by the maximum number of stations supported by FDDI.

5.2 FDDI LAYERS AND SERVICES

The FDDI functions are grouped into four layers.

- Physical Medium Dependent layer (PMD)
- Physical layer (PHY)
- Medium Access Control layer (MAC)
- Station Management layer (SMT)

The relationship of each layer to the OSI Reference Model [91] is shown in

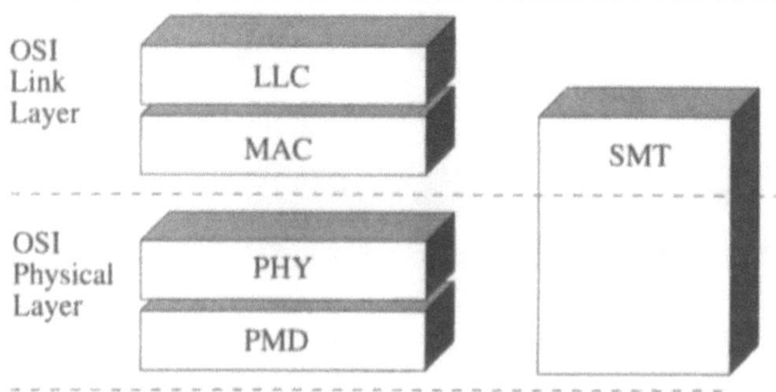

Figure 5.4: Relationships between the FDDI layers and the OSI Reference Model

Figure 5.4. Although the main focus of this chapter will be on the FDDI MAC layer, some information on the other layers is given below. For more details about the FDDI layers, see [2] and [92].

THE PMD LAYER. The PMD layer provides all the necessary functions to transport a suitably coded, digital bit stream from node to node. The PMD layer defines and characterizes transmitters, receivers, cables, connectors, power budgets, and other physical medium- and hardware-related characteristics. In order to include existing media installations in the standard, as well as to provide for flexibility in the selection of media, several options are specified in the PMD layer:

- Multimode fiber
- Single-mode fiber
- Low-cost fiber
- Shielded twisted pair
- Unshielded twisted pair

- FDDI on SONET

REMARK. The Synchronous Optical NETwork (SONET) is a high-speed digital transport system with bit rates in multiples of 51.48 Mbps up to 2488.32 Mbps, and has recently been adopted as a standard for use primarily in long-haul communication networks.

◊

THE PHY LAYER. The PHY layer provides medium-independent functions to the upper layer, i.e., the MAC layer and the SMT layer. The PHY layer decodes an incoming bit stream into a symbol stream which is passed to the MAC layer. The PHY layer encodes the data and control symbols (see below) provided by the MAC layer for transmission via the PMD layer (Figure 5.5). Additional functions carried out by the PHY layer include delineating octet boundaries as required by the MAC layer, and establishing clock synchronization with the incoming bit stream.

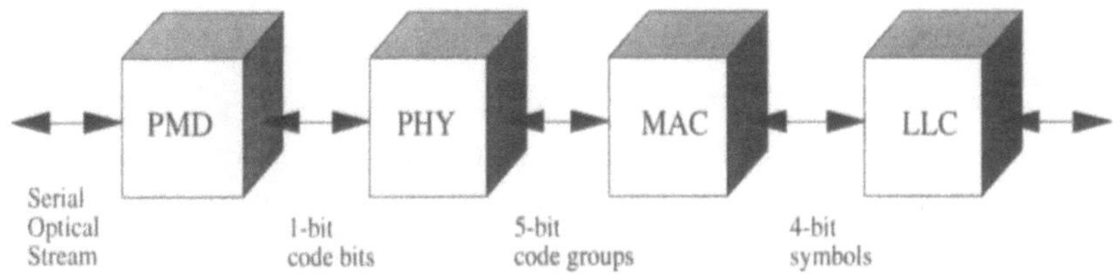

Figure 5.5: Bit coding across the layer interfaces

MAC LAYER. The MAC layer includes the functions necessary for accessing the medium. In FDDI, the data are transmitted in units of 4 bits, called *symbols*. Hence, there are 16 data symbols, labelled 0, 1, 2, ..., A, B, ..., F. Although a data symbol consists of 4 data bits, it is transmitted on the fiber as a code consisting of 5 code bits (Figure 5.5). For the purpose of this book, it is enough to know that 5 code bits result in 32 symbols, 16 of which are used as *data symbols* (Table 5.1). The remaining 16 symbols are available for use as *control symbols*, and are labelled Q, I, H, J, K, T, R, S and so on. Control symbols consist of bit patterns that cannot occur in user data, and their use is shown in Table 5.2.

Table 5.1 4B5B code for data symbols

Code Bits	Symbol	Assignment
11110	0	0000
01001	1	0001
10100	2	0010
10101	3	0011
01010	4	0100
01011	5	0101
01110	6	0110
01111	7	0111
10010	8	1000
10011	9	1001
10110	A	1010
10111	B	1011
11010	C	1100
11011	D	1101
11100	E	1110
11101	F	1111

SMT LAYER. The SMT layer provides those necessary functions which make cooperation between the nodes on the ring possible. Using the services provided by the PMD, PHY, and MAC layers, the SMT layer provides services such as control configuration management (i.e., the coordination of node insertion and removal), fault isolation and recovery. For example, for an FDDI station to join the ring, its SMT layer initiates the so-called *Claim Token Process* (see Section 5.3.6). If this fails, the station's SMT layer activates the *Beacon Process* (see Section 5.3.7).

Table 5.2 4B5B code for control symbols

Code Bits	Symbol	Assignment
00000	Q	Quiet
11111	I	Idle
00100	H	Halt
11000	J	First symbol of starting delimiter
10001	K	Second symbol of starting delimiter
01101	T	End
00111	R	Reset
11001	S	Set

5.3 MAC PROTOCOL

The FDDI MAC standard [4] describes the protocol which allows stations to share the medium. The functions performed by the protocol are located in the MAC layer. In FDDI, the MAC protocol functions are partitioned and carried out by two cooperating asynchronous processes, the MAC Receiver and the MAC Transmitter. These processes interact, and they are controlled, enabled, and monitored by the SMT.

The FDDI MAC protocol supports two types of service, referred to as *synchronous* and *asynchronous* (see Figure 5.6). The former type receives guaranteed throughput and guaranteed delay, thus making FDDI suitable for packetized voice and video, real-time control, and other applications which require a guaranteed Quality of Service (*QoS*). The asynchronous type of service does not receive any guaranteed QoS, and is typically used to support data applications for which bandwidth and response time requirements are not critical.

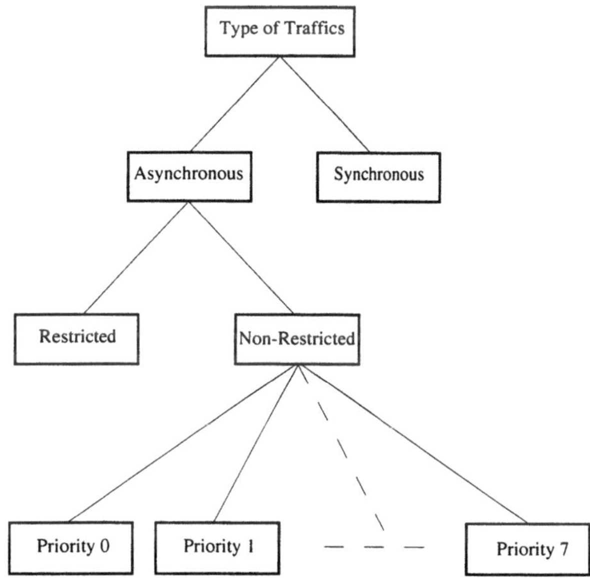

Figure 5.6: Taxonomy of FDDI traffic types

The asynchronous traffic is managed by a *restricted token mode* and *non-restricted token mode* of operations. The former is available for extended asynchronous dialogues between two stations, while the latter allows each station to partition the asynchronous traffic into eight priority levels.

5.3.1   Token Structure

The token structure is shown in Figure 5.7. A description of the various fields follows.

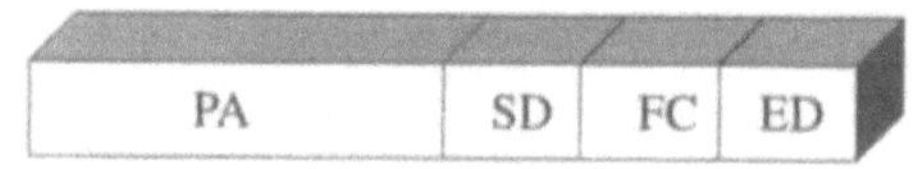

Figure 5.7: Token structure

PREAMBLE. The preamble (PA) consists of idle symbols, which are used for synchronizing the receiver clock to the upcoming frame. The functions of adding and deleting preambles are located in the PHY layer.

STARTING DELIMITER. The starting delimiter (SD) sequence consists of the uniquely recognizable symbol sequence JK.

FRAME CONTROL. The 1-byte frame control (FC) field identifies the type of token (i.e., non-restricted or restricted). Values of 1000-0000 and 1100-0000 indicate non-restricted and restricted tokens, respectively.

TOKEN ENDING DELIMITER. The token ending delimiter (EDT) consists of two consecutive T symbols

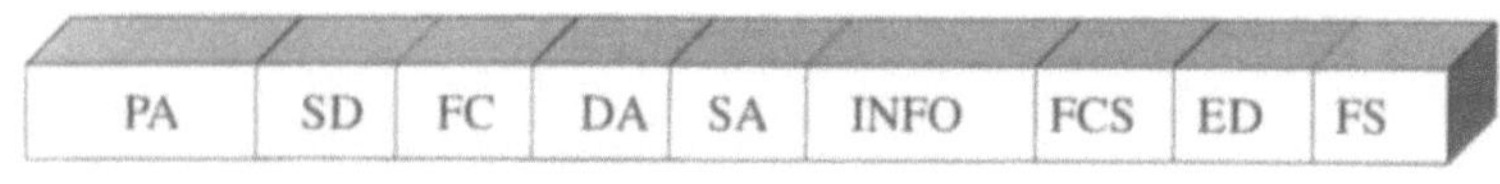

Figure 5.8: FDDI frame structure

5.3.2 Frame Structure

The structure of a frame is shown in Figure 5.8. Like a token, it includes a preamble (PA) and a starting delimiter (SD) followed by a 1-byte frame-control (FC) field. The FDDI frame structure contains further fields: these and the frame control field will be discussed, in turn, below.

Frame Control Field. As stated above, the values 1000-0000 and 1100-0000 of the FC field identify a token; other patterns in this field are used to identify the class of a frame and the length of the address field. The format of the frame control byte is shown in Figure 5.9, while Table 5.3 reports the

Figure 5.9: Frame control byte

FC coding for the frames which will be referred to throughout this chapter. The 8 bits of the frame control field are denoted by CLFF-ZZZZ. The frame class bit (C), is set to zero for asynchronous frames, and to one for synchronous frames. The address length bit (L) is set to zero for 16-bit addresses,

Table 5.3 FDDI FC coding

CLFF	ZZZZ	
1L00	0011	MAC claim frame
1L00	0010	MAC beacon frame
0L00	0001	SMT frame
....	....	....
0L00	1111	SMT frame
0L01	0PPP	LLC asynchronous frame
1L01	0000	LLC synchronous frame

and to one for 48-bit addresses (Figure 5.10). For tokens, this bit is used to distinguish restricted from non-restricted tokens.

The two format bits FF, together with the CL and four control bits ZZZZ, provide further subdivision of frame types. These include:

1. *LLC Frames*. These are the frames that have been passed by the LLC entity to the MAC layer for transmission. Protocol data units coming from higher layers of a network architecture are generally encapsulated into LLC frames. The FF bits in the frame control field of LLC frames are 01. For asynchronous frames, the last three bits indicate frame priority (P); zero is the lowest priority and seven is the highest.

2. *MAC Frames*. These frame types relates to the MAC layer only, and therefore are not passed to higher layers. The frame control field for these frame types uses the pattern 1L00-ZZZZ, where the ZZZZ bits define the

type of MAC frame. Claim frames and beacon frames are two examples of MAC frames belonging to this class. These are identified by frame control values of 1L00-0011 and 1L00-0010, respectively.

3. *SMT Frames.* The frame control field of station management frames is coded 0L00-ZZZZ.

DESTINATION AND SOURCE ADDRESSES. Each FDDI frame shall contain two address fields: the destination MAC address (DA) and the source MAC address (SA). The destination address is used to identify the MAC(s) for which the frame is intended. The leftmost bit (I/G) of the destination address distinguishes individual from group addresses, i.e., the destination of a frame can be either an individual MAC or a group of MACs. Both types of addresses, individual and group, are supported in FDDI in both a 16-bit and 48-bit addressing mode (see the description of the L bit, above). The two addressing modes are shown in Figure 5.10. In the 48-bit address mode, further distinction is made with a universal vs. local (U/L) bit which is used for frame-routing purposes [4].

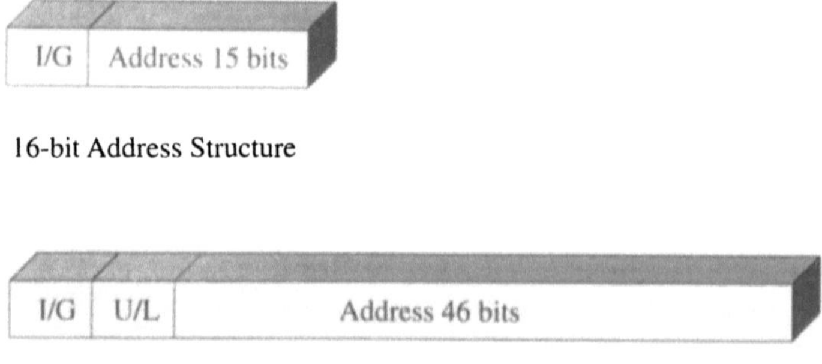

16-bit Address Structure

48-bit Address Structure

Figure 5.10: 16-bit and 48-bit addressing structures

FRAME DATA. The frame data (INFO) field is not mandatory. However, frames with no data are, in general, used for control purposes. The maximum frame length is 9,000 symbols or 4,500 bytes, including the four symbols of the preamble.

FRAME CHECK SEQUENCE. The frame check sequence (FCS) is the standard cyclical redundancy checker (CRC) using a degree 32 polynomial for the generator and checking function. The fields covered by the FCS include FC, DA, SA, DATA and the FCS itself. All frames except tokens have FCS protection.

FRAME ENDING DELIMITER. The ending delimiter (ED) field of a frame shall consist of a single T symbol.

FRAME STATUS. The Frame Status (FS) field shall consist of a sequence of Control Indicator symbols (R and S) of arbitrary length which follows the ending delimiter of a frame (i.e., the T symbol).

Together, the ED and FS fields form the End of Frame Sequence (EFS), as shown in Figure 5.11. The first three Control Indicators of the Frame Status field are mandatory, indicating Error Detected (E), Address Recognized (A), and Frame Copied (C).

At the transmitting station the E, A and C indicators start out as R. Each

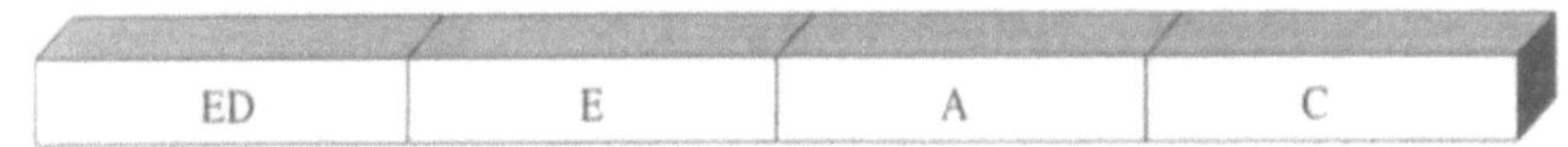

Figure 5.11: End of Frame Sequence

station on the ring checks each frame for validity and, if any errors are detected, it sets the E indicator in the corrupted frame to S. Once set to S, other stations on the ring cannot reset the E indicator even if they find the frame valid. Also, other indicators (A and C) have no reliable meaning if the E indicator is set, since the source and destination addresses may be corrupted.

If a station on the ring recognizes the destination address of a frame as its own address, it sets the A indicator. Thus, the source station can find out if the destination station is on the ring. If a station on the ring copies the frame, it sets the C indicator. Generally, this station is one whose address matches that specified in the destination address field of the frame. If so, both A and C will be set. The use of additional trailing Control Indicators in the FS field is optional and implementor-defined.

5.3.3 FDDI Token Passing Mechanism

As it was mentioned several times in previous sections, in FDDI, access to the medium is controlled by a timed token. The token gives the node that receives it the right to transmit frames. The other nodes repeat these frames, and the destination node copies the frames addressed to it in addition to repeating them. The MAC layer of the node that originated the frames has the responsibility of removing them and of issuing the token to the next downstream node. The MAC layer generates the frame check sequences for outgoing frames and verifies those of incoming frames. Whenever a frame has reached its destination node, the corresponding MAC layer passes the frame to the LLC entity, which has the task of processing it.

To explain the FDDI token passing mechanism, an example employing a three-station FDDI network, whose stations are denoted by S_1, S_2, and S_3 respectively, is considered. As shown in Figure 5.12, the initial status of the network is assumed to be characterized as follows:

(i)   the token is in transit between stations S_1 and S_2;

(ii)   stations S_2 and S_3 both have a queued frame to send to station S_1.

As mentioned earlier (see Section 5.2), a station wishing to transmit frames

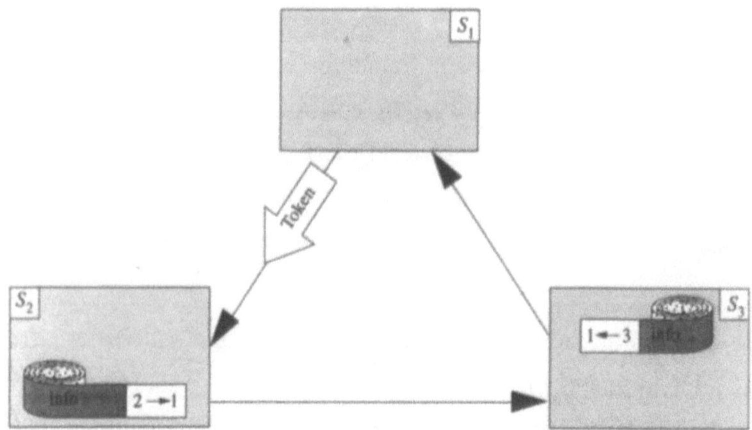

Figure 5.12: Snapshot of the initial status of the network

must first wait for the (timed) token to arrive; the token is then *captured*, as represented in Figure 5.13.

After the token is seized by station S_2, that station can then transmit its

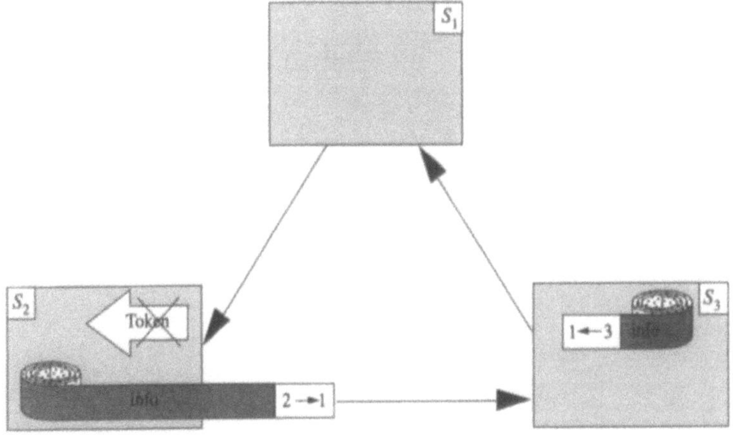

Figure 5.13: Token seized by station S_2

queued frame(s) of data (Figure 5.13), after which it issues a new token which provides other stations with the opportunity to access the medium.

The frames of data transmitted by the originating station are regener-

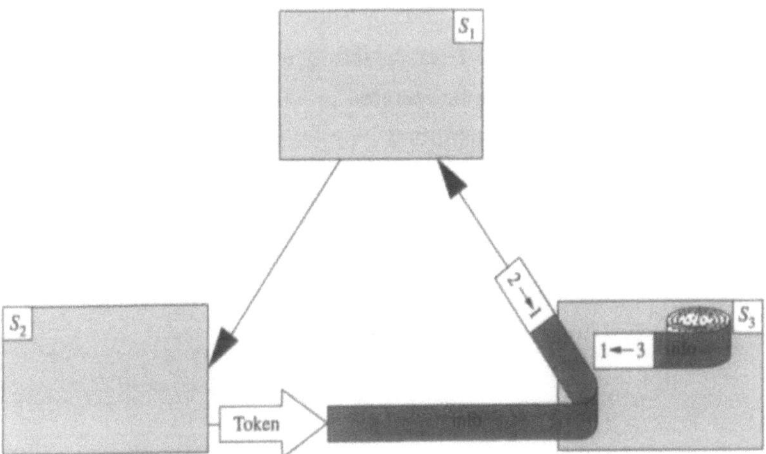

Figure 5.14: Token released by station S_2

ated and repeated by the other active stations on the ring. In Figure 5.14, station S_3 is performing these operations. While repeating incoming frames, a station also examines the destination address for a match with its own address. If a match occurs, the station copies the frame contents into its receiving buffer (Figure 5.15) as it transmits the frame onwards.

Figure 5.16 highlights the process which leads to the coexistence of

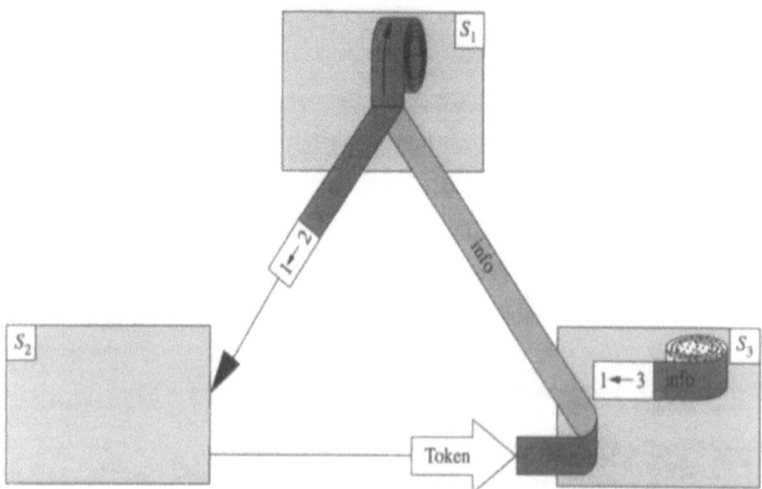

Figure 5.15: The frame previously sent by station S_2 reaches its destination

multiple circulating frames along the ring. Specifically, by the time the header of the frame previously sent by station S_2 is in transit between stations S_1 and S_2, the token has been seized by station S_3 and the transmission of a new frame along the ring has begun.

Each frame of data transmitted eventually comes back to the station

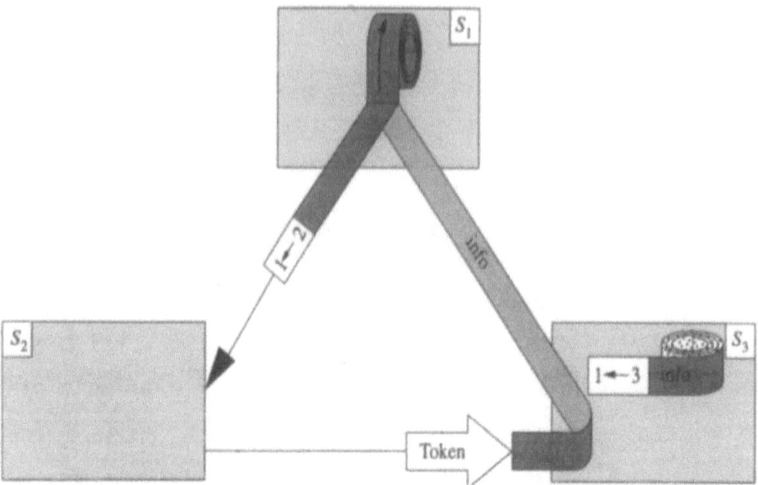

Figure 5.16: Multiple frames circulating along the ring

which originated it, and which is then responsible for stripping the frame from the ring. The originating station (station S_2, in the current example) recognizes the source address contained in the frame as its own, and so stripping takes place (Figure 5.17). The stripping process leaves frame frag-

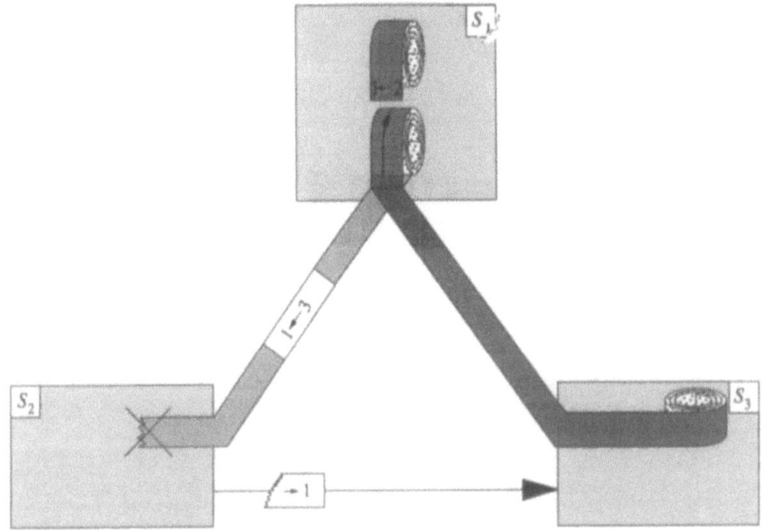

Figure 5.17: Frame fragment left by station S_2

ments, consisting of the starting delimiter, frame control field, destination and source addresses, and some additional parts of each frame, circulating along the ring (Figure 5.17). Frame fragments are ultimately removed when they arrive at a station that is transmitting its own frames onto the medium (Figure 5.18).

Hence, multiple frames originated by different stations can coexist along the ring. However, according to the FDDI MAC protocol, there are still intervals of time during which no data can be placed on the medium. These are the token walking times from each active station to the next active one downstream which, depending upon the number of active stations and the length of the ring, can be significantly less than the ring latency. This point is highlighted in Figure 5.19, in which several snapshots of the ring are taken from the time a frame (F) has just been transmitted onto the medium by a generic active station$\{i\}$, up to the time the token (T) is captured by the next active downstream station$\{i+1\}$.

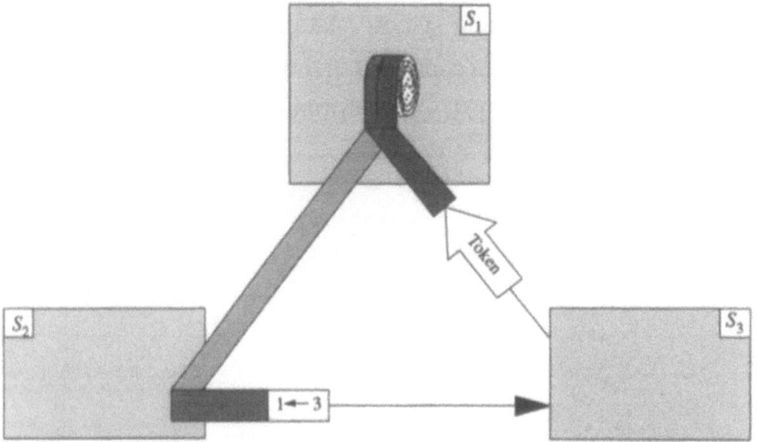

Figure 5.18: Frame fragment removal by station S_3

5.3.4 FDDI Timed Token Protocol

In this subsection, the protocol mechanisms implemented by FDDI for transmitting synchronous and asynchronous frames are described. These mechanisms are based on the fact that each station times the interval between successive token arrivals. The timing operation is performed by means of a timer and a counter, called the *Token Rotation Timer* (*TRT*) and the *Late_Counter* (*Late_Ct*), respectively.

TRT values vary from zero up to a threshold called the *Target Token Rotation Time* (*TTRT*). The *TTRT* value is negotiated as part of ring initialization during the so-called *claim token process* (see Section 5.3.6), at the end of which each station selects the same value for the *TTRT*. A station's *TRT* begins timing upon token arrival, and expires if it reaches the *TTRT* value. Whenever a *TRT* expires, it is restarted from zero and the *Late_Ct* value is increased by one. Furthermore, the *Late_Ct* value is cleared to zero on the next token arrival (see Figure 5.20). Hence, the number of *TRT* expirations since a token was last received are counted by the *Late_Ct*.

When the token arrives at a station with *Late_Ct* = 1, the token is called a *Late Token*. In this case the *Late_Ct* value is cleared to zero and the *TRT* is not restarted.

When the token arrives at a station with *Late_Ct* = 0, the token is called an *Early Token*. In this case, the *Late_Ct* value remains zero, and the

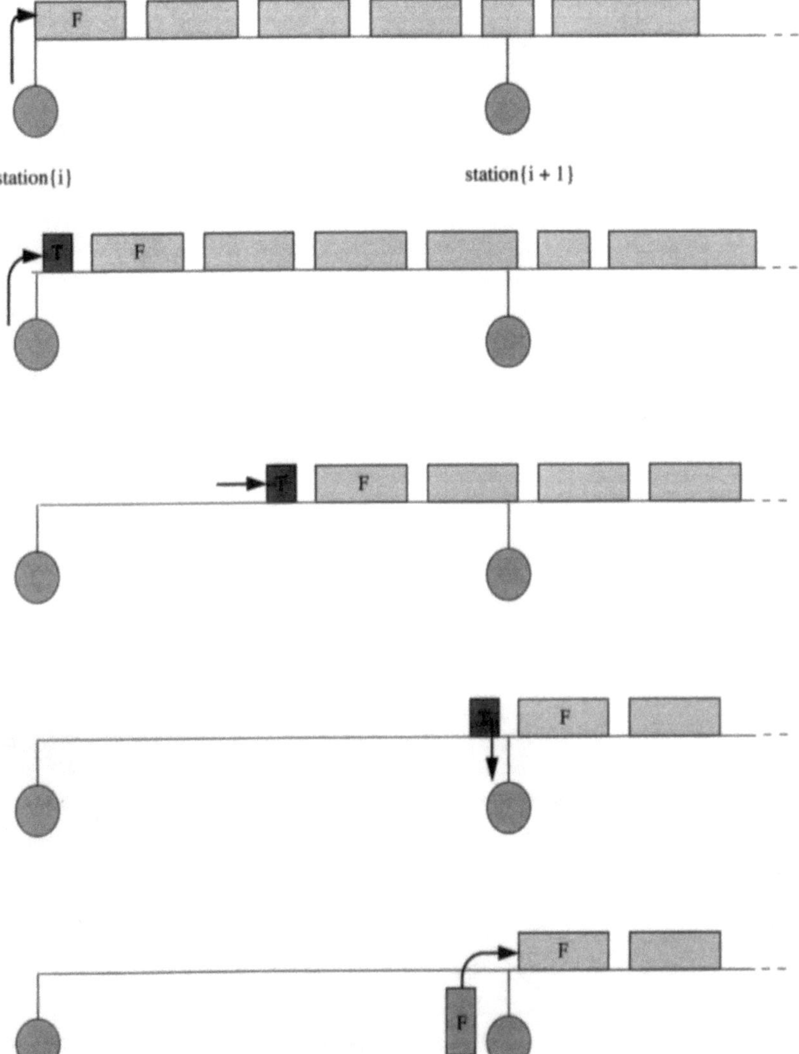

Figure 5.19: FDDI token propagation between two adjacent stations

TRT is restarted (see Figure 5.21).

The behavior of the FDDI MAC protocol depends on whether the token is early or late, according to the following rules:

(i)   synchronous frames can be transmitted by the station whenever the token arrives; while

(ii)   asynchronous frames can be sent only if the token is an Early Token.

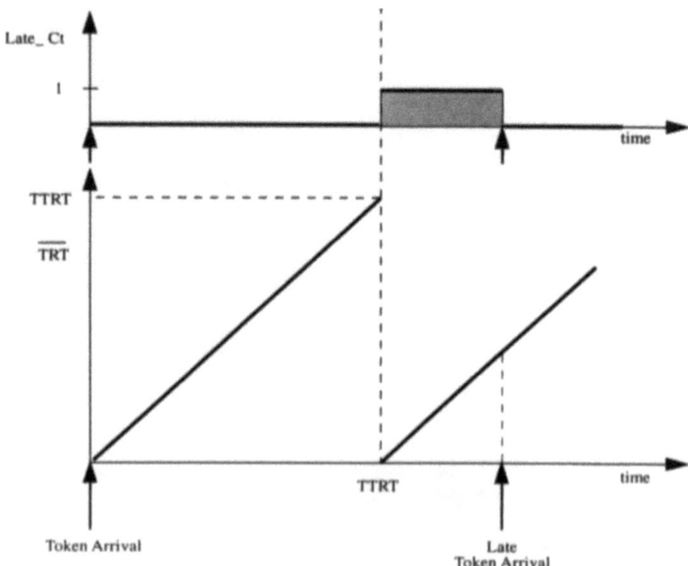

Figure 5.20: Late Token scenario

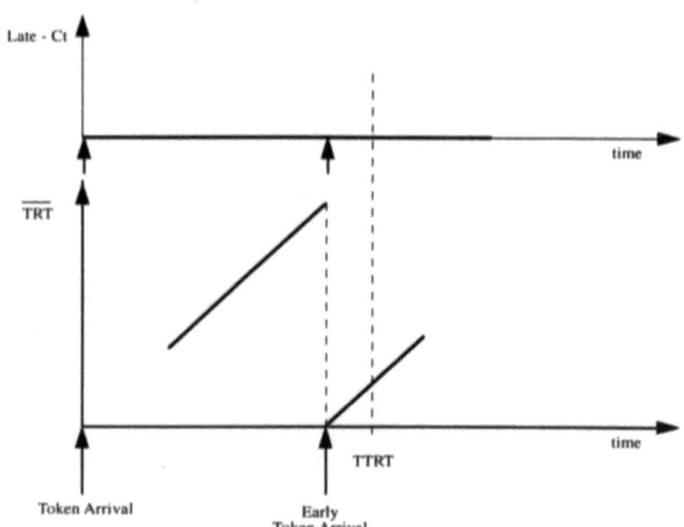

Figure 5.21: Early Token scenario

Rules (i) and (ii) are supplemented by rules which establish the maximum quota of synchronous traffic that can be transmitted by any station on token arrival, and the portion of asynchronous traffic that can be transmitted by a station on an Early Token arrival. In the following discussion, synchronous traffic transmission will be analyzed first.

SYNCHRONOUS TRANSMISSION. To provide guaranteed QoS to synchronous traffic, FDDI enforces a limit on how much synchronous traffic each station can send per token received. Specifically, in FDDI, the synchronous bandwidth allocated to station$\{i\}$, $i \in \{1, 2, ..., K\}$, is a fraction f_i of $TTRT$: that is, $TTRT \times f_i$. The aggregate synchronous allocation must be less than $TTRT$, and therefore $f = \sum_{i=1}^{K} f_i < 1$. More specifically, the aggregate synchronous bandwidth can never exceed $TTRT - Ring\_Latency$, where $Ring\_Latency$ is a term which depends on the maximum ring latency, on the maximum frame length, and on the time it takes to transmit a token (see Section 5.3.8).

Hence, according to point (i) on page 217, every time station$\{i\}$, $i \in \{1, 2, ..., K\}$, captures the token, it can transmit synchronous frames for a time up to $TTRT \times f_i$. If, on token arrival, each station transmits synchronous frames for its allowed fraction, the inter-token time at all stations is equal to $TTRT$. In the following examples, unless explicitly stated, it is assumed that all nodes may transmit the same amount S of synchronous traffic.

Under the assumption that the TRT restarts from zero when the token arrives at station S_1 at time 0, the behavior of the FDDI MAC protocol is shown in Figure 5.22 which plots the TRT trajectory observed by station S_1. In addition, the left half of Figure 5.22 shows the token departure from station S_1 and the time instants at which the token visits the other stations. From the same figure it is evident that after an interval of time equal to $TTRT$ since the last visit to station S_1, the token returns to station S_1. The right half of Figure 5.22 shows the operational scenario in which the token comes back to station S_1 before TRT has reached the $TTRT$ threshold. This can be due, for example, to station S_K not having any synchronous traffic to transmit. In this case, FDDI allows S_1 to transmit of asynchronous frames, as will be shown below.

ASYNCHRONOUS TRANSMISSION. According to rule (ii) on page 217, a station can transmit asynchronous frames only when it captures an Early Token. For

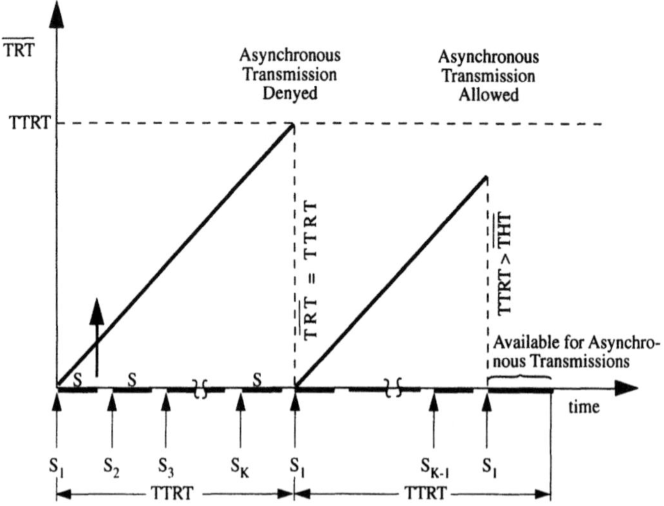

Figure 5.22: Synchronous frame transmission

example, the right half of Figure 5.22 depicts the scenario in which station
S_1 receives an Early Token. Whenever an Early Token is captured by a station, FDDI performs the following actions:

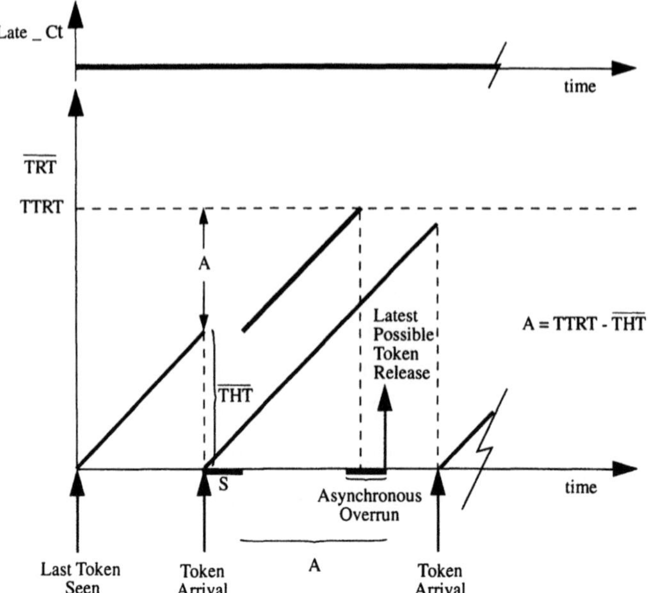

Figure 5.23: Asynchronous frame transmission

1. the current value of *TRT* is transferred to an additional timer, called the *Token Holding Timer* (*THT*);

2. the *TRT* is reset to zero so that it will time the next rotation of the token;

3. synchronous frames are transmitted according to the rules listed above; then

4. the *THT* is enabled for timing asynchronous frame transmission, which can last, at most, until the *THT* value reaches the *TTRT* threshold; i.e., for a time interval equal to $A = TTRT - \overline{THT}$.

REMARK. At the end of action 1, the *THT* contains a value which is denoted by $\overline{THT}$. The *THT* is kept disabled during synchronous frame transmission, after which asynchronous frame transmission can begin (Figure 5.23).

$\Diamond$

Any unused time remaining in the *THT* at the end of asynchronous frame transmission is lost by the current station (since it cannot be retained until the next token arrival) but, if needed, it can be utilized by any downstream station.

If during the transmission of an asynchronous frame the *THT* timer expires (since it has reached the *TTRT* threshold), the transmission of that frame will be completed before releasing the token. This may cause an additional delay in the release of the token; this delay is called *asynchronous overrun* (Figure 5.23).

There is now enough background information to understand how *Late Token* scenarios can develop. Figure 5.24 shows the *TRT* and *Late_Ct* trajectories of a specific scenario in which a *Late Token* event occurs. In that figure the FDDI behavior is observed from a generic station, which will be called the *tagged station* throughout the following discussion, under the assumptions that:

(i)   an Early Token (*Late_Ct* contains zero) arrives at the tagged station at time zero (arrow (1) in Figure 5.24);

(ii)   all stations are "empty" during the first rotation of the token around the ring, and

(iii) by the time the token has rotated around the ring, the tagged station has become full of synchronous and asynchronous frames.

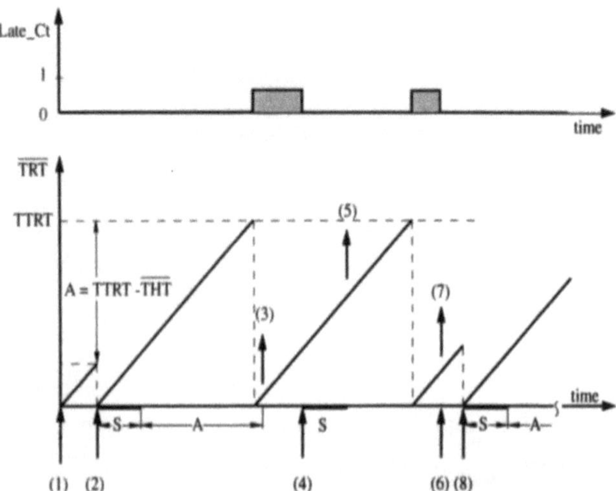

Figure 5.24: Example of a scenario triggering a Late Token event

According to the previous assumptions, the token comes back to the tagged station after a time equal to the ring latency (arrow (2) in Figure 5.24). It can easily be verified that the token is still early, and thus actions (1) to (4) on page 221 will be performed. Specifically, the *THT* of the tagged station is equal to the ring latency and this, in the example in Figure 5.24, is negligible compared to the *TTRT*. Hence, the $A = TTRT - \overline{THT}$ value is very close to *TTRT*. From arrow (3) in Figure 5.24, it can be seen that the token is released by the tagged station after the *TRT* has expired. Therefore, when the token leaves the tagged station, that station's *Late_Ct* content is one. When the token comes back to the tagged station (arrow (4) in Figure 5.24), its *Late_Ct* content is one, and thus the token is a Late Token which is managed by FDDI according to the algorithm explained below.

Whenever a Late Token is captured by a station (arrow (4) in Figure 5.24), FDDI performs the following actions:

(i)   the *Late_Ct* is set to 0;

(ii)   only synchronous frame transmission can occur, at the end of which the token is released (arrow (5) in Figure 5.24);

(iii) the *TRT* is not restarted.

Hence, the choice made by FDDI on Late Token arrival is very clear: the *TRT* is left running to count both the amount of time by which the token arrived late, plus the next cycle time. Obviously, the token remains late until the aggregate quota of synchronous frames transmitted in a cycle offsets the *accumulated lateness*. For example, continuing the analysis of Figure 5.24, arrow (6) shows that when the token arrives at the tagged station, the *Late_Ct* once again contains one, and thus the token is still a Late Token. However, this time the tagged station has nothing to transmit and, hence, the token is immediately released, as indicated by arrow (7). As shown by arrow (8), when the token comes back again to the tagged station, the *Late_Ct* contains zero. Thus, the token becomes an Early Token and the tagged station can now perform both synchronous and asynchronous frame transmission.

When FDDI is functioning correctly, the accumulated lateness should never exceed $TTRT$, so that the TRT should never expire twice during a single token rotation, i.e., the token rotation time is bounded up by $2 \times TTRT$. This property will be formally proved in Section 5.5.

5.3.5 Additional FDDI Features

In FDDI, asynchronous frame transmission can be controlled by two additional mechanisms: the *restricted token* and *priorities*, which allow MAC users to establish how to share the available asynchronous bandwidth. While the former mechanism involves two stations at a time, the latter involves all the stations visited during the rotation of the token.

RESTRICTED-TOKEN COMMUNICATION. Restricted-token mode permits two asynchronous stations to share all the available asynchronous bandwidth. Once this mode is entered, all the other active stations can only use synchronous bandwidth. The restricted-token mode can be used, for example, when a powerful workstation needs to initiate an exchange of a large quantity of data with a computer requiring all of the unallocated ring bandwidth. The management of the extended dialogue (e.g., decisions to initiate, continue and terminate the dialogue) shall be under the responsibility of higher-level protocols. Two stations enter restricted-token mode by running the following algorithm

1. the initiating station captures a non-restricted token, transmits its data frames, then issues a restricted token;

2. the addressed station receives the above data frames, enters restricted mode, may send data frames to the initiating station, and issues the restricted token.

The exchange of restricted tokens between the initiating and addressed stations can proceed for the duration of the dialogue (potentially many times $TTRT$). This algorithm is fair at the dialogue level, since all stations on the ring will have the opportunity to enter the restricted-token mode in a round-robin fashion. When two stations interact in the restricted-token mode, they exchange data frames according to the asynchronous data exchange algorithm. To detect potential "hang" conditions, SMT monitors the duration of the dialogue. Whenever the duration exceeds the maximum time limit allowed, SMT aborts the restricted mode and asynchronous frames can be exchanged among all stations. The restricted-token mode can be resumed again by going through steps 1 and 2, above.

PRIORITIES. The asynchronous traffic can be divided into eight priority levels. For each priority level (n), a threshold value $(T\_Pri\,(n))$ is established, forming a set of threshold values.

An early, non-restricted token may be captured by a station for transmission of asynchronous traffic of priority n when the current TRT (which is moved to the THT) is less than the associated priority threshold value $T\_Pri(n)$. If priorities are not implemented at a station, only the $TTRT$ threshold is in use. If multiple priority levels are implemented, they are commonly (but not mandatorily) ordered in such a way that the highest and lowest priority levels are $T\_Pri\,(7)$ and $T\_Pri\,(0)$, respectively. Hence,

$$Ring\_Latency < T\_Pri\,(0) < T\_Pri\,(1) < ... < T\_Pri\,(7) \leq TTRT \quad .$$

Therefore, by setting lower threshold values for lower priority levels, transmission of lower priority frames is deferred when the ring is heavily loaded (i.e., when the token rotation time exceeds the threshold).

As was shown earlier (see Section 5.3.4), the amount of time available to a station for asynchronous transmission is equal to $A = TTRT - \overline{THT}$. With priorities implemented, the asynchronous data a station can transmit is limited by the greatest of the $T\_Pri\,(n)$ values.

REMARK. Since the *TRTs* at successive stations are different, lower priority traffic at one station may be transmitted, while higher priority traffic at another station may be denied access during the same token rotation.

5.3.6 Claim Token Process

Any station detecting a requirement of ring (re-)initialization shall initiate the *claim token process* which develops according to the following rules.

In FDDI, every station has its own requirement, denoted by T_Req, on the maximum inter-token time. The claim token process is designed to satisfy the following specification. The station with the most stringent requirement (i.e., the shortest T_Req value) is granted a claim to create the token. If the shortest T_Req value is shared by two or more stations, then the station with the longest address (48 bit or 16 bit) wins the claim. If these two rules do not result in a unique winner, the station with the numerically highest address wins the claim.

The claim winner is selected via a distributed bidding process which is carried out according to the following steps.

1. Every station on the ring will declare its bid via continuously generating *claim frames* (see Table 5.3). When issuing a claim frame, a station stores its bid value, which coincides with its T_Req value, in the first four bytes of the information field of the claim frame (T_Bid). It should be noted that a token is not necessary to transmit claim frames.

2. When a station receives a claim frame, it compares the bid carried by the frame (T_Bid) with its own required target (T_Req). If $T\_Bid < T\_Req$, the station stops bidding and forwards the received claim frames unchanged.

3. If $T\_Bid > T\_Req$, the station removes the incoming claim frames from the ring and generates its own claim frames with its own T_Bid value.

4. If $T\_Bid = T\_Req$, the rules (previously mentioned) based on the address length and address value are applied. If the station is the winner of the claim, it keeps generating its own claim frames, otherwise it stops bidding.

5. The claim token process ends when one station receives its own claim frame and is, therefore, the claim winner.

It should be noted that the claim token process works even if more than one station starts sending claim frames simultaneously.

The claim winner issues a non-restricted token. During the first rotation, stations align their timers. Specifically, all stations clear their $TRTs$ and store the negotiated T_Req in a register called T_Neg. Furthermore, on the first rotation the stations set $Late\_Ct$ to one. Therefore, only synchronous traffic can be transmitted on the second rotation, and $Late\_Ct$ gets reset on this rotation. Thus, asynchronous traffic can be transmitted on the third and subsequent rotations.

REMARK. The FDDI standard documents make use of several variables, namely, T_Req, T_Opr, T_Bid, and T_Neg, that relate to $TTRT$. The proper variable (and hence the register) used by FDDI depends on the state of the ring as follows

- T_Req contains the value *requested* by a station. The T_Req value is local to each station, and so may differ from station to station. Each station also maintains its own local minimum and maximum limits, T_Min and T_Max. $TTRT$ must be between these limits. The default maximum value of T_Min is 4 msec, whereas the default minimum value of T_Max is 165 msec.
- T_Bid contains the value declared by a station in its claim frame. Since stations bid their T_Req values, the T_Bid value of each station is equal to its T_Req value.
- T_Neg contains the value *negotiated* during the claim token process. The winning station's T_Bid becomes T_Neg for all stations, and is therefore a network-wide variable.
- T_Opr contains the *operational* value of the $TTRT$. The value of T_Opr is T_Max before initialization, and T_Neg afterwards.

Figure 5.25 shows the relationship among the above variables.

5.3.7 Beacon Process

A station initiates the *beacon process* if it detects that the claim token process has failed, or if it receives a request to do so from the SMT. In these cases, the ring has probably been physically interrupted, and may have been globally reconfigured. Upon entering the beacon process, a station continu-

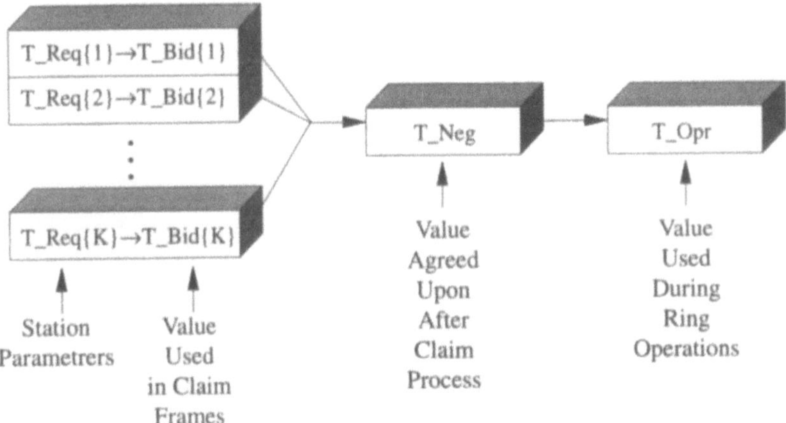

Figure 5.25: Relationship among the various *TTRT* values

ously transmits beacon frames (see Table 5.3). The beacon process terminates when the station receives its own beacon frame. At this point the claim
process is re-entered. If the beacon process does not terminate within a prespecified time limit the station concludes that the ring is broken.

5.3.8 Examples of Parameter Calculations

In this section, several previously defined parameters are calculated to give
an idea of their order of magnitude.

RING LATENCY. The ring latency ($Ring\_Latency$) is mainly affected by the
values of the following parameters: the *fiber delay* ($Prop\_Delay$), which is
proportional to the total fiber length ($Fiber\_Length$) ; the *node delay*
(SD_Max) [4], i.e., the delay introduced by each active station; the number
of active stations, K; and the token transmission time ($Token\_Time$). By
taking into consideration that in the optical fiber the light speed is
1.967×10^8 *m/s* or, equivalently, the propagation delay is 5.085 $\mu s/Km$, the
following relations hold

$$Prop\_Delay = Fiber\_Length \times 5.085\,\mu s/Km \quad , \tag{5.1}$$

$$MD\_Stations = K \times SD\_Max \quad , \tag{5.2}$$

$$Token\_Time = 11\ bytes \times 80\ \frac{ns}{byte} = 880\ ns = 0.00088\ ms \quad , \qquad (5.3)$$

$$Ring\_Latency = Prop\_Delay + MD\_Stations + Token\_Time \quad , \qquad (5.4)$$

where $MD\_Stations$ denotes the maximum delay introduced by the active stations. Therefore, to calculate the maximum values for these parameters, it is necessary to evaluate the above formulae using the following values: $K = 1,000$, $SD\_Max = 600 ns/station$ (the value provided by the standard), and $Fiber\_Length = 100 Km$. Hence

$$Prop\_Delay = 100\ Km \times 5.085\ \mu s/Km = 0.508\ ms \quad , \qquad (5.5)$$

$$MD\_Stations = 1,000 \times 600 ns/station = 0.6\ ms \quad , \qquad (5.6)$$

$$Ring\_Latency = 0.508\ ms + 0.6\ ms + 0.00088\ ms = 1.10888\ ms \quad , \quad (5.7)$$

SYNCHRONOUS BANDWIDTH ALLOCATION. This section presents the computation of the maximum quota of synchronous bandwidth allocated to each station under the assumption that the same bandwidth is assigned to each station. The computation is performed assuming a negotiated $TTRT$ value of $25 msec$, an optical fiber length of 10 Km, and $K = 20$ stations.

First, the contributions from fiber, stations, and token are

$$Prop\_Delay = 10\ Km \times 5.085\ \mu s/Km = 50.85\ \mu s = 0.051\ ms \quad , \quad (5.8)$$

$$MD\_Stations = 20\ stations \times 600\ \frac{ns}{station} = 0.012\ ms \quad , \qquad (5.9)$$

$$Token\_Time = 11\ bytes \times 80\ \frac{ns}{byte} = 0.00088\ ms \quad , \qquad (5.10)$$

from which $Ring\_Latency = 0.0639\ ms$. Thus, the time available per cycle for synchronous frame transmission (T_{Synch}) is given by the expression

$$T_{Synch} = TTRT - Ring\_Latency = 24.93\ ms \quad ,$$

allowing each station to transmit synchronous frames for up to

$$S = \frac{24.93\ ms}{20} = 1.246\ ms \quad . \qquad (5.11)$$

The number of synchronous, maximum-size frames that each station can transmit can be calculated as follows:

$$T_{Frame} = 4,500 \frac{bytes}{frame} \times 80 \frac{ns}{byte} = 0.36 \frac{ms}{frame} \quad , \tag{5.12}$$

So,

$$Number\ of\ Synchronous\ Frames = \left\lfloor \frac{S}{T_{Frame}} \right\rfloor = 3\ frames \quad . \tag{5.13}$$

5.4 FDDI MAC PROTOCOL CAPACITY

This section is devoted to the derivation of the FDDI MAC protocol capacity. To proceed in a tutorial fashion, first the *Synchronous Capacity*, and then the *Asynchronous Capacity*, is introduced. At the end of the section, the *Global Capacity* of the FDDI MAC protocol, handling a combination of synchronous and asynchronous traffic types is given.

SYNCHRONOUS CAPACITY. Assuming that the K stations are saturated with synchronous traffic, the FDDI protocol capacity for synchronous frame transmission, denoted throughout by $\rho_{max}^{(S)}$, can be easily calculated, recalling that the maximum total transmission time over a token rotation, which lasts $TTRT$, can never exceed $TTRT - Ring\_Latency$. Hence,

$$\rho_{max}^{(S)} \doteq \frac{TTRT - Ring\_Latency}{TTRT} \quad . \tag{5.14}$$

By assuming that the length of synchronous frames is constant and equal to F, and that the K stations can transmit exactly h synchronous frames during a token rotation, i.e., $hF = TTRT - Ring\_Latency$, formula (5.14) can be rewritten as

$$\rho_{max}^{(S)} = \frac{1}{1 + \dfrac{a}{h}} \quad , \tag{5.15}$$

where $a = Ring\_Latency/F$.

REMARK. As shown by Equation (5.14) the synchronous capacity changes with the value of $TTRT$. The trade-off in $TTRT$ selection is basically the following: the longer the $TTRT$, the smaller the influence of $Ring\_Latency$ and, hence, the more efficient the utilization of channel bandwidth. On the other hand, the shorter the $TTRT$, the shorter the inter-token time becomes.

$\Diamond$

ASYNCHRONOUS CAPACITY. The maximum FDDI asynchronous capacity, denoted throughout by $\rho_{max}^{(A)}$, was first derived by Dykeman and Bux [61]. The model they use to calculate $\rho_{max}^{(A)}$ is based on the following assumptions:

1. all stations are saturated[1] by asynchronous frames to send, i.e., they always have asynchronous frames waiting for transmission;

2. asynchronous priority levels are in use;

3. frame transmission times have constant duration F, and

4. asynchronous overruns are of constant length R.

Under these assumptions, Dykeman and Bux show that the network behavior can be represented by a regenerative process. By applying Theorem 2.4, they derive the expression for the maximum aggregate asynchronous throughput, $\gamma_{max}^{(A)}$, that FDDI can manage. Specifically, $\gamma_{max}^{(A)}$ is obtained through the formula

$$\gamma_{max}^{(A)} = \frac{E\,[tot\_tx\_time]}{E\,[Reg\_Cycle\_Length]} \times C \quad, \tag{5.16}$$

where

- $E\,[tot\_tx\_time]$ is the (average) amount of time stations transmit asynchronous frames in a regenerative cycle;
- $E\,[Reg\_Cycle\_Length]$ is the (average) length of a regenerative cycle; and
- C is the nominal channel rate (100 Mbps for FDDI).

$\rho_{max}^{(A)}$ is then obtained by making the ratio between $\gamma_{max}^{(A)}$ and C. The derivation of $\rho_{max}^{(A)}$ presented in [61] is quite complex. In this book a simpler approach is presented, which is valid if the asynchronous overrun can be

---

1. In the literature, such an offered load condition is often referred to as the *asymptotic condition*.

neglected and priorities are not implemented. Before showing this simpler derivation, it is necessary to introduce the following notations

- $t_i^{(n)}$ denotes the value of the *TRT* timer when the token arrives at station $\{i\}$ during the n-th cycle;
- $A_i^{(n)}$ denotes the quota of asynchronous traffic transmitted by station $\{i\}$ during the n-th cycle.

Starting from any given initial state, station$\{1\}$ measures an inter-token time equal to $Ring\_Latency + \varepsilon$ where ε is the amount of time in which asynchronous traffic was transmitted in the previous cycle. In the most general case, ε can be written as $\varepsilon = \sum_{i=1}^{K} \varepsilon_i$, where ε_i is the amount of the asynchronous traffic transmitted by station$\{i\}$ in the previous cycle. The system evolves as shown in Table 5.4. The table indicates that the system regenerates after $K+1$ token rotations. Specifically, the table starts with the initial status in which, on token arrival at station$\{1\}$, station$\{1\}$ can transmit asynchronous frames for the maximum possible time, i.e., $TTRT - \tau - \varepsilon$, where $\tau = Ring\_Latency$. Station$\{2\}$ then observes a cycle length equal to $TTRT - \varepsilon_1$, and thus transmits an asynchronous quota equal to ε_1. In general, in cycle 1, a station$\{i\}$, with the exception of station$\{1\}$, observes a cycle length equal to $TTRT - \varepsilon_{i-1}$, and thus transmits an asynchronous quota equal to ε_{i-1}. The system evolves in the deterministic way shown in Table 5.4. Finally, when the token comes back to station $\{1\}$, after station$\{K\}$ has transmitted in the K-th cycle, the system repeats the same behavior again and again, starting from Cycle 1 of Table 5.4.

From the evolution shown in Table 5.4, it can be shown that

1. $E\,[Reg\_Cycle\_Length] \;=\; (K \times TTRT) + Ring\_Latency$;

2. $E\,[tot\_tx\_time] \;=\; K \times (TTRT - Ring\_Latency)$;

3. Each station transmits the same amount of asynchronous traffic in a regenerative cycle, i.e., $TTRT - Ring\_Latency$.

Hence, by substituting 1 and 2 in Formula (5.16), it follows that

$$\rho_{max}^{(A)} \;=\; \frac{K \times (TTRT - Ring\_Latency)}{(K \times TTRT) + Ring\_Latency} \; . \tag{5.17}$$

In addition, point 3 indicates that in asymptotic conditions, FDDI fairly subdivides the asynchronous bandwidth among its stations. This means that the

Table 5.4 Structure of the regenerative cycle for asynchronous transmission

Cycle 0	Cycle 1		Cycle 2		...	Cycle K	
$A_1^{(0)} = \varepsilon_1$	$t_1^{(1)} = \tau + \varepsilon$	$A_1^{(1)} = TTRT - \tau - \varepsilon$	$t_1^{(2)} = TTRT - \varepsilon_K$	$A_1^{(2)} = \varepsilon_K$	...	$t_1^{(K)} = TTRT - \varepsilon_2$	$A_1^{(K)} = \varepsilon_2$
$A_2^{(0)} = \varepsilon_2$	$t_2^{(1)} = TTRT - \varepsilon_1$	$A_2^{(1)} = \varepsilon_1$	$t_2^{(2)} = \tau + \varepsilon$	$A_2^{(2)} = TTRT - \tau - \varepsilon$	...	$t_1^{(K)} = TTRT - \varepsilon_3$	$A_1^{(K)} = \varepsilon_3$
...	...	...	...	...	...	...	...
$A_K^{(0)} = \varepsilon_K$	$t_K^{(1)} = TTRT - \varepsilon_{K-1}$	$A_K^{(1)} = \varepsilon_{K-1}$	$t_K^{(2)} = TTRT - \varepsilon_{K-2}$	$A_K^{(2)} = \varepsilon_{K-2}$	...	$t_K^{(K)} = \tau + \varepsilon$	$A_K^{(K)} = TTRT - \tau - \varepsilon$

FDDI MAC protocol is *fair* as far as asynchronous transmission is concerned.

To analyze the dependency of the protocol capacity on the a parameter values (see Chapter 1), Formula (5.17) needs to be manipulated slightly. Because of the assumption $R = 0$, $TTRT$ and F must be adjusted in such a way that

$$kF = TTRT - Ring\_Latency \quad , \tag{5.18}$$

where k is an integer greater than or equal to one.

Substituting (5.18) into (5.17) and remembering the definition of the a parameter, i.e., $a = Ring\_Latency/F$, the following relation holds

$$\rho_{max}^{(A)} = \frac{1}{1 + \dfrac{a}{k}\left(1 + \dfrac{1}{K}\right)} \quad . \tag{5.19}$$

As was stated before, (5.19) is valid when the asynchronous overrun is zero[1] (i.e., $R = 0$). This was generalized by Dykeman and Bux [61] to the case in which the asynchronous overrun is of constant length, R. Their approach starts from the observation that in the asymptotic condition, on token arrival at any active station on the ring, the associated *transmission state* is specified by the token rotation timer value $(\overline{TRT})$ and the (asynchronous) frame transmission time (tx_time) value. Furthermore, as was shown for the case with $R = 0$, Dykeman and Bux pointed out that the ring converges to steady-state operation, so that after a fixed number of token rotations (which depends upon the number of active stations), the set of active stations cycle through a finite number of transmission states.

To show this in the simplest case, let us consider the FDDI configuration of only one active station which starts from the initial scenario in which the FDDI ring is completely empty during the first token rotation. Assume that the priorities are not implemented and that by the time the token circulates around the ring for the first time, the station gets an infinite number of asynchronous frames to transmit. By defining the time instant at which the token is observed by the station as the beginning of the token rotation, the sequence of the station transmission states evolves as follows:

---

1. In other words, the asynchronous frame transmission stops immediately as soon as THT reaches $TTRT$.

1. beginning of second token rotation

$$\overline{TRT} = Ring\_Latency$$

$$tx\_time = TTRT - Ring\_Latency + R$$

2. beginning of third token rotation

$$\overline{TRT} = TTRT + R \qquad (\text{token is late by R})$$

$$tx\_time = 0$$

3. beginning of fourth token rotation

$$\overline{TRT} = R + Ring\_Latency$$

$$tx\_time = TTRT - (R + Ring\_Latency) + R = TTRT - Ring\_Latency$$

4. beginning of fifth token rotation

$$\overline{TRT} = TTRT$$

$$tx\_time = 0$$

Obviously, during the fifth token rotation, the station does not transmit any frames. When the token comes back to the station at the end of the fifth rotation, the previous four transmission states are again experienced, and in the same order, by the active station. This behavior repeats over and over on successive token rotations. Hence, the regenerative process which represents the behavior of the FDDI network just analyzed, cycles through the previous four transmission states. The regenerative cycle has a constant length (Reg_Cycle_Length) of

$$Reg\_Cycle\_Length = (TTRT - Ring\_Latency + R) + \qquad (5.20)$$
$$(TTRT - Ring\_Latency) + 4 \times Ring\_Latency \ ,$$

whereas the total transmission time (tot_tx_time) of the station during a regenerative cycle is

$$tot\_tx\_time = (TTRT - Ring\_Latency + R) +$$
$$(TTRT - Ring\_Latency) \ .$$

Therefore, the maximum throughput achieved by the (only) active station is

$$\gamma_{max}^{(A)} = \qquad\qquad\qquad (5.21)$$

$$\frac{tot\_tx\_time}{Reg\_Cycle\_Length} \times C = \frac{(tot\_tx\_time)}{(tot\_tx\_time + 4 \times Ring\_Latency)} \times C \ .$$

In [61], by following the same line of reasoning used with one active station, a scenario with two active stations is analyzed. The resulting expression for the aggregate throughput achieved by both stations has a structure similar to (5.21):

$$\gamma_{max}^{(A)} = \frac{tot\_tx\_time}{tot\_tx\_time + 9 \times Ring\_Latency} \times C \quad . \tag{5.22}$$

By examining additional scenarios (not reported in [61]) with more than two active stations, Dykeman and Bux derive the $\gamma_{max}^{(A)}$ expression for an FDDI network configuration with an arbitrary number, K, of active stations, from which $\rho_{max}^{(A)}$ can be obtained

$$\rho_{max}^{(A)} = \frac{K \times tot\_tx\_time + K^2 \times tx\_window}{K \times tot\_tx\_time + K^2 \times tx\_window + \Delta} \quad , \tag{5.23}$$

where

- $\Delta = (K^2 + 2K + 1) \times Ring\_Latency$,

- $tx\_window = TTRT{-}Ring\_Latency$,

- $tot\_tx\_time = tx\_window + R = \left\lceil \dfrac{tx\_window}{F} \right\rceil \times F$.

Since

$$R = \left\lceil \frac{TTRT{-}Ring\_Latency}{F \cdot} \right\rceil \times F{-}TTRT{-}Ring\_Latency \quad , \tag{5.24}$$

Equation (5.23) can be written as

$$\rho_{max}^{(A)} = \frac{K \times (TTRT{-}Ring\_Latency) + \dfrac{K}{K+1} \times R}{(K \times TTRT) + Ring\_Latency + \dfrac{K}{K+1} \times R} \quad . \tag{5.25}$$

REMARK. As expected, when $R = 0$, (5.25) coincides with (5.17).

◊

Taking the limit of $\rho_{max}^{(A)}$ as the number of active stations, K, approaches infinity, a bound for the maximum asynchronous protocol capacity is obtained:

$$\lim_{K \to \infty} \rho_{max}^{(A)} = \frac{tx\_window}{tx\_window + Ring\_Latency} = \frac{TTRT - Ring\_Latency}{TTRT} . \tag{5.26}$$

Keeping into consideration that $0 \le R \le F$, $\rho_{max}^{(A)}$ can vary within the range

$$\frac{K \times (TTRT - Ring\_Latency)}{(K \times TTRT) + Ring\_Latency} \le \rho_{max}^{(A)} \le \tag{5.27}$$

$$\frac{K \times (TTRT - Ring\_Latency) + \dfrac{K \times F}{K + 1}}{(K \times TTRT) + Ring\_Latency + \dfrac{K \times F}{K + 1}} .$$

After some algebraic manipulations, (5.27) can be expressed in terms of a:

$$\frac{1}{1 + \dfrac{a}{k}\left(1 + \dfrac{1}{K}\right)} \le \rho_{max}^{(A)} \le \frac{1}{1 + \dfrac{a}{k}\left(1 + \dfrac{1}{K}\right)\left[\dfrac{k(K+1)}{k(K+1)+1}\right]} . \tag{5.28}$$

It can be observed that when $K \to \infty$, the upper and lower bounds coincide, and are equal to

$$\lim_{K \to \infty} \rho_{max}^{(A)} = \frac{1}{1 + \dfrac{a}{k}} . \tag{5.29}$$

It is interesting to compare (5.29) with (1.11) in Chapter 1 which gives the Token Ring protocol capacity when $a > 1$. From Formula (1.11), as the a parameter values increase, the protocol capacity decreases. On the other hand, the presence of the k factor in (5.29), i.e., of the multiple frame mechanism, reduces the negative effect on the protocol capacity due to an increase in the a parameter values. Specifically, if a/k is kept much lower than one, the resulting FDDI protocol capacity can be close to one.

Figure 5.26 highlights this behavior for several k values. For a given value of a, the greater the k value, the greater the value of $\rho_{max}^{(A)}$.

Figure 5.27 shows the influence of $TTRT$ on $\rho_{max}^{(A)}$ for three FDDI network configurations (small, medium, and large) characterized by the param-

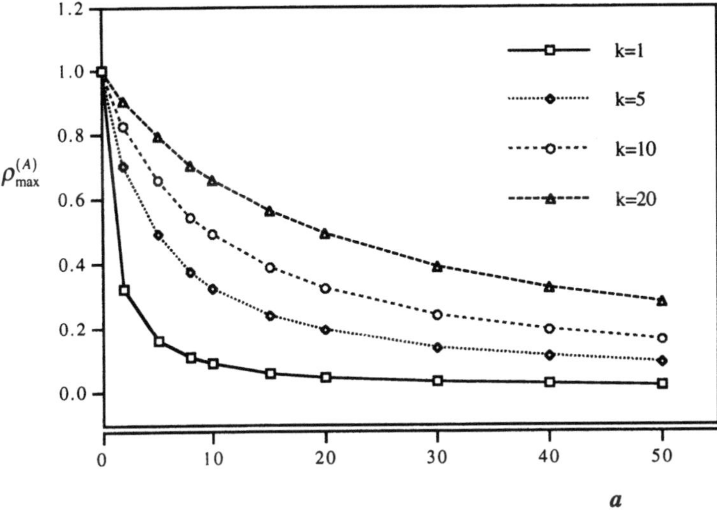

Figure 5.26: $\rho_{max}^{(A)}$ versus a for $K=20$

eter values listed in Table 5.5.

Table 5.5 FDDI configurations

Configuration	small	medium	large
ring length	1 Km	10 Km	100 Km
K	20	50	100
Ring_Latency	17.085 μsec	254.250 μsec	568.5 μsec

The values in the last row of Table 5.5 are calculated while keeping in mind that the ring latency value $(Ring\_Latency)$ is the sum of the signal propagation delay on the fiber $(5.085 \ \mu sec/Km)$ and of the station's latencies $(600 \ nsec/station)$. Figure 5.27 shows that, for each Table 5.5 FDDI configuration, an increase in the *TTRT* results in an increase in the protocol capacity. This can be easily understood from the analysis of formula (5.19) since, for a given FDDI configuration, increasing the *TTRT* results in an increase in k, i.e., more frames are transmitted during each token rotation. Furthermore, from the analysis of (5.19), it can be understood why increasing the number of active stations increases the protocol capacity.

Figure 5.28 shows the influence of the *Ring_Latency* on the protocol capacity for an FDDI configuration of 50 active stations. The curves were

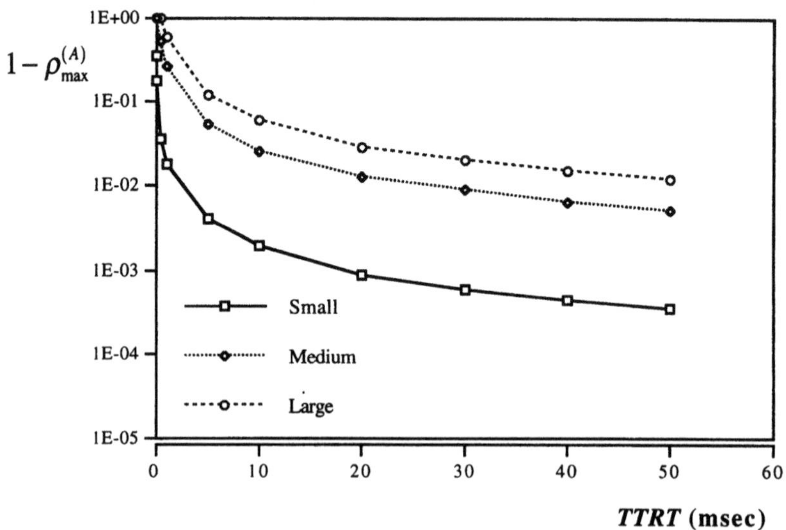

Figure 5.27: $1 - \rho_{max}^{(A)}$ versus *TTRT* for the three FDDI configurations

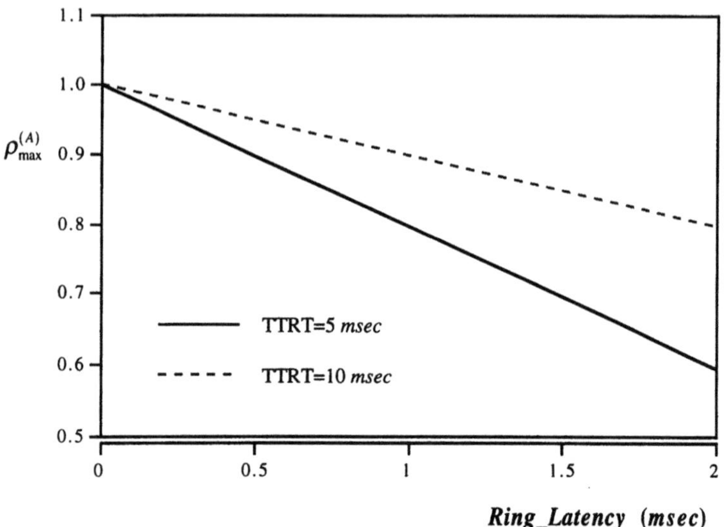

Figure 5.28: $\rho_{max}^{(A)}$ versus *Ring_Latency* for an FDDI configuration of 50 active stations, *TTRT*=5 msec and *TTRT*=10 msec

obtained by applying (5.23). The range of the *Ring_Latency* values varies from 0.25 *msec*, representing an FDDI ring with 50 connected stations and a

44 Km fiber, to 1.62 *msec,* which represents an FDDI ring with 1000 connected stations and a 200 Km fiber (i.e., the maximum configuration allowed by FDDI). Figure 5.28 shows that, with a *TTRT* of 10 *msec*, the protocol capacity remains high (i.e., 0.82), even in the case of the maximum configuration allowed by FDDI. When the *TTRT* is reduced to 5 msec, the protocol capacity corresponding to the largest FDDI configuration drops to 0.65.

REMARK. Similarly to the synchronous capacity, increasing the *TTRT* value leads to higher asynchronous capacity, but this degrades access delay performance figures.

◊

BIBLIOGRAPHIC NOTES. The equation for $\rho_{max}^{(A)}$ was generalized in [61] to produce the maximum total throughput for FDDI when multiple asynchronous priority levels are being used.

◊

GLOBAL CAPACITY. Formulae (5.19) and (5.15) give the FDDI protocol capacities when active stations operate in saturated conditions with one traffic type only, i.e., asynchronous and synchronous, respectively. However, an FDDI network can operate with both types of traffic. Thus, the FDDI protocol capacity should be derived under the assumptions that the K stations

- operate in asymptotic conditions as far as the asynchronous traffic is concerned; and
- use a fraction f, $f < 1$ of their maximum allowed synchronous allocation on every token rotation, i.e., $f \times TTRT$, and thus the synchronous bandwidth left over will be used for asynchronous transmission.

Under the above assumptions, the behavior of the FDDI network can be represented, neglecting the asynchronous overrun, by a regenerative process which regenerates after every $K + 1$ token rotations. Specifically, during any given token rotation, the K stations transmit synchronous frames for a total duration of $f \times TTRT$, while the amount of the asynchronous traffic is computed by following the same line of reasoning used in the analysis of the FDDI asynchronous capacity (see Table 5.4). Starting from a generic initial state in which station$\{1\}$ measures an inter-token time equal to $Ring\_Latency + \varepsilon + f \times TTRT$,[1] the system has an evolution similar to that shown in Table 5.4. Hence, the regenerative cycle is the sum of the following times

1. total asynchronous transmission time
$$K \times (TTRT - Ring\_Latency - (f \times TTRT));$$
2. total synchronous transmission time $(K + 1) \times (f \times TTRT)$;

3. total latency due to token rotations $(K + 1) \times Ring\_Latency$.

Thus, the duration of the regenerative cycle is
$$K \times TTRT + Ring\_Latency + f \times TTRT \quad .$$

Therefore, the computation of the asynchronous $(\rho_{max}^{(A)})$, synchronous $(\rho_{max}^{(S)})$, and global $(\rho_{max}^{(FDDI)})$ capacities proceeds in a straightforward manner with the following results

$$\rho_{max}^{(A)} = \frac{K \times (TTRT - Ring\_Latency - (f \times TTRT))}{(K \times TTRT) + Ring\_Latency + f \times TTRT} \quad , \tag{5.30}$$

$$\rho_{max}^{(S)} = \frac{(K + 1) \times (f \times TTRT)}{(K \times TTRT) + Ring\_Latency + f \times TTRT} \quad , \tag{5.31}$$

and

$$\rho_{max}^{(FDDI)} = \rho_{max}^{(A)} + \rho_{max}^{(S)} = \tag{5.32}$$

$$\frac{K \times (TTRT - Ring\_Latency) + (f \times TTRT)}{(K \times TTRT) + Ring\_Latency + f \times TTRT} \quad .$$

It must be noted that when $f = 0$ (i.e., there is no synchronous traffic), (5.32) reduces to (5.19), and when only synchronous traffic is transmitted, i.e., $f \times TTRT = TTRT - Ring\_Latency$, (5.32) reduces to (5.15).

5.5 FDDI CYCLE PROPERTIES

This section is devoted to prove the two most important properties of the token rotation time:

---

1. As before, $\varepsilon = \sum_{i=1}^{K} \varepsilon_i$ is the amount of the asynchronous traffic transmitted in the previous cycle.

1. the maximum token rotation time cannot exceed twice the $TTRT$, and
2. the average token cycle time cannot exceed the $TTRT$.

The bounds on the cycle length were formally proved by Sevcik & Johnson in [138]. Here, by exploiting a different approach, a simpler derivation of these properties is presented.

5.5.1 Maximum Cycle Length

The upper bound of the cycle length is derived by observing that the maximum quota of asynchronous traffic which can be transmitted in a cycle is bounded by $TTRT - Ring\_Latency$. In fact, by focusing on the cycle observed by a station (say station{1}, without any loss of generality), and by denoting with A_i the quota of asynchronous traffic transmitted by station{i}, the following inequalities hold

$$0 \le A_1 \le TTRT - Ring\_Latency \quad ,$$
$$0 \le A_2 \le TTRT - Ring\_Latency - A_1 \quad ,$$

$$\ldots\ldots$$

$$0 \le A_K \le TTRT - Ring\_Latency - \sum_{j=1}^{K-1} A_j \quad .$$

Hence, the quota of asynchronous traffic in a cycle satisfies the following inequalities

$$0 \le \sum_{j=1}^{K} A_j \le TTRT - Ring\_Latency \quad . \tag{5.33}$$

By considering that the maximum quota of synchronous traffic is $TTRT - Ring\_Latency$, it follows that the token rotation time has an upper bound of $2 \times TTRT$.

Figure 5.29 shows a scenario in which the bound on the cycle length is reached. Arrows (1) and (2) in that figure[1] delimit a cycle, observed by station{1}, in which no transmission occurs. Assume that, when the token arrives again at station{1}, all the stations are full of synchronous and asynchronous frames. Hence, station{1} transmits its maximum synchronous quota (S) and the maximum quota of asynchronous traffic it is authorized to transmit, $A = TTRT - Ring\_Latency$. When the token (arrow (3)) leaves the tagged station, all the remaining stations transmit their maximum quota

---

1. Arrows (1) and (2) are separated by an interval time equal to $Ring\_Latency$.

of synchronous frames and, eventually, the token comes back to the tagged station (arrow (4)).

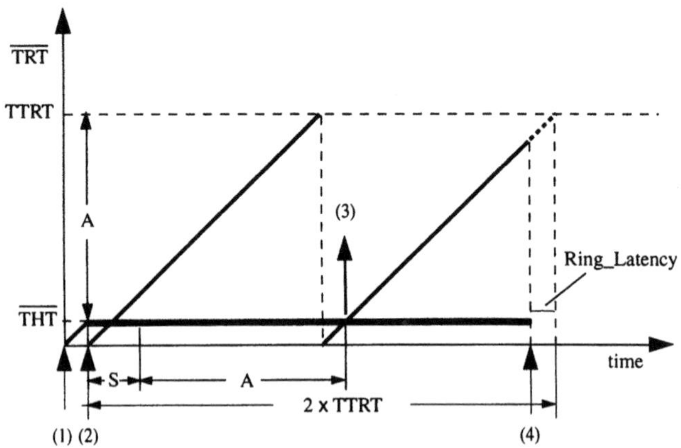

Figure 5.29: FDDI worst case scenario

By using geometric arguments, it can be seen that the time interval between arrows (2) and (4), which measures the inter-token arrival time at the tagged station, lasts approximately $2 \times TTRT$. From Figure 5.29 it also appears that the inter-token arrival time approaches $2 \times TTRT$ as $Ring\_Latency$ approaches zero.

5.5.2 Average Cycle Length

To derive the bound on the average cycle length it is useful to observe that FDDI behaves as a polling system with total switchover time equal to $Ring\_Latency$. Hence its average cycle length satisfies the following relationship (see Section 2.1.3)

$$E[C] = \frac{Ring\_Latency}{1-\rho} . \tag{5.34}$$

From (5.34) it follows that the maximum average cycle length is obtained when ρ takes its maximum value, i.e., $\rho_{max}^{(FDDI)}$. Hence, by substituting (5.32) in (5.34)

$$E[C] = \frac{Ring\_Latency}{1 - \dfrac{K \times (TTRT - Ring\_Latency) + (f \times TTRT)}{(K \times TTRT) + Ring\_Latency + f \times TTRT}} , \tag{5.35}$$

which after some algebraic manipulations can be rewritten as

$$E[C] = \frac{(K \times TTRT) + Ring\_Latency + f \times TTRT}{(K + 1)} \; . \qquad (5.36)$$

Finally, by observing that

$$Ring\_Latency + f \times TTRT \leq TTRT \; , \qquad (5.37)$$

it results that $E[C] \leq TTRT$.

5.6 REMARKS ON THE IEEE 802.4 TOKEN BUS PROTOCOL

The IEEE 802.4 Token Bus standard [88] specifies a token passing protocol on a bus with an optional priority mechanism. Specifically, this MAC protocol identifies four priority classes, called access classes, termed 0, 2, 4, and 6, with 6 being the highest priority and 0 the lowest.The priority mechanism is optional; if it is not used, only priority 6 is active. Each access class acts as a virtual substation in that the token is passed, internally, from the highest access class downward, in the order 4, 2, 0. A time parameter, called the *hi_pri_token_hold_time*, is assigned to class 6 while each of the other three classes is assigned a parameter, called *target token rotation time* (abbreviated as $TTRT_i$, $i = 4, 2, 0$). Hence, for the IEEE 802.4 standard, there can be at most three different values for the $TTRT_i$ parameters. Each station has three rotation timers for the three lower access classes and each access class has a queue where frames to be transmitted are queued. When a station receives the token, it is allowed to transmit data frames of class 6 until the station becomes empty, or until a period of time equal to *hi_pri_token_hold_time* has elapsed, whichever comes first. For each of the other access classes, the corresponding virtual substation measures the time it takes the token to circulate around the logical ring. If the token returns in less than $TTRT_i$, then the substation transmits frames of that class until such frames are transmitted or until $TTRT_i$ has elapsed, whichever comes first. Otherwise, if the token returns later than $TTRT_i$, the station cannot send frames of that priority on this pass of the token and forwards the token immediately. Hence, the priority 6 access class supports the time-constrained service while priorities 4, 2, and 0 support no-time-constrained services.

Obviously, if the total transmission time of class 6 data frames in a token cycle exceeds all the $TTRT$s, then no lower class can transmit at all. The intended effect of the cycle-dependent timing mechanism is that, as the aggregate offered load of class 6 traffic decreases, lower classes are allowed to access the channel in succession, starting from the access class with the largest $TTRT_i$ down to the one with the smallest $TTRT_i$.

The IEEE 802.4 token bus and FDDI MAC protocols are almost equivalent [119]. The main difference between them is the handling of negative residual values, or "accumulated latency" as it is called in FDDI. While the FDDI MAC protocol takes accumulated latency into account, the other MAC protocol does not. In other words, the IEEE 802.4 MAC protocol loses the memory of the accumulated latency by resetting the appropriate token rotation timer. FDDI, on the other hand, keeps track of the accumulated latency (by setting the $Late\_Ct$ counter to one) until it has been recovered.

5.7 CURRENT USE OF FDDI

FDDI is now mainly used for private data communication networks both as backbone technology for interconnecting existing LANs (via FDDI bridges and routers) and to connect directly powerful workstations (via FDDI adapters and concentrators).

FDDI has experienced a slow market penetration due to the high cost per attachment. Initially the cost was around $10,000 (second mid of 80s); at the beginning of the 90s it was still in the order of $5,000 and only recently the cost per attachment dropped to about $1000. This last figure is still five times higher than an Ethernet attachment.

Note that initially FDDI standards were designed for multimode fiber, extensions to the more widely used single-mode fiber were approved in 1992. In addition to these fiber-optic-based standards in the 90s vendor-based products utilizing the twisted-pair technology were also introduced. These systems, called now Copper Distributed Data Interface ($CDDI$), are cheaper because they both don't require a new fiber-based cabling system, and personal computers can be connected with low-cost adapters.

6 FDDI Models

6.1 INTRODUCTION

Although many FDDI analytical studies have already been published, they contain simplifying assumptions. The main difficulty for the analysis of the FDDI MAC protocol is the high degree of complexity and interdependence of the various processes that describe the operations of the protocol itself. In fact, when a station has seized the token, synchronous frames (if any) are always transmitted, whereas asynchronous frames are only transmitted if the token is early. This implies that there are interdependencies between the cycle length, the service period at one station, and the service periods at subsequent stations. Therefore, exact analytically-tractable solutions for an FDDI network are very difficult to formulate. Simplifying assumptions thus have to be made in order to obtain analytically-tractable solutions. Furthermore, even when synchronous traffic alone is transmitted, only approximate solutions are known for the resulting model, i.e., a polling system with an *exhaustive time-limited* (visit) service discipline (see Section 2.1).

To provide a structured overview of the FDDI analytical studies it is useful to introduce the taxonomy shown in Figure 6.1, which preliminarily classifies the FDDI models into two categories: the first category contains *Network-wide models*, while the second contains *Station-in-isolation models* (see Figure 6.2).

The analytical models in the first category characterize the overall FDDI structure by considering all the stations spread around the ring. All the models in this category belong to the class of multiqueue systems with cyclic service (see Chapter 4).

The analytical models in the second category (Station-in-isolation mod-

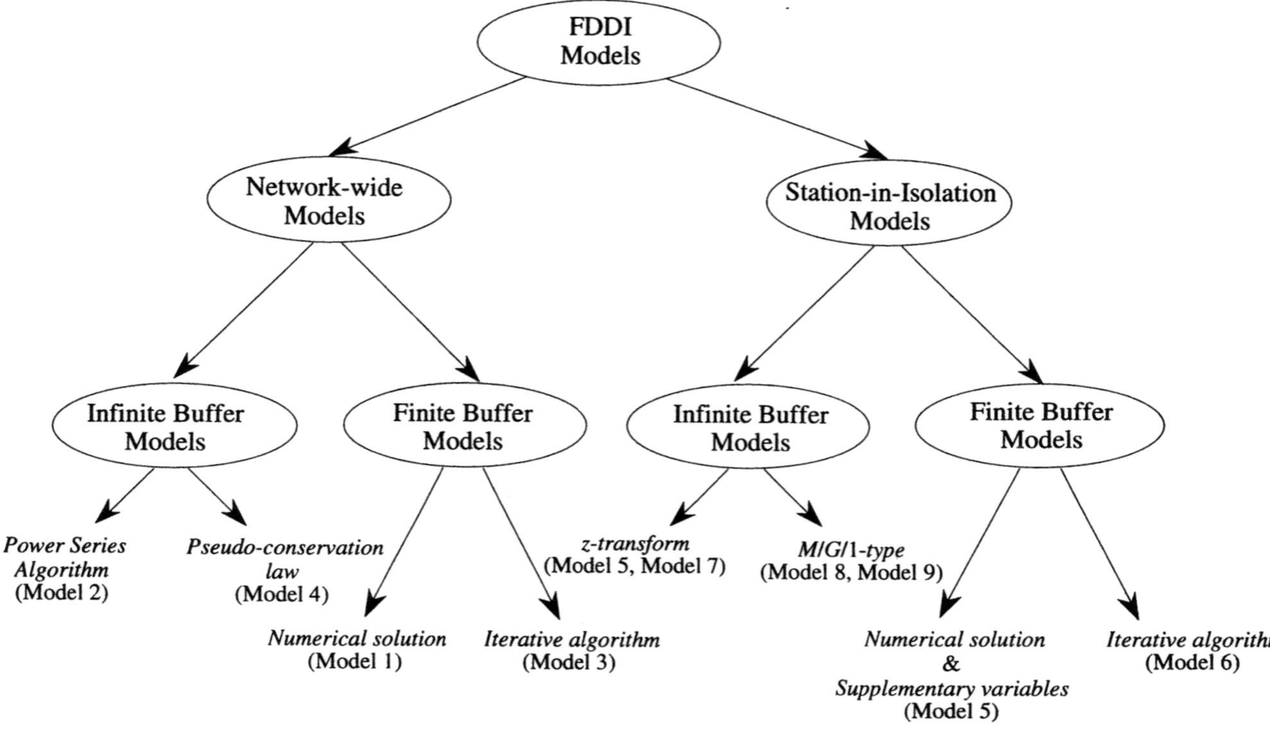

Figure 6.1   FDDI models taxonomy

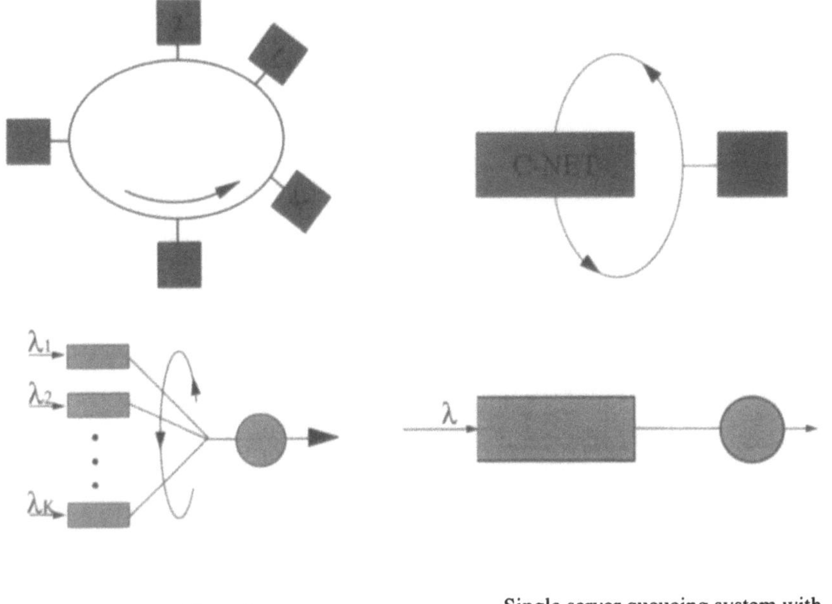

Multiqueue system with cyclic
service

Single server queueing system with
server vacations

Figure 6.2: Network-wide and station-in-isolation models

els) tag the station under study. With respect to the tagged station, the FDDI
network is partitioned into the tagged station itself, and the complementary
part of the network (C–NET) which aggregates all the other stations. In
this approach the tagged station is modeled as a single server queueing system with server vacations (see Section 3.2). The server vacation time represents the period between the token's departure from the tagged station and
its subsequent arrival at this station, i.e., the time it takes the token to cross
C–NET.

Models belonging to both classes can be further subdivided on the basis
of buffer size (i.e., $M=1$, M finite and M infinite). Figure 6.1 shows the computational methods used to solve the resulting models. A list of significant
models is associated to each leaf of the taxonomy, presented in Figure 6.1.
The list will be described later in this chapter.

REMARK. Although in the FDDI standards the unit of transmission is named
frame, in the description of the various FDDI model the words frame and

packet are used interchangeably. This is because in most of the FDDI modeling literature, and also in several chapters of this book, the word packet is used to identify the unit of transmission.

6.2 NETWORK-WIDE MODELS

This section outlines four special cases of the network-wide models. All the models in this category are derived from the following abstract representation of the system.

- The system has K queues (stations) and a single server. The queues can have either finite (one or more) or infinite buffers.
- Synchronous and asynchronous packets arrive at the queues in accordance with Poisson arrival processes.
- The server (token) walks from queue to queue in a fixed order. The time needed to switch the server from queue to queue is modelled by a delay (switchover time), which can be zero (e.g., Model 2).
- When the server reaches a queue, synchronous frames are always transmitted (if any), whereas asynchronous frames are only transmitted if the preceding token rotation time does not exceed $TTRT$. Once the server has served a queue it goes to the next queue. Packets can either be of constant or of variable length.

An FDDI network-wide model has a more complex behavior than a polling system with an exhaustive-limited service discipline (see Chapter 4). FDDI models have thus been solved by imposing some simplifying assumptions on the system behavior, or by providing an approximate solution of the model.

In the first model (Model 1 [148]), Takagi provides an exact analysis of an FDDI network, with synchronous and asynchronous traffic, under the assumption that a maximum of one packet can be queued at each node (i.e., a single buffer model).

Under several assumptions on system behavior (e.g., TRT evaluation, zero switchover time), Altman [3] provides an exact analysis of the resulting model (Model 2) by applying the PSA algorithm (see Section 4.2.2 on page 168).

In Model 3, Chang & Shandu [33] provide an analysis of an FDDI network with synchronous traffic only. On the other hand, Tangeman & Sauer

Table 6.1 Performance figures of station Q_i derived with network-wide models

	Traffic Types	Buffer size	Queue indices	Delay indices	Other indices
Model 1	$S + A$	$M = 1$	$p_i(\bullet)$, $P_L(i)$ $E[N_i^S]$, $E[N_i^A]$	$W_i^{S^*}(s)$ $E[W_i^S]$, $E[W_i^A]$	F_{C_i} γ_i^S, γ_i^A
Model 2	$S + A$	$1 \leq M < \infty$	$p_i(\bullet)$	$E[W_i^S]$, $E[W_i^A]$	
Model 3	S	$M = \infty$	$E[N_i^S]$	$E[W_i^S]$	
Model 4	A	$1 \leq M < \infty$	$p_i(\bullet)$, $P_L(i)$	$E[W_i^A]$	F_{C_i}, γ_i^A

LEGEND:

S =Synchronous; A =Asynchronous;

$p_i(\bullet) = Q_i$ queue length distribution at a random point in time;

$P_L(i) = Q_i$ packet-loss probability;

N_i^S (N_i^A) = number of synchronous (asynchronous) packets queued at Q_i;

W_i^S (W_i^A) = waiting time of synchronous (asynchronous) packets at Q_i;

γ_i^S (γ_i^A) = Q_i synchronous (asynchronous) traffic throughput;

F_{C_i} = distribution of the Q_i polling-cycle length.

[152], in Model 4, analyze an FDDI network with asynchronous traffic subdivided into several priority levels. Only approximate solutions have been devised for both Models 3 and 4. The main differences between these models is in the buffer size (infinite vs. finite), and hence in the solution technique used to solve them: approximation based on a pseudo-conservation law vs. approximation based on an iterative scheme.

To clarify the information provided by the different models Table 6.1 reports, for each model presented in this chapter, the type of traffic (Synchronous (S), Asynchronous (A)) analyzed, the buffer size, and the performance measures characterizing the QoS experienced by each class of traffic. Specifically, the performance figures are subdivided into three classes: those describing the queue occupancy (*Queue indices*), the indices describing the delays experienced by the packets (*Delay indices*), and other performance measures (*Other indices*) such as the throughput and the statistics on the token rotation time.

6.2.1 Single-buffer Model with Synchronous and Asynchronous Traffic (Model 1)

This model was proposed and analyzed by Takagi [148]. To perform a delay analysis of an FDDI network which supports both synchronous and asynchronous traffic, Takagi proposes a simplified network model based on the following assumptions

- a single-packet buffer for each station;
- packets are of constant length;
- the *TRT* timer at a station always measures the elapsed time from the last token arrival at that station.[1]

Following these hypotheses, the resulting network model is a polling system with K single-buffer queues $\{Q_1, Q_2, ..., Q_K\}$ served in a cyclic order (see Figure 6.3). Synchronous and asynchronous packets arrive at station$\{i\}$ according to a Poisson process with rate λ_i^S and λ_i^A, respectively.[2] All packets have the same constant length b.

---

1. This means that the token's accumulate delay does not propagate to subsequent cycles.
2. Packets that arrive when the buffer is occupied are lost.

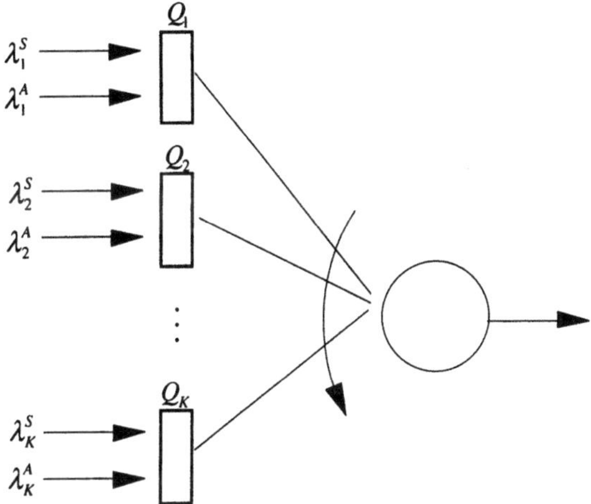

Figure 6.3: FDDI single-buffer model

In the model the TTRT is set equal to the ring latency (τ) plus $m \cdot b$, where m is an integer number in the range $[0, K]$. Therefore, when a station receives the token it is authorized to transmit an asynchronous packet only if in the previous cycle less than m packets have been transmitted (i.e., less than m stations have transmitted in the last cycle).

By describing the system behavior with a Markov chain, for each station, the throughput and the mean waiting times of synchronous and asynchronous packets, the mean token rotation time, and the buffer utilization can be obtained from the Markov chain solution.

NETWORK MARKOV MODEL. In this model, on token arrival a station can be in state 0, 1, 2, or 3; where

- state 0 means that no packet is queued at this station;
- state 1 means that an asynchronous packet is queued at the station and the token is late;[1]
- state 2 means that an asynchronous packet is queued at the station and the token is not late;
- state 3 means that a synchronous packet is queued at the station.

---

1. This means that the last cycle is greater than $\tau + m \cdot b$.

Hence, by denoting with $t_i^{(n)}$ the time instant, during the n-th cycle, at which the servers arrive at Q_i, the state of Q_i at this time is described by the r.v. $X_i(t_i^{(n)})$, where $X_i(t_i^{(n)}) \in \{0, 1, 2, 3\}$.

Note that, for each station$\{i\}$, the value of $X_i(t_i^{(n)})$ only depends on the state of Q_i at the previous token arrival at this station (i.e., $t_i^{(n-1)}$), and on the number of packets transmitted by the other stations in the last cycle. It is thus easy to verify that a Markovian description of the system can be provided by defining a Markov chain embedded at the time instants at which the server arrives at a station (i.e., $t_i^{(n)}$, $i = 1, \ldots, K$, $n = 1, 2, \ldots$). Specifically, the state of the Markov chain, $s(t_i^{(n)})$, is

$$s(t_i^{(n)}) = \{\xi^{(n)} = i, X_1(\xi^{(n)}), \ldots, X_i(\xi^{(n)}), \ldots, X_K(\xi^{(n)})\} \quad , \qquad (6.1)$$

where

- the r.v. $\xi^{(n)}$ denotes the index of the station visited by the token during the n-th cycle;
- the r.v. $X_i(\xi^{(n)})$, $i = 1, \ldots, K$, describes the state of station$\{i\}$ the last time it received the token, given that the token is now visiting the station with index $\xi^{(n)}$.

As the Markov chain has a finite number of states, and by observing that the chain is irreducible, it follows that all states are positive recurrent [29] and the steady-state probabilities satisfy the following equations

$$P(\xi = i, X_1(\xi) = x_1, X_2(\xi) = x_2, \ldots, X_K(\xi) = x_K) = \qquad (6.2)$$
$$\lim_{n \to \infty} P(\xi^{(n)} = i, X_1(\xi^{(n)}) = x_1, X_2(\xi^{(n)}) = x_2, \ldots, X_K(\xi^{(n)}) = x_K) \quad .$$

Hereafter, to simplify the presentation, $s(i)$ is used to denote the state $\{\xi = i, X_1(\xi), \ldots, X_K(\xi)\}$, and the notation $P(i, x_1, x_2, \ldots, x_K)$ will be used instead of $P(\xi = i, X_1(\xi) = x_1, \ldots, X_K(\xi) = x_K)$.

To solve (6.2) it is useful to exploit the relationships between the steady-state probabilities at two consecutive embedding points i-1 and i. To establish these relationships it must be noted that in the transition between $i-1$ and i only the r.v. $X_i(\xi)$ may change its value, and its new value depends on

- the state of Q_i at the previous visit of the token, which is described by the r.v. $X_i(i-1)$, i.e. the value of $X_i(\xi)$ while the token is visiting Q_{i-1};

- the length, $\tilde{C}_i$, of the last cycle observed by Q_i which the event "late token" or "early token" depends on. $\tilde{C}_i$ includes the last intervisit period at Q_i, i.e., $\tilde{I}_i$, plus the length of the service period at Q_i in the previous cycle, $\tilde{Sp}_i$;
- the distribution of the number of packets which arrive at Q_i during the intervisit period $\tilde{I}_i$.

Note that given $s(i-1) := \{i-1, x_1, x_2, ..., x_K\}$ the random variables $X_i(i-1)$, $\tilde{C}_i$, $\tilde{I}_i$ and $\tilde{Sp}_i$ take a constant value (see (6.3)-(6.6)), and therefore, $P\{s(i)\}$ can be derived by conditioning on the $s(i-1)$.

$$X_i(i-1) = x_i \tag{6.3}$$

$$\tilde{I}_i = \tau + b \cdot \sum_{j \neq i} I_{\{x_j \in \{2,3\}\}} \tag{6.4}$$

$$\tilde{Sp}_i = b \cdot I_{\{x_i \in \{2,3\}\}} \tag{6.5}$$

$$\tilde{C}_i = \tilde{Sp}_i + \tilde{I}_i = \tau + b \cdot \sum_j I_{\{x_j \in \{2,3\}\}} \tag{6.6}$$

As stated before, $X_i(\xi)$ is the only r.v. which changes its value while the server switches from Q_{i-1} to Q_i. Hence, by conditioning on $s(i-1)$, four possible cases exist depending on th value of $X_i(i)$.

CASE $X_i(i) = 0$. Q_i is empty when it receives the token if its queue was empty when the token departs from Q_i at the previous visit (i.e., $X_i(i-1) \neq 1$), and no packets arrive at Q_i during $\tilde{I}_i$, hence

$$P(i, x_1, ..., x_{i-1}, 0, x_{i+1}, ..., x_K) = \tag{6.7}$$

$$\sum_{x_i \neq 1} e^{-(\lambda_i^S + \lambda_i^A)\tilde{I}_i} \cdot P(i-1, x_1, ..., x_{i-1}, x_i, x_{i+1}, ..., x_K) \quad,$$

where, given $s(i-1)$, $\tilde{I}_i$ is a constant and its value is given by (6.4).

$\Diamond$

CASE $X_i(i) = 1$. The event "Q_i contains an asynchronous packet which is not authorized to transmit since the last cycle length exceeds $\tau + b \cdot m$" occurs in two possible cases depending on the value of $X_i(i-1)$.

(i) $X_i(i-1) \neq 1$; in this case $X_i(i) = 1$ if (a) $\tilde{C}_i > \tau + b \cdot m$; (b) at least one packet arrives during the interval $\tilde{I}_i$; and (c) the first packet which arrives during $\tilde{I}_i$ belongs to the asynchronous class;

(ii) $X_i(i-1) = 1$ in this case $X_i(i) = 1$ if $\tilde{C}_i > \tau + b \cdot m$.

Taking into consideration (i) and (ii) above, it follows that

$$P(i, x_1, \ldots, x_{i-1}, 1, x_{i+1}, \ldots, x_K) = \tag{6.8}$$

$$P(i-1, x_1, \ldots, x_{i-1}, 1, x_{i+1}, \ldots, x_K) \cdot I_{(\tilde{C}_i > \tau + m \cdot b)} +$$

$$\sum_{x_i \neq 1} I_{(\tilde{C}_i > \tau + m \cdot b)} \cdot \left(1 - e^{-(\lambda_i^S + \lambda_i^A)\tilde{I}_i}\right) \frac{\lambda_i^A}{\lambda_i^S + \lambda_i^A} \cdot$$

$$P(i-1, x_1, \ldots, x_{i-1}, x_i, x_{i+1}, \ldots, x_K) \quad .$$

◊

CASE $X_i(i) = 2$. This case occurs when Q_i contains an asynchronous packet which is authorized to be transmitted as the last cycle length does not exceed $\tau + m \cdot b$. Hence this case differs from the previous one (i.e., $X_i(i) = 1$) in term of the condition $\tilde{C}_i \leq \tau + m \cdot b$, and

$$P(i, x_1, \ldots, x_{i-1}, 2, x_{i+1}, \ldots, x_K) = \tag{6.9}$$

$$P(i-1, x_1, \ldots, x_{i-1}, 1, x_{i+1}, \ldots, x_K) \cdot I_{(\tilde{C}_i \leq \tau + m \cdot b)} +$$

$$\sum_{x_i \neq 1} I_{(\tilde{C}_i \leq \tau + m \cdot b)} \cdot \left(1 - e^{-(\lambda_i^S + \lambda_i^A)\tilde{I}_i}\right) \frac{\lambda_i^A}{\lambda_i^S + \lambda_i^A} \cdot$$

$$P(i-1, x_1, \ldots, x_{i-1}, x_i, x_{i+1}, \ldots, x_K) \quad .$$

◊

CASE $X_i(i) = 3$. This event (i.e., at the token arrival instant, Q_i contains a synchronous packet) occurs if $X_i(i-1) \neq 1$; at least one packet arrives during the interval $\tilde{I}_i$; and the first packet which arrives during $\tilde{I}_i$ belongs to the synchronous class. Hence

$$P(i, x_1, \ldots, x_{i-1}, 3, x_{i+1}, \ldots, x_K) = \qquad (6.10)$$

$$\frac{\lambda_i^S}{\lambda_i^S + \lambda_i^A} \cdot \sum_{x_i \neq 1} P(i-1, x_1, \ldots, x_{i-1}, x_i, x_{i+1}, \ldots, x_K) \cdot \left(1 - e^{-(\lambda_i^S + \lambda_i^A)\bar{t}_i}\right).$$

$\Diamond$

Equations (6.7), (6.8), (6.9) and (6.10) define the steady-state probabilities of $s(i)$ in terms of the steady-state probability of $s(i-1)$. If $(i = 1) \rightarrow (i-1 = K)$, and by exploiting (6.7), (6.8), (6.9) and (6.10), with $1 \leq i \leq K$, a set of $K \cdot 4^K$ equations is established among the $K \cdot 4^K$ unknowns $P(i, x_1, x_2, \ldots, x_K)$, $1 \leq i \leq K$, $0 \leq x_i \leq 3$. The steady-state probabilities can be derived by numerically solving these equations under the normalization condition $\sum_i \sum_{\mathbf{x}} P(i, \mathbf{x}) = 1$ $(\mathbf{x} = [x_1, x_2, \ldots, x_K])$.

REMARK. For a symmetric polling system (i.e., for each Q_i, $\lambda_i^S = \lambda^S$, $\lambda_i^A = \lambda^A$, and $b_i = b$) the computational complexity can be reduced. In fact, in this case all stations are statistically identical. Since,

$$P(i, x_1, \ldots, x_{i-1}, x_i, x_{i+1}, \ldots, x_K) = P(i-1, x_2, \ldots, x_i, x_{i+1}, x_{i+2}, \ldots, x_K, x_1),$$

equations (6.7), (6.8), (6.9) and (6.10) define a set of 4^K equations from which the steady-state probabilities of the Q_i state when it captures the token can be derived.

$\Diamond$

Once the steady-state probabilities have been derived each node can be analyzed in isolation to derive its performance figures. In the following, for a given station$\{i\}$, it is shown how to derive the throughput and the delay performance figures, for each class of traffic queued at this station.

THROUGHPUT ANALYSIS. For each station$\{i\}$, the throughput achieved by both synchronous and asynchronous traffic, denoted by γ_i^S and γ_i^A, respectively, can be derived by applying classical renewal/regenerative arguments (see Section 2.3). It can be verified that the sequence $\{s(t_i^{(n)}), n \in \mathbb{N}\}$, where

$$s(t_i^{(n)}) = \{\xi^{(n)} = i, X_1(\xi^{(n)}), \ldots, X_i(\xi^{(n)}), \ldots, X_K(\xi^{(n)})\},$$

is a regenerative process with respect to the sequence S of time instants (renewal instants) at which the server arrives at Q_i when all the queues in the polling systems are empty. Hence, the throughput of the synchronous (asynchronous) traffic is

$$\gamma_i^S = \frac{\sum_{j=1}^{M} E\,[N_{tr}^S(i, j)]}{\sum_{j=1}^{M} E\,[C_{i, j}]} \quad , \qquad \gamma_i^A = \frac{\sum_{j=1}^{M} E\,[N_{tr}^A(i, j)]}{\sum_{j=1}^{M} E\,[C_{i, j}]} \quad , \qquad (6.11)$$

where:

- $E\,[N_{tr}^S(i, j)]$ and $E\,[N_{tr}^A(i, j)]$ are the mean numbers of synchronous and asynchronous packets transmitted by Q_i in the j-th polling cycle of a renewal cycle, respectively;
- $E\,[C_{i, j}]$ is the average length of the j-th polling cycle of a renewal cycle; and
- M is the number of polling cycles in a renewal cycle (i.e., the number of times the token arrives at Q_i in a renewal cycle).

By applying Theorem 2.4, after some routine algebraic manipulations, (6.11) can be rewritten as

$$\gamma_i^S = \frac{E\,[N_{tr}^S(i)]}{E\,[C_i]} \quad , \qquad \gamma_i^A = \frac{E\,[N_{tr}^A(i)]}{E\,[C_i]} \quad , \qquad (6.12)$$

where $E\,[N_{tr}^S(i)]$ $(E\,[N_{tr}^A(i)])$ is the average number of synchronous (asynchronous) packets transmitted in a polling cycle, and $E\,[C_i]$ is the average length of a polling cycle[1].

From the steady-state probabilities it immediately follows that

$$E\,[N_{tr}^S(i)] = \sum P(i, x_1, x_2, \ldots, x_{i-1}, 3, x_{i+1}, \ldots, x_K) \quad , \qquad (6.13)$$

$$E\,[N_{tr}^A(i)] = \sum P(i, x_1, x_2, \ldots, x_{i-1}, 2, x_{i+1}, \ldots, x_K) \quad , \qquad (6.14)$$

and

$$E\,[C] = \tau + b \cdot \sum_{j=1}^{K} (E\,[N_{tr}^S(j)] + E\,[N_{tr}^A(j)]) \quad . \qquad (6.15)$$

By denoting with $P_L^S(i)$ $(P_L^A(i))$ the packet-loss probability for the synchronous (asynchronous) traffic the following relationships hold

$$\gamma_i^S = \lambda_i^S \cdot (1 - P_L^S(i)) \,, \qquad \gamma_i^A = \lambda_i^A \cdot (1 - P_L^A(i)) \,.$$

---

1. As already mentioned in Section 2.1.3, the average length of a polling cycle does not depend on the station index; hence $E\,[C_i] = E\,[C]$.

from which

$$P_L^S(i) = 1 - (\gamma_i^S / \lambda_i^S) , \qquad P_L^A(i) = 1 - (\gamma_i^A / \lambda_i^A) . \tag{6.16}$$

Due to the PASTA property (see Theorem 2.3), $P_L^S(i)$ and $P_L^A(i)$ are equal to the probability that the station buffer is full at a random point in time, $p_i(1)$,

$$P_L^S(i) = P_L^A(i) = p_i(1) . \tag{6.17}$$

From (6.16) and (6.17) it immediately follows that the average number of packets queued at Q_i ($E[N_i]$) is

$$E[N_i] = 1 - \frac{\gamma_i^S}{\lambda_i^S} = 1 - \frac{\gamma_i^A}{\lambda_i^A} . \tag{6.18}$$

DELAY ANALYSIS. The computation of the average waiting time is more complex. First, the average waiting time of the synchronous packets, $E[W_i^S]$, must be computed. Then, the average waiting time for asynchronous packets, $E[W_i^A]$, can be derived by exploiting (6.18) and Little Theorem (see Section 2.2.1).

By applying Little's formula

$$E[N_i^S] = \gamma_i^S (E[W_i^S] + b) , \qquad E[N_i^A] = \gamma_i^A (E[W_i^A] + b) ,$$

where $E[N_i^S]$ and $E[N_i^A]$ are the average number of synchronous and asynchronous packets queued at Q_i, respectively.

Hence, once $E[W_i^S]$ is known, from (6.18) also $E[W_i^A]$ can be derived by noting that $E[N_i] = E[N_i^S] + E[N_i^A]$.

To compute the waiting time of synchronous packets, W_i^S, only the packets which arrive at Q_i when the queue is empty should be considered. This event (empty queue) may occur only if Q_i is empty when the token departs from this queue, i.e., the state of Q_i at the token arrival is $\{0, 2, 3\}$. Hereafter, $T(I_i)$ denotes the type of Q_i intervisit period, which indicates the Q_i state at the previous token arrival. Furthermore, $L(I_i)$ denotes the length of a Q_i intervisit period, and A_i indicates the sub-interval $(y, y + dy]$ of I_i, in which the first arrival occurs during an intervisit period. The probability that a packet is accepted at Q_i, $P\{Acc\}$, is

$$P\{Acc\} = \tag{6.19}$$

$$\sum_{t \in \{0, 2, 3\}} \sum_{\mathbf{x}(t) = l} \int_0^l P\{y < A_i \le y + dy, Acc, T(I_i) = t, L(I_i) = l\}\, dy \quad,$$

where

- Acc denotes the event "one synchronous packet is accepted";
- $\mathbf{x}(t)$ denotes a vector $\mathbf{x} = [x_1, x_2, ..., x_K]$, such that $x_i = t$; and
- the event $\{\mathbf{x}(t) = l\}$ indicates that $\{L(I_i) = l\}$.

After some routine manipulations (6.19) becomes

$$P\{Acc\} = \sum_{t \in \{0, 2, 3\}} \sum_{\mathbf{x}(t) = l} \left[1 - e^{-(\lambda_i^S + \lambda_i^A) l} \right] \cdot \tag{6.20}$$

$$\frac{\lambda_i^S}{\lambda_i^S + \lambda_i^A} \cdot P(i, x_1, x_2, ..., x_{i-1}, t, x_{i+1}, ..., x_K) \quad.$$

To compute the average waiting time it is useful to exploit the LST $W_i^{S^*}(s)$ of W_i^S. By conditioning $W_i^{S^*}(s)$ first on the type and length of the Q_i intervisit period (I_i) in which the arrival occurs, and then on the sub-interval $(y, y + dy]$ of I_i in which the arrival occurs it follows that

$$E[W_i^{S^*}(s)] = \tag{6.21}$$

$$\sum_{t \in \{0, 2, 3\}} \sum_{l} \int_0^l E[W_i^{S^*}(s) | Acc, T(I_i) = t, L(I_i) = l, y < A_i \le y + dy]$$

$$P\{y < A_i \le y + dy, Acc | T(I_i) = t, L(I_i) = l\}$$

$$P\{T(I_i) = t, L(I_i) = l\} / P\{Acc\} \quad.$$

Since

$$\int_0^l E[W_i^{S^*}(s) | Acc, T(I_i) = t, L(I_i) = l, y < A_i \le y + dy] =$$

$$\frac{\lambda_i^S}{\lambda_i^S + \lambda_i^A - s} \cdot \left[e^{-sl} - e^{-(\lambda_i^S + \lambda_i^A) l} \right] \quad,$$

it follows that

$$E\,[\,W_i^{S^*}(s)\,] \;=\; \frac{1}{P\{Acc\}} \cdot \left\{ \sum_{t \in \{0, 2, 3\}} \sum_{x(t) = t} \frac{\lambda_i^S}{\lambda_i^S + \lambda_i^A - s} \cdot \right. \tag{6.22}$$
$$\left. \left[e^{-st} - e^{-(\lambda_i^S + \lambda_i^A)t} \right] \cdot P(i, x_1, x_2, ..., x_{i-1}, t, x_{i+1}, ..., x_K) \right\} \quad .$$

Finally, from (6.20) and (6.22), with routine manipulations the average waiting time can easily be computed.

BIBLIOGRAPHIC NOTES.   The model presented above was developed by Takagi [148] who applied this model to the analysis of a symmetric FDDI network with both synchronous and asynchronous traffic. Nakamura et al. (see [123]) applied this model to study an asymmetric network. In [123] Takagi's model was also extended to consider different priority levels for the asynchronous traffic.

6.2.2 An FDDI Model with Zero Switchover Time (Model 2)

The FDDI model considered in [3] consists of K queues. Q_j has a waiting room equal to $M_j > 1$ which may also be infinite. Packets arrive at Q_j according to a Poisson process with arrival rate λ_j. Here and below it might be useful to read queue K for queue 0. Thus for any vector $\mathbf{V}$ with s components (one for each queue) notations $\mathbf{V} = (V_1, ..., V_K)$ and $\mathbf{V} = (V_0, ..., V_{K-1})$ are used interchangeably. In the latter representation, it is understood that $V_j = V_{(j \bmod K)}$ for $j \geq K$ or $j < 0$. Service times (i.e., transmission times) of frames arriving at Q_j are assumed to be i.i.d. exponentially distributed with rate μ_j. The switchover times are assumed to be negligible. The local offered load at Q_j is $\rho_j = \lambda_j / \mu_j$, and the total offered load to the system is $\rho = \sum_{j=1}^{K} \rho_j$. Queue j has a fixed Target Token Rotation Time $TTRT_j$ which depends upon Q_j. By using this modeling feature, with $TTRT_j = \infty$, it is possible to model a station transmitting synchronous traffic only.[1] In [3] two service disciplines are considered for the transmission of asynchronous traffic: 1-*limited* and *exhaustive up to the TTRT limit*. For both service disciplines, an asynchronous frame starts to be transmitted at Q_j, if, upon token arrival at Q_j, $\overline{TRT}_j < TTRT_j$ and Q_j is not empty.[2]

---

1. In a real FDDI network the *TTRT* value is equal for all stations.

2. As stated in Chapter 5, a timer symbol with a bar indicates its value.

The behavior of the above two service disciplines differs when transmission of a frame ends at Q_j. Specifically, in the 1-*limited* service discipline the token passes to the next downstream queue, while in the *exhaustive* service discipline the transmission of a new frame at Q_j begins if $\overline{THT_j} < TTRT_j$ and Q_j is not empty. If on completion of frame transmission Q_j becomes empty or $\overline{THT_j} \geq TTRT_j$, then the token passes to the next downstream queue.

Throughout, PSA will be applied to the *FDDI* model that implements the 1-limited service discipline. Details on the exhaustive service discipline can be found in [3].

Before proceeding to the analysis of the 1-limited service discipline it is necessary to comment on the approximations of the model proposed in [3].

1. TRT_j is always cleared and starts counting up again upon token arrival at Q_j. In other words, the *accumulated delay* feature (see Section 5.3.4) of the FDDI MAC protocol is not modeled.

2. Upon token arrival at Q_j, $\overline{TRT_j}$ is computed by summing up the *average* frame transmission times at the various queues during the last cycle as observed by Q_j. On the other hand, in the *FDDI* MAC protocol upon token arrival at Q_j, $\overline{TRT_j}$ is computed by summing up the frame transmission times at the various queues during the last cycle as observed by Q_j.

3. The switchover times between stations are assumed to be negligible.

Under the above approximations and for the 1-limited service discipline the state of the system includes:

- the number of frames in the different queues described by the vector $\mathbf{N} = (N_0, N_1, ..., N_{K-1})$ where the N_i component represents the number of frames in the i-th queue,
- the location of the token described by the random variable $H \in \{0, ..., K-1\}$, and
- the values of the timers TRT_j (i.e., $\overline{TRT_j}$) in the various queues $j = 0,1, ..., K-1$.

In order to simplify the representation an equivalent state description is used in [3] which, in addition to the random variables $\mathbf{N}$ and H, includes the vector $\mathbf{U} = (U_0, U_1, ..., U_{K-1})$; where U_j is the r.v. which denotes the number (which is at most one) of frames transmitted at Q_j during the last cycle as observed by queue H. Therefore, the state of the system is described by the vector $(\mathbf{N}, \mathbf{U}, H)$. With the two supplementary variables $\mathbf{U}$ and H, the queue length process

$$\Theta = \{\, (\mathbf{N}, \mathbf{U}, H) = (\mathbf{n}, \mathbf{u}, h), h = 0, 1, \ldots, s - 1; 0 \leq n_h \leq K_h; u_h \in \{0, 1\} \,\}$$

becomes Markovian. Specifically it becomes a multidimensional QBD process (see Section 4.2.2).

To show the use of vector $\mathbf{U}$, assume that the token arrives to queue $H = h$. Due to assumption 2 above, $\overline{TRT}_h$ can be reconstructed for that queue from vector $\mathbf{U}$ by means of the following relationship

$$\overline{TRT}_h = \sum_{i=0}^{K-1} \frac{u_i}{\mu_i} \quad .$$

The state probabilities of the QBD process will be denoted by $p(\mathbf{n}, \mathbf{u}, h)$.

Following assumption 3, $\mathbf{U}$ and H are not defined when the system is empty. Thus, the empty system is fully defined by $\mathbf{N} = \mathbf{0}$. Hence

$$p(\mathbf{n}, \mathbf{u}, h) = 0, \ \text{if } n_h = 0 \quad . \tag{6.23}$$

REMARK. In this model when all queues become empty the system regenerates, i.e., its future evolution does not depend on the past.

$$\Diamond$$

For $h = 0, 1, \ldots, K - 1$, $0 < n_h \leq M_h$, and $\sum_{i=h}^{h+K-1} u_i / \mu_i < TTRT_h$, by denoting with $Z = \{0, 1\}^K$, the global balance equations are:

$$\left[\sum_{j=0}^{K-1} \lambda_j I_{\{n_j < M_j\}} + \mu_h\right] p(\mathbf{n}, \mathbf{u}, h) \tag{6.24}$$

$$= \sum_{j=0}^{K-1} \lambda_j p(\mathbf{n} - \mathbf{e}_j, \mathbf{u}, h) I_{\{n_j > 0, \ if \ (j \neq h)\,; \ n_h > 1\}} \tag{6.25}$$

$$+ \lambda_h p\,(\mathbf{0})\, I_{\{\mathbf{n} = \mathbf{e}_h\}} I_{\{\mathbf{u} = \mathbf{0}\}} \tag{6.26}$$

$$+ \sum_{j=1}^{K} \mu_{h-j} I_{\{n_{h-j} < M_{h-j}\}} \tag{6.27}$$

$$\times \sum_{\mathbf{v} \in Z} p(\mathbf{n} + \mathbf{e}_{h-j}, \mathbf{v}, h-j) \tag{6.28}$$

$$\times \left[I_{\{u_i = 0 \,:\, i = h-j+1, \,...,\, h-1\}} I_{\{u_{h-j} = 1\}} \right. \tag{6.29}$$

$$\times \prod_{r=1}^{j-1} I_{\left\{ \sum_{i=h}^{h+K-j} \frac{u_i}{\mu_i} + \sum_{i=h-r}^{h-1} \frac{v_i}{\mu_i} \geq TTRT_{h-r} \text{ or } n_{h-r} = 0 \right\}} \tag{6.30}$$

$$\times I_{\{v_i = u_i \,:\, i = h, h+1, \,...,\, h+K-j-1 \text{ or } j = K\}} \tag{6.31}$$

$$\times I_{\{\mathbf{u} = \mathbf{0}\}} \tag{6.32}$$

$$\times I_{\{n_{h-r} = 0, \, r = 1, 2, \,...,\, j-1\}} \tag{6.33}$$

$$\times I_{\left\{ n_{h+r} = 0 \text{ or } \frac{1}{\mu_{h-j}} + \sum_{i=h+r}^{h-j+K-1} \frac{v_i}{\mu_i} \geq TTRT_{h+r}, \, (r = 0, 1, \,...,\, K-j) \right\}} \right] \,, \tag{6.34}$$

and

$$\sum_{j=0}^{K-1} \lambda_j p(\mathbf{0}) = \sum_{j=0}^{K-1} \sum_{\mathbf{u} \in Z} p(\mathbf{e}_j, \mathbf{u}, j) \,. \tag{6.35}$$

As already seen in a more general setting, equation (6.24) represents the probability flow out of state $(\mathbf{n}, \mathbf{u}, h)$, whereas (6.25) to (6.35) represent the probability flow into the same state. In the following hints on the terms contributing to the flow into the state $(\mathbf{n}, \mathbf{u}, h)$, are given.

The term (6.25) corresponds to arrivals to the various queues and the indicator function

$$I_{\{n_j > 0, \text{ if } (j \neq h)\,;\, n_h > 1\}}$$

takes into account of the fact that the state from which the transition occurs is $(\mathbf{n} - \mathbf{e}_j, \mathbf{u}, h)$. Hence, $n_j > 0$, for $j \neq h$ whereas $n_h > 1$ because the server is already serving at Q_h.

When the state of the system is $(\mathbf{0})$ (i.e., the system is empty) an arrival at queue h will cause a transition to the state $(\mathbf{e}_h, \mathbf{0}, h)$ with a rate given by (6.26).

The events which will be considered next are those which lead the server from queue $h - j$ $(j = 1, 2, ..., K)$ to Q_h in one transition with a rate given by (6.27). It must be noted that these types of transitions are possible due to assumption 3 (i.e., there is no switchover time). The number of frames at Q_{h-j} before the end of service was $n_{h-j} + 1$ which cannot be larger than the buffer size of Q_{h-j}. This justifies the presence of the indicator function in (6.27). Summation in (6.28) is taken on the values assumed by a new vector, denoted by $\mathbf{v}$, the components of which describe the server activity at the various queues (i.e. whether or not these stations have transmitted one packet) while serving Q_{h-j}. The relationship between $\mathbf{u}$ and $\mathbf{v}$ can be explained as follows: while the $\mathbf{u}$ components describe the server activity right at the beginning of service of Q_h, the $\mathbf{v}$ components describe the server activity before the end of service in Q_{h-j}

The components $u_h, u_{h+1}, ..., u_{h+K-j-1}$ of the $\mathbf{u} \in \mathbf{U}$ vector may have been changed since the last time the server visited Q_h due to the fact that frames from queues $h, h+1, ..., h+K-j-1$ may have been served. On the other hand, the components $u_{h-j+1}, ..., u_{h-1}$ have still not been changed since the server is serving Q_{h-j}.

To derive the probabilities of the events which lead the server from Q_{h-j} to Q_h in one transition, it is useful to partition these events into the following two mutually exclusive classes

1. class including the events for which the $\overline{TRT}_h$ is smaller than $TTRT_h$ when the server leaves Q_{h-j},
2. class including the events for which the $\overline{TRT}_h$ is greater than $TTRT_h$ when the server leaves Q_{h-j}.

For the events belonging to class 1 it follows that

$$\begin{cases} u_i = 0, \ i = h - j + 1, ..., h - 1 \text{ for } j \neq K \\ u_i = 0, \ i \neq h \qquad\qquad\qquad\quad \text{for } j = K \end{cases} \qquad (6.36)$$

and this justifies the indicator function in (6.29). If $j \neq 1$, the fact that the last queue served was Q_{h-j} and that the server is now attending Q_h rather than queue $h - j + r, r = 1, \ldots, j - 1$, implies that these queues were either empty or that $\overline{TRT}_{h-j+r}$ exceeded $TTRT_{h-j+r}$ when the server visited them. This justifies the product of indicator functions in (6.30). The indicator function in (6.31) puts a constraint on the values of $v_{h-j+1}, \ldots, v_{h-1}$ on which the sum in (6.28) is taken. Specifically, (6.31) states that when the server moves from Q_{h-j} to Q_h in one transition, some components of the $\mathbf{U}$ and $\mathbf{V}$ random vectors overlap. If Q_h was the last queue served, i.e., the one which is currently served (i.e., $j = K$), the overlap may not exist.

For the events belonging to class 2 it follows that more than a whole cycle has elapsed since the server served Q_{h-j}, which explains the indicator function in (6.32).

If $j \neq 1$ the fact that the last queue served was Q_{h-j} and that the server is now attending Q_h rather than queue $h - j + r, r = 1, \ldots, j - 1$, implies that these queues were empty when the server visited them immediately after the server completed its service at Q_{h-j} and they will remain empty up until the round which led the server to Q_h. This justifies the indicator functions in (6.33).

The fact that after service ended at Q_{h-j} the server jumped to Q_h rather than to queues $h + r, r = 0, 1, \ldots, K - j$, means that those queues were either empty, or that $\overline{TRT}_{h+r}$ exceeded $TTRT_{h+r}$ when the server visited them. This is expressed by (6.34).

By exploiting the PSA methodology introduced in 4.2.2 the steady-state probabilities of the global balance equations

$$p(\mathbf{n}, \mathbf{u}, h) = \chi^{n_1 + n_2 + \ldots + n_s} \sum_{k=0}^{\infty} \chi^k b(k; (\mathbf{n}, \mathbf{u}, h)), \quad \mathbf{n} \neq 0 \quad , \tag{6.37}$$

$$p(0) = \sum_{k=0}^{\infty} \chi^k b(k;0) \quad , \tag{6.38}$$

can be derived.

THROUGHPUT ANALYSIS. For each station$\{i\}$, the throughput is given by the probability that Q_i is not empty, times μ_i. Obviously, for an infinite queue, the throughput of station$\{i\}$, is equal to λ_i.

DELAY ANALYSIS. The queueing delay experienced by a packet at Q_i can be computed by evaluating first the expected number of packets $(E[N_i])$ at this queue, and then by applying Little's theorem, i.e.,

$$E[W_i] = E[N_i] / \lambda_i - 1 / \mu_i \quad . \tag{6.39}$$

The first and higher moments of the length of Q_i are given by

$$E[N_i^v] = \sum_{k=1}^{\infty} \chi^k f_v(k;i), \quad v = 1, 2, \ldots \quad , \tag{6.40}$$

where $f_v(k;i)$ are given by

$$f_v(k;i) = \sum_{|n| \le k} \sum_{h=0} \sum_{u \in V} n_j^v b(k-|n|;n, u, h) \quad . \tag{6.41}$$

6.2.3 An FDDI Network with Synchronous Traffic (Model 3)

An FDDI network in which the stations transmit synchronous traffic only can be modeled with a polling system with an exhaustive time-limited service discipline. Time-limited service disciplines limit the length of the stations' service-periods. Hence, with an exhaustive time-limited discipline, the station which holds the token can transmit its packets as soon as its queue becomes empty, or the ongoing service period exceeds a threshold, whichever of the two events occurs first.

Depending on the way the event "the ongoing service period exceeds a threshold" is defined, two variants of the exhaustive time-limited service discipline exist: *non-preemptive (NP)* and *look-ahead (LH)*.

- The NP discipline enables a station to start a packet transmission if the length of the service period (up to this time) does not exceed a given threshold. A packet transmission is always completed even though this may mean that the service period exceeds the maximum length.
- The LH discipline enables a station to start a packet transmission only if by adding this packet transmission time to the service period (already elapsed) the threshold is not exceeded.

The LH service discipline can be used to model an FDDI network with synchronous traffic only; while the NP variant is suitable for modeling a Token Ring network when the network traffic belongs to the highest priority level.

There are some small differences in the analysis of these two models. An approximate analysis of both models will be presented in this section following the approach proposed by Chang & Sandhu in [33].

When service times are constant, the NP and LH disciplines enable a network to transmit a fixed number of packets. Thus the basic idea behind Chang & Sandhu's approach is to reduce the study of a polling system with an NP or LH service discipline to the analysis of a polling system with an exhaustive l-limited service discipline (see Section 4.1.5). Specifically, this entails approximating the maximum number of packets a station can transmit with a constant number $\tilde{l}_i$.

The main contribution of [33] is therefore the definition of a function which for each station Q_i translates the time-limit, H_i, of its service period[1] into the maximum number of packets the station can transmit in an l-limited polling system, $\tilde{l}_i$. To define the relationship between H_i and $\tilde{l}_i$, Q_i is analyzed in *saturated conditions* (i.e., under the hypothesis that its queue is never empty). In this case, let S_i denote the elapsed time between the server's arrival at Q_i and the end of the transmission of the i-th packet in this service period; in addition, let $S_0 = 0$ denote the server arrival instant at Q_i. Under the assumption that service times at Q_i are independent random variables all distributed as the r.v. B_i, the process $\{S_n, n \geq 0\}$, where

$$S_0 = 0, \text{ and } S_n = \sum_{j=1}^{n} B_{i,j}$$

is a renewal process (see Section 2.3).

The maximum number of packets which Q_i may transmit in a service period is therefore dependent on the r.v. $N_r(H_i)$ which denotes the number of renewals in the interval $(0, H_i]$, where

$$N_r(H_i) = \sum_{n=1}^{\infty} I_{\{S_n \leq H_i\}} . \tag{6.42}$$

---

1. H_i is the maximum length of the Q_i service period in the polling system with a time-limited service discipline.

By denoting with $L_i^{NP}(H_i)$ $(L_i^{LH}(H_i))$ the maximum number of packets that Q_i may transmit in a service period under the non-preemptive (look-ahead) service discipline

$$L_i^{LH}(H_i) = N_r(H_i) \quad , \tag{6.43}$$

$$L_i^{NP}(H_i) = 1 + N_r(H_i) \quad . \tag{6.44}$$

Equations (6.43) and (6.44) indicate that the maximum number of packets Q_i may transmit in a service period is a random variable. In order to translate these variable limits in a constant number $\tilde{l}_i$, it is assumed (see [33]) that $\tilde{l}_i$ is equal to the integer nearest to the r.v. average value

$$\tilde{l}_i^{LH} \approx E[L_i^{LH}(H_i)] = E[N_r(H_i)] \quad , \tag{6.45}$$

$$\tilde{l}_i^{NP} \approx E[L_i^{NP}(H_i)] = 1 + E[N_r(H_i)] \quad . \tag{6.46}$$

The problem is therefore reduced to the computation of the renewal function $m(t)$ (see Section 2.3)

$$m(t) = E[N_r(t)], \quad t \geq 0 \quad . \tag{6.47}$$

From (6.42) and (6.47)

$$m(t) = E\left[\sum_{n=1}^{\infty} I_{\{S_n \leq t\}}\right] = \sum_{n=1}^{\infty} E\left[I_{\{S_n \leq t\}}\right] = \sum_{n=1}^{\infty} P\{S_n \leq t\} \quad , \tag{6.48}$$

and by exploiting the LST transform

$$m^*(s) = \int_0^{\infty} e^{-st} \cdot dm(t) = \sum_{n=1}^{\infty} [B_i^*(s)]^n = \frac{B_i^*(s)}{1 - B_i^*(s)} \quad . \tag{6.49}$$

For exponentially distributed service times, $B_i^*(s) = \mu_i/(s + \mu_i)$, where $\mu_i = 1/E[B_i]$, and hence from (6.49) it follows that $m(t) = t/E[B_i]$. For other service time distributions, $m(t)$ can be obtained by inverting its Laplace-Stieltjes transform $m^*(s)$.

DELAY ANALYSIS. As mentioned before, if packets have a constant length the time-limited policy is equivalent to an l-limited policy for which, as shown in Section 2.6.4, the pseudo-conservation law (6.50) holds.

$$\sum_{i=1}^{K} \rho_i \left(1 - \frac{\lambda_i s}{\tilde{l}_i(1 - \rho_i)}\right) E[W_i] = A + \frac{s}{(1 - \rho_i)} \cdot \sum_{i=1}^{K} \rho_i \Gamma_i \quad , \tag{6.50}$$

where

$$A = \frac{\sum_{i=1}^{K} \lambda_i b_i^{(2)}}{\tilde{l}_i (1 - \rho_i)} + \rho \frac{s^{(2)}}{2s} + \frac{s}{2(1-\rho)} (\rho^2 - \sum_{i=1}^{K} \rho_i^2) \quad ,$$

and

$$\Gamma_i = \frac{1}{2\tilde{l}_i} \left[(1-\rho_i) \cdot \sum_{j=1}^{\tilde{l}_i - 1} j (\tilde{l}_i - j) \, p_{i,0,j} - [\tilde{l}_i - 1 - (\tilde{l}_i + 1) \rho_i] \right] \quad .$$

As shown in Section 4.1.5, Γ_i is estimated by using an *M/G/1* vacation model.

Unfortunately, a formula equivalent to (6.50) for the case of service times sampled from a general distribution does not exist. In addition, numerical results (see [33]) indicate that by using (6.50) in the general case, the r.h.s. of (6.50) underestimates the weighted sum of the stations' average delays. To overcome this problem Chang and Sandhu (see [33]) proposed to substitute to the r.h.s. of (6.50) with an upper bound for the weighted sum

$$\sum_{i=1}^{K} \rho_i \left(1 - \frac{\lambda_i s}{\tilde{l}_i (1 - \rho_i)} \right) E[W_i] \approx A + \frac{s}{(1-\rho_i)} \cdot \sum_{i=1}^{K} \rho_i \tilde{\Gamma}_i \quad , \qquad (6.51)$$

where $\tilde{\Gamma}_i = \rho_i / \tilde{l}_i$.

Simulative results presented in [33] have shown that for exponential service times the approximation (6.51) is accurate. In [33] Chang and Sandhu also show results obtained by applying the approach presented in this section to derive the average performance figures of a Token Ring network.

6.2.4 An FDDI Network with Asynchronous Traffic (Model 4)

The model of an FDDI network with K stations transmitting asynchronous traffic is analyzed by Tangeman & Sauer in [152]. Specifically, in this model it is assumed that the asynchronous traffic is subdivided into multiple-priority levels. Each station transmits traffic of one priority level only, and the priority level of a station$\{i\}$ is characterized by its threshold $T_{Pri}(i)$[1]. Buffers are assumed to be of a finite length, and up to M_i packets can be buffered in

---

1. In this model the threshold of a priority level may vary from station to station. For this reason the notation used in this model to denote the threshold is slightly different from the notation used in Chapter 5.

a station$\{i\}$. Packets arrive at station$\{i\}$ according to a Poisson distribution with rate λ_i, and require i.i.d. transmission times sampled from a general distribution.

NETWORK MODEL. The network is modeled with a polling system with K finite-capacity queues $\{Q_1, Q_2, ..., Q_K\}$ served in a cyclic order, where a service period at Q_i represents the time station$\{i\}$ holds the token.

According to the polling-model notation introduced in previous chapters, the r.v. B_i denotes the transmission times at Q_i, while the r.v. S_i represents the time it takes the token to switch from Q_i to Q_{i+1}, i.e., the switchover time. In addition, $C_{i,n}$ denotes the length of the n-th polling cycle observed by Q_i, i.e., the elapsed time between the $(n-1)$th and n-th token arrival at Q_i. Specifically, as shown in Figure 6.4, $C_{i,n+1}$ includes

(i)   the transmission times of the packets transmitted by stations with an index greater than or equal to i at their n-th service period (i.e., $\sum_{j=i}^{K} Sp_j^{(n)}$), in [152] the service period is referred to as a scanning epoch;

(ii)   the transmission times of the packets transmitted by stations with an index lower than i at their $(n+1)$th scanning epoch (i.e., $\sum_{j=1}^{i-1} Sp_j^{(n+1)}$); and

(iii)   the switchover times.

As pointed out in the description of the FDDI MAC protocol (see Chapter 5), the amount of asynchronous traffic a station$\{i\}$ can transmit depends on the value of the $TRT(i)$ timer when the token arrives at station$\{i\}$ ($\overline{TRT(i)}$). To represent the time-token protocol behavior, in the model the service discipline is *Gated-Limited with Limit variation*: when a station$\{i\}$ captures the token it can use the transmission media for a maximum time equal to

$$max\{0, T_{Pri}(i) - \overline{TRT(i)}\} \quad . \tag{6.52}$$

In addition, when the station$\{i\}$ holds the server, it can only transmit the packets that were in its buffer at the token arrival (i.e., gated service). The transmission period of a station ends when there are no more packets to transmit or the maximum transmission time for this cycle has been reached,

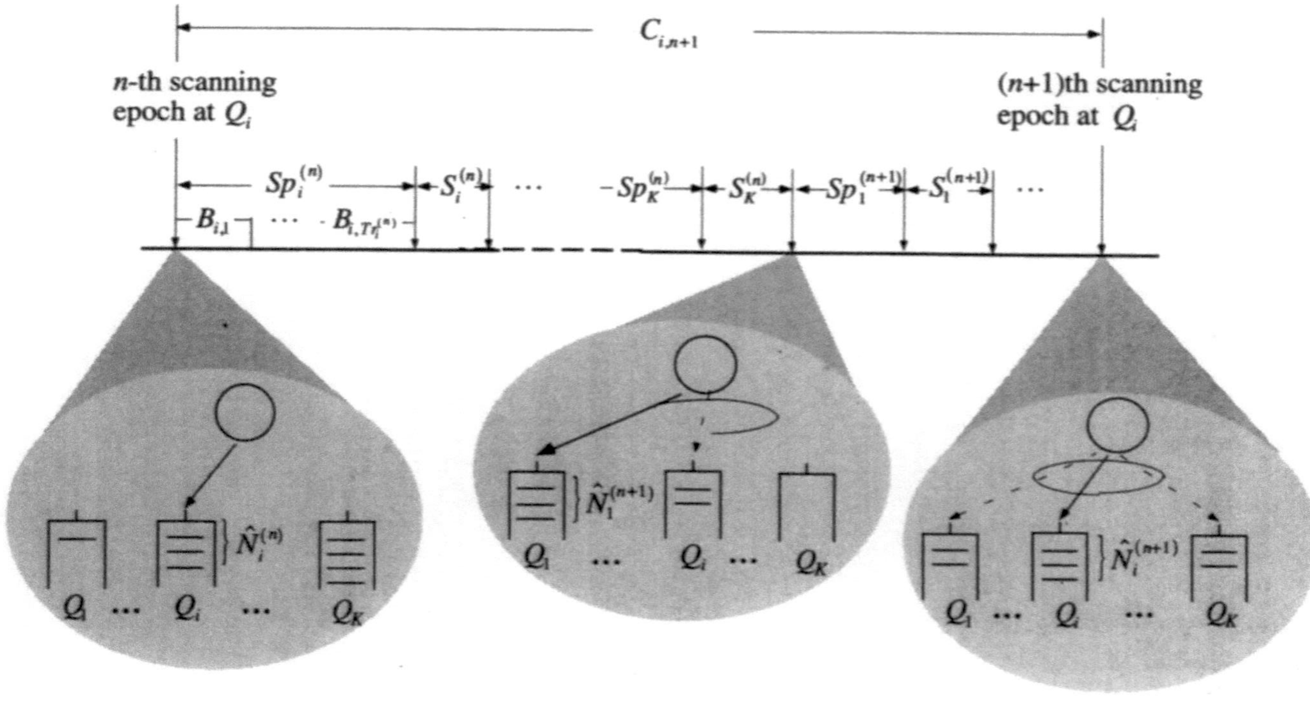

Figure 6.4 Relationship between the n-th and $(n+1)$th scanning epoch at Q_i

whichever occurs first. In the latter case the transmission of the packet in service (if any) is completed.

To model the service discipline, the value of the timer of a station$\{i\}$ when the token arrives at this station should be computed. This computation may be extremely difficult as, when the token is late, the $\overline{TRT(i)}$ may depend on the length of several consecutive cycles. To simplify the computation, it is assumed (see [152]) that the TRT timer is always reset when the token arrives at Q_i, and hence $\overline{TRT(i)}$ coincides with $C_{i,n}$. By exploiting this hypothesis and equation (6.52), the token holding time at Q_i at the n-th scanning epoch is

$$max\,\{0,\, T_{Pri}(i) - C_{i,n}\}\quad .\tag{6.53}$$

MODEL ANALYSIS. The analysis is divided into two main parts. In the first part, the dependencies among the network stations are analyzed by deriving, for each station$\{i\}$, the joint distribution of the cycle length and the queue length at the token arrival instant. In the second part, each station is analyzed in isolation to derive its packet-loss probability and the average waiting time experienced by its packets.

By denoting with $\hat{N}_i^{(n)}$ the number of packets queued at Q_i at the n-th scanning epoch, the first part of the analysis is devoted, for each station$\{i\}$, to the computation of the steady-state probabilities of the processes $\{\hat{N}_i^{(n)}\}$ and $\{C_{i,n}\}$, embedded at token arrival instants at Q_i. Specifically, by denoting with $P_j^{(n)}(k)$, $k = 0, 1, ..., M_j$ the probability mass function of $\hat{N}_i^{(n)}$, and with $C_{i,n}(t)$ the density function of the r.v. $C_{i,n}$, the target of the computation is the estimate

$$P_j(k) = \lim_{n \to \infty} P_j^{(n)}(k),\ \ k = 0, 1, ..., M_j,\, j = 1, 2, ..., K\quad ,\tag{6.54}$$

$$C_i(t) = \lim_{n \to \infty} C_{i,n}(t),\quad t \geq 0\quad .\tag{6.55}$$

To perform this task, the method developed by Tran-Gia & Raith for non-exhaustive polling systems with finite capacity (see Section 4.2.1) is exploited. In principle, to derive the limiting probabilities (6.54) and (6.55) the method entails building up for each Q_j, with an iterative algorithm, the n-sequences $\{P_j^{(n)}(k),\ k = 0, 1, ..., M_j\}$ and $\{C_{i,n}(t)\}$. However, to simplify the computation, instead of explicitly deriving the n-sequence of density functions $\{C_{i,n}(t)\}$, only the n-sequence of vectors $\{E[C_{i,n}], E[(C_{i,n})^2]\}$ is computed, where $E[C_{i,n}]$ and $E[(C_{i,n})^2]$ are the first and second moments of $C_{i,n}$, respectively. The distribution function of $C_{i,n}$ is then approximated by applying a two-moment fitting procedure (see Section 4.2.1).

The limiting probabilities (6.54) and (6.55) are approximated by exploiting the vector $\{P_j^{(\tilde{n})}(k), \ k = 0, 1, ..., M_j\}$, $\{E[C_{i,n}], E[(C_{i,n})^2]\}$ where $\tilde{n}$ is the step of the iterative algorithm such that

$$\sum_{j=1}^{K} \frac{E[\hat{N}_j^{(\tilde{n})}] - E[\hat{N}_j^{(\tilde{n}-1)}]}{E[\hat{N}_j^{(\tilde{n}-1)}]} < \varepsilon \quad , \tag{6.56}$$

and ε is the selected approximation level.

REMARK. The condition (6.56) for the termination of the iterative computation was proposed by Tangeman & Sauer [152]. On the other hand, in the iterative algorithm presented in Section 4.2.1 the condition to stop the iterative process is

$$\left| P_j^{(\tilde{n})}(k) - P_j^{(\tilde{n}+1)}(k) \right| \le \varepsilon, \quad \forall (j, k) \quad . \tag{6.57}$$

$\Diamond$

To construct the sequences $\{P_j^{(n)}(k)\}$ and $\{C_{i,n}(t)\}$, relationships are established between the couple of random variables $(C_{i,n}, \hat{N}_i^{(n)})$ and $(C_{i,n+1}, \hat{N}_i^{(n+1)})$. Specifically, as shown in Figure 6.5, in the first step (denoted in the graph with label (1)) from the knowledge of

(i)   the length of the n-th cycle observed by Q_i $(C_{i,n})$, and

(ii)   the threshold for the transmission of Q_i asynchronous traffic $(T_{Pri}(i))$,

the mass function of the r.v. $L_i^{(n)}$ is estimated. $L_i^{(n)}$ denotes the limit on the number of packets Q_i can transmit at the n-th scanning epoch. Hence, from $L_i^{(n)}$ and $\hat{N}_i^{(n)}$ the number of packets, $Tr_i^{(n)}$, transmitted by Q_i at the n-th scanning epoch[1] can be immediately derived (labels (2) in the graph)

---

1. These transmissions will contribute to $C_{i,n+1}$.

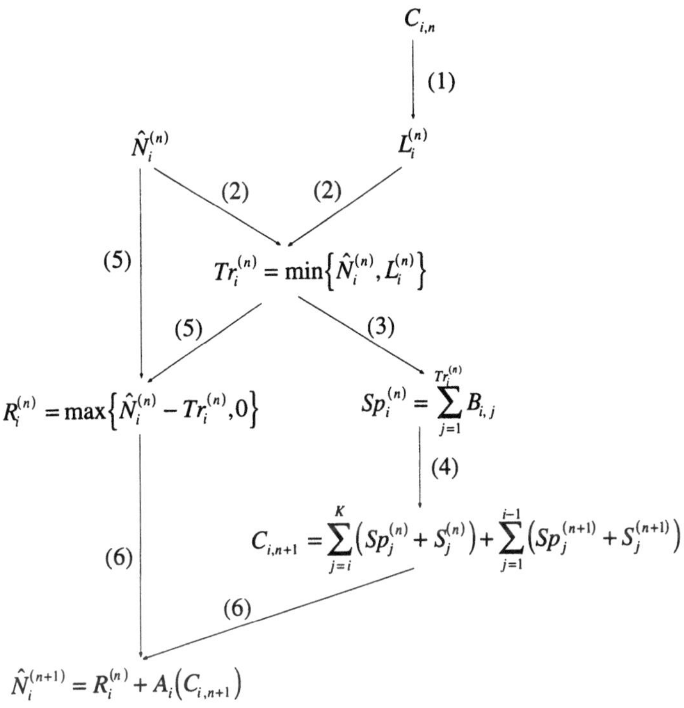

Figure 6.5: Flow chart of the computation

$$Tr_i^{(n)} = \text{Min}\{\hat{N}_i^{(n)}, L_i^{(n)}\} \ . \tag{6.58}$$

Finally, the number of packets transmitted by Q_i at the n-th scanning epoch determines the length of the n-th service period at Q_i (label (3)), from which the cycle length observed by Q_i at the $(n+1)$th scanning epoch ($C_{i,n+1}$) can be derived (label (4)).

The computation of the number of packets queued at Q_i at the $(n+1)$th scanning epoch requires two additional steps. First (labels (5)), the residual number of packets at Q_i after the n-th scanning epoch, $R_i^{(n)}$, is computed, and then by adding to $R_i^{(n)}$ the number of arrivals which occur in the $(n+1)$th cycle, $A_i(C_{i,n+1})$, the queue length at the $(n+1)$th scanning epoch, $\hat{N}_i^{(n+1)}$, is derived[1].

Details on each step of the computation can be found in Section 6.4.

---

1. The symbol $A_i(C_{i,n+1})$ indicates that the number of arrivals depends on the cycle length $C_{i,n+1}$.

REMARK. To start the computation, for each Q_i, the distributions of the r.v. $C_{i,0}$, $\hat{N}_i^{(0)}$ must be initialized. In [152] Tangeman and Sauer suggested to set these distributions by assuming that the system is empty. Then, by applying the relationships between two consecutive scanning epochs (see Figure 6.5) to each station (starting from Q_1 and increasing the node index) at the end of the first step of the iterative algorithm, for each station $\{i\}$, the distributions of the r.v. $C_{i,1}$, $\hat{N}_i^{(1)}$ are obtained. The iterative steps are repeated until the condition (6.56) is satisfied.

$$\Diamond$$

Once the limiting probabilities (6.54) and (6.55) have been derived, each node can be analyzed in isolation. By exploiting the distribution $\{P_i(k),\ k = 0, 1, ..., M_i\}$ the steady-state distributions of the queue length at an arbitrary point in time, denoted by $\{p_i(0), p_i(1), ..., p_i(M_i)\}$, are derived. Then from $\{p_i(0), p_i(1), ..., p_i(M_i)\}$ the packet-loss probability,[1] $P_L(i)$, and the average response time $(E[R_i])$ are easily obtained. The packet-loss probability at Q_i is derived by noting that, due to the PASTA property (see Theorem 2.3),

$$P_L(i) = p_i(M_i) \quad , \tag{6.59}$$

while, from Little's theorem (see Theorem 2.1) it follows that

$$E[R_i] = \frac{\sum_{k=1}^{M_i} k \cdot p_i(k)}{\lambda_i \cdot (1 - P_L(i))} \quad . \tag{6.60}$$

The computation of the queue length distribution at an arbitrary point in time follows the same line of reasoning applied in Section 4.2.1. For this reason only the description of the main steps of the derivation is reported here. Specifically, $\{p_i(0), p_i(1), ..., p_i(M_i)\}$ are derived by conditioning on the cycle type (i.e., on the number of transmissions the station performs in the cycle) in which the observation instant falls. By denoting with $p_i^{(l)}$ the probability that a random point in time falls in a cycle which contains l services at Q_i

---

1. In [152] this quantity is referred to as the blocking probability.

(hereafter type l *cycle*), and by exploiting (6.61) the computation of $p_i(k)$ is reduced to derive

- $p_i^{(l)}$, $l = 0, ..., M_i$, and
- the steady-state probabilities, at an arbitrary point in time, in a type l cycle (i.e., $p_i(k|\ type\ l\ cycle)$).

$$p_i(k) = \sum_{l=0}^{M_i} p_i(k|\ type\ l\ cycle) p_i^{(l)} \tag{6.61}$$

By applying the same arguments used in Section 4.2.1, $p_i^{(l)}$ is derived from equation (6.62), where $E[C_{i|l}]$ is the average length of a cycle which contains l packet transmissions at Q_i, and $E[C]$ is the unconditional cycle length.

$$p_i^{(l)} = \frac{P\{Tr_i = l\}\, E[C_{i|l}]}{E[C]} \tag{6.62}$$

To derive $p_i(k|\ type\ l\ cycle)$ it is useful to note that a random point in time may fall either during the service period at Q_i or during the vacation. By denoting with $E[V_{i|l}]$ the average length of a vacation in a type l cycle, the probability that a random point in time falls in the vacation period is equal to $E[V_{i|l}]/E[C_{i|l}]$, while $E[B_i]/E[C_{i|l}]$ is the probability that the random point occurs during the j-th $(j = 1, 2, ..., l)$ service time at Q_i. Hence,

$$p_i(k|\ type\ l\ cycle) = \frac{E[B_i]}{E[C_{i|l}]} \sum_{j=1}^{l} \dot{p}_{j|l}(k) + \frac{E[V_{i|l}]}{E[C_{i|l}]} \dot{p}_{v|l}(k) \quad , \tag{6.63}$$

where $\dot{p}_{j|l}(k)$ and $\dot{p}_{v|l}(k)$ denote the probability of having k packets at Q_i at a random point in time, given that the random point falls in the j-th service time or in the vacation period of a type l cycle, respectively.

Finally, $\dot{p}_{j|l}(k)$ and $\dot{p}_{v|l}(k)$ are derived through the following steps:

1. Compute the steady-state distribution of the number of packets at Q_i at a scanning epoch given that l packets will be transmitted in that service period (i.e., type l cycle) $\{P_{i|l}(k),\ k = 0, 1, ..., M_j\}$. These probabilities are simply derived from the unconditional steady-state probabilities by conditioning on l transmissions in a cycle.

2. Compute the steady-state distribution of the number of packets at Q_i just after the j-th service time in a type l cycle $(j = 1, 2, ..., l)$ $\{\pi_{i|l}^{(j)}(k), \ k = 0, 1, ..., M_j\}$. This computation can be performed by observing that the number of packets after the j-th transmission is equal to the number of packets in the queue before this transmission (i.e., after the $(j-1)$ th service time[1]) plus all the packets which arrive during the service time, and minus the packet which is served during the j-th transmission.

3. The number of packets at an arbitrary point in time during the j-th service (vacation) time in a type l cycle, is easily obtained by adding to the number of packets in the system immediately after the $(j-1)$ th service time (l-th service time) the number of packets that arrive during the elapsed service (vacation) time which corresponds to the backward recurrence time of the r.v. B_i $(V_{i|l})$.

This algorithm was validated via simulation. Results on this validation process can be found in [152] together with some experimental results on the convergence speed of the algorithm. In the network configurations analyzed, the number of iteration steps required to achieve the convergence when $\varepsilon = 10^{-5}$ is always less than 70, and the complexity of a single iteration is $O(K(M^2 + 2M))$, where M is the station buffer size.[2]

6.3 STATION-IN-ISOLATION MODELS

This section outlines four station-in-isolation models. All the models falling in this category are variants of the $M/G/1$ queueing model with vacation (see Section 3.2). As outlined in Section 6.1, the token is a server that provides a service during its visit to the tagged station and is on vacation (i.e., within C-NET) when it is away from it. Thus, with respect to the tagged node, C-NET can also be viewed as a generator of vacation periods. In some of the station-in-isolation models described in this section, the queue has a finite capacity and this is the reason for the use of $M/G/1/K$ queueing model with vacation, where the queue capacity is equal to K.

---

1. If $j - 1 = 0$ this point coincides with the scanning epoch, i.e., token arrival at the station.
2. This formula is derived by assuming the same buffer size in each station.

REMARK. As already stated in Chapter 3, when station-in-isolation models are considered, K is used to denote the size of the buffer, rather than the number of stations in a MAN. This should not cause any problems since node-in-isolation models only focus on a tagged station.

$\Diamond$

To represent the behavior of the tagged station the limited service disciplines already introduced in Section 2.1.2 are commonly used. In some models, to better represent the protocol behavior, the *limited with limit variation* service disciplines are used. These service disciplines extend the previous ones, since a dynamic limit (chosen upon server arrival) is placed either on the maximum number of frames that can be served, or on the amount of time the server serves the queue before going on vacation.

The $M/G/1$ and $M/G/1/K$ queueing systems with vacation and limited (or limited with limit variation) service disciplines are analyzed with the embedded Markov chain technique. The state of the Markov chain includes a random variable to denote the number of frames in the system and other random variables whose semantics vary from model to model. If the tagged station is modelled with a gated limited (or gated limited with limit variation) service discipline the embedding points correspond to the vacation termination instant (i.e., the arrival instants of the server at the tagged node); otherwise, if the service discipline is exhaustive limited (or exhaustive limited with limit variation), apart from the vacation termination instants the service completion instants are also included among the embedding points.

Two techniques have been used to solve the models presented in this section: z-transform and matrix-analytic.

The first model (Model 5 [106]) performs a delay analysis of the asynchronous traffic alone. The performance figures of an FDDI tagged station are studied by solving, with the z-transform technique, an $M/G/1$ queueing system with exhaustive limited and limit variation service discipline.

Model 6 ([41], [45]) was proposed and solved to analyze the QoS achieved by the synchronous frames. Since the synchronous service of FDDI is designed to deliver time critical packets, the emphasis is on the development of a model which provides, under the heavy-load assumption, estimates of the probability distribution of the number of packets queued in the tagged station, and of the delay experienced by these packets in the buffer.

Model 7 ([35]) and Model 8 ([112]) used the matrix-analytic technique (see Section 3.3) to analyze an FDDI network with asynchronous traffic.

To better organize the information provided by the different models Table 6.2 reports, for each model presented in this chapter, the type of traffic (Synchronous (S), Asynchronous (A)) analyzed, and the performance measures characterizing the QoS experienced by each class of traffic.

6.3.1 M/G/1 with Vacation and Exhaustive Limited with Limit Variation Service Discipline (Model 5)

This model was proposed by LaMaire ([106]). In this model the tagged station, which is fed with asynchronous traffic belonging to one priority level, is modeled as an $M/G/1$ single-server queueing system with server vacations. The service discipline is such that the server provides service until either the system is emptied or a randomly chosen limit of l frames has been served (exhaustive limited with limit variation or ELV service discipline). The server then goes on a vacation before returning to service the queue again.

In the analysis performed in [106], the limit L is assumed to be a bounded discrete random variable with probability mass function, p_l, $l \in \{0, 1, ..., l_{max}\}$. The value of L for a service interval, l, is determined at the last vacation termination instant.

REMARK. In a real FDDI network, the value of l will differ from one token arrival at the next, depending upon the $Late\_Ct$ and the TRT values at the time the token is seized by the tagged station.

◊

To simplify the analysis, the following approximations are introduced

1. the sequence of limits chosen at the vacation termination instants are i.i.d. random variables;
2. the vacation time is independent of all system variables except for the limit l of the previous token visit.

Table 6.2 Performance figures of the tagged station derived with the station-in-isolation models

	Traffic Types	Buffer size	Queue indices	Delay indices	Other
Model 5	A	$M = \infty$	$E[N^A]$	$E[W^A]$,	
Model 6	S+A	$M = \infty$	$E[N^S]$		$\pi(\bullet)$
Model 7	S, A	$M = \infty$	$E[N^S], E[N^A], p(\bullet)$	$E[W^S], E[W^A]$	
Model 8	S, A	$M = \infty$	$E[N^S]\ E[N^A], p(\bullet)$		

LEGEND:

S=Synchronous; A=Asynchronous;

$p(\bullet)$ queue length distribution at a random point in time

$\pi(\bullet)$ queue length distribution at tthe embedding points

$N^S (N^A)$ = number of synchronous (asynchronous) packets queued in the tagged station;

$W^S (W^A)$ = waiting time of synchronous (asynchronous) packets at the tagged station;

The frame arrival process is Poisson with arrival rate λ and the service times (i.e., the frame transmission times) are assumed to be independent of any process in the system. The conditional pdf of the vacation time, given the limit l, is denoted by $v_l(t)$ with LST $V_l^*(s)$ and first and second moments v_l and $v_l^{(2)}$, respectively. Similarly, the unconditional pdf of the vacation time is denoted by $v(t)$ with LST $V^*(s)$ and the first and second moments v and $v^{(2)}$, respectively.

The $M/G/1$ vacation model described above is analyzed by using a similar approach to the one used by Lee for an E-limited (i.e., with a fixed limit) service [110] (see also [149]).

STABILITY ANALYSIS. To perform this type of analysis it is necessary to calculate the maximum rate (μ_{max}) at which the server of the tagged station can serve frames in steady state. If M denotes the mean number of frames that are served between two consecutive visits of the server at the queue (i.e., during a cycle) of the tagged station, the mean cycle time is given by $Mb + v$. Hence, in steady state, the service rate (μ) of the server is given by the ratio

$$\mu = \frac{M}{Mb + v} \ . \tag{6.64}$$

Since this ratio is an increasing function of M, the maximum service rate can be obtained by noting that $M \le E[L]$ where $E[L] = \sum_{l=0}^{l_{max}} l p_l$. Hence,

$$\mu_{max} = \frac{E[L]}{E[L]\, b + v} \ . \tag{6.65}$$

For system stability, the frame arrival rate must be less than μ_{max}, that is,

$$\lambda < \frac{E[L]}{E[L]\, b + v} \ ,$$

or, by recalling that $\rho = \lambda b$ (see Section 2.1.1).

$$\rho < 1 - \frac{\lambda v}{E[L]} \ .$$

DELAY ANALYSIS. This type of analysis employs the embedded Markov chain technique where the embedding points are the vacation termination and service completion instants. The state of the system at an embedding point during the n-th service period is described by a triplet, $\{\xi, N, L\}_{(n)}$. $\xi = 0$ indicates that an embedded point is a vacation termination instant, while $\xi = k$, $(k = 1, 2, ..., l_{max})$ indicates that the embedding point is the service completion instant of the k-th frame in the n-th service period. The random variable N, is the number of frames in the system at the embedding point. Finally, L, is the limit on the number of frames that can be served during the (possibly zero length) n-th service period. It is worth remembering that the L value is sampled at the vacation termination instant from probability mass function p_l.

The resulting Markov chain is irreducible and aperiodic. Assuming that the system is stable, the following steady-state joint probabilities exist

$$f_{n,l} = \tag{6.66}$$
$$\lim_{m \to \infty} P\{\xi = 0, N = n, L = l\}_{(m)}, \; n = 0, 1, ...; \; l = 0, 1, ..., l_{max} \quad,$$

$$\pi_{n,l}^{[k]} = \lim_{m \to \infty} P\{\xi = k, N = n, L = l\}_{(m)}, \tag{6.67}$$
$$n = 0, 1, ...; \quad l = 1, ..., l_{max}; \quad k = 1, 2, ..., l \quad,$$

where,

- $f_{n,l}$ is the steady state probability that, at vacation termination instants, there are n frames in the system and that a limit l is chosen for the next service period;
- $\pi_{n,l}^{[k]}$ is the steady state probability that at the k-th embedding point in a service period in which at maximum l frames can be transmitted (i.e., $0 < k \leq l$), there are n frames in the system.

Since the number of frames in the system at the vacation termination instant is independent of the limit l that is chosen at this instant for the next service period, $f_{n,l}$ can be expressed as:

$$f_{n,l} = f_n p_l, \quad n = 0, 1, ...; \; l = 0, 1, ..., l_{max} \quad, \tag{6.68}$$

where f_n is the steady-state probability that an embedded point is a vacation termination instant and that there are n frames in the system.

Rather than using the joint probabilities $\pi_{n,l}^{[k]}$, the steady-state equations for the embedded Markov chain are written in terms of $\pi_{n|l}^{[k]}$, i.e., steady-state conditional probabilities that, given the limit l, an embedded point is the k-th service completion in this service period and that there are n frames in the system.

Once the model has been solved for $f_{n|l}$ and $\pi_{n|l}^{[k]}$, the joint probabilities $f_{n,l}$ and $\pi_{n,l}^{[k]}$ can be calculated by means of their relationships, i.e.:

$$\pi_{n,l}^{[k]} = \pi_{n|l}^{[k]} p_l, \quad n = 0, 1, \ldots; \ l = 1, 2, \ldots, l_{max}; \ k = 1, 2, \ldots, l \quad . \tag{6.69}$$

With the above definitions, for $n = 0, 1, 2, \ldots$, the following steady state equations hold

$$f_n = p_0 \left[\sum_{j=0}^{n} f_j g_{n-j|0} \right] + \sum_{l=1}^{l_{max}} p_l \left\{ \left[\sum_{k=1}^{l-1} \pi_{0|l}^{[k]} + f_0 \right] g_{n|l} + \sum_{j=0}^{n} \pi_{j|l}^{[l]} g_{n-j|l} \right\} , \tag{6.70}$$

where

$$g_{i|l} = \int_0^{\infty} \frac{(\lambda t)^i}{i!} e^{-\lambda t} v_l(t) dt, \quad i = 0, 1, \ldots; \ l = 0, 1, \ldots, l_{max} \quad , \tag{6.71}$$

is the probability that i frames arrive during a vacation time, given that the previous service period had limit l.

Furthermore, for $n = 0, 1, 2, \ldots; \ l = 1, 2, \ldots, l_{max}$

$$\pi_{n|l}^{[k]} = \begin{cases} \displaystyle\sum_{j=1}^{n+1} f_j h_{n-j+1} & k = 1, \\[4mm] \displaystyle\sum_{j=1}^{n+1} \pi_{j|l}^{(k-1)} h_{n-j+1} & k = 2, 3, \ldots, l; \ l \geq 2 \quad , \end{cases} \tag{6.72}$$

where

$$h_i = \int_0^{\infty} \frac{(\lambda t)^i}{i!} e^{-\lambda t} b(t) dt, \quad i = 0, 1, \ldots \quad , \tag{6.73}$$

is the probability that i frames arrive during a service time.

From (6.72) it follows that

$$\pi_{n|l}^{[k]} = \pi_{n|m}^{[k]}, \quad k \leq min\,(l, m) \quad . \tag{6.74}$$

Therefore, once the model has been solved for $\pi_{n|l_{max}}^{(k)}$, all the other $\pi_{n|l}^{(k)}$, for $l = 1, 2, ..., l_{max} - 1$ can be determined from (6.74). Hereafter, for simplicity of notation, the following definition is introduced

$$\pi_n^{[k]} \equiv \pi_{n|l_{max}}^{[k]}, \quad k = 1, 2, ..., l_{max}; \, n = 0, 1, 2, ... \quad . \tag{6.75}$$

By defining

$$F(z) = \sum_{n=0}^{\infty} f_n z^n \quad , \tag{6.76}$$

$$\Pi(z) = \sum_{l=1}^{l_{max}} p_l \sum_{k=1}^{l} \left(\sum_{n=0}^{\infty} \pi_n^{[k]} z^n \right) \quad , \tag{6.77}$$

after extensive algebraic manipulation the following expressions are derived

$$F(z) = \tag{6.78}$$

$$\frac{z^{l_{max}} \sum_{l=1}^{l_{max}} p_l V_l^*(\lambda - \lambda z) \cdot \left[\left(1 - \left[\frac{B^*(\lambda - \lambda z)}{z} \right]^l \right) f_0 + R_l(1) - R_l\left(\frac{B^*(\lambda - \lambda z)}{z} \right) \right]}{z^{l_{max}} - \sum_{l=0}^{l_{max}} p_l V_l^*(\lambda - \lambda z) \left[B^*(\lambda - \lambda z) \right]^l z^{l_{max} - l}}$$

$$\Pi(z) = \frac{B^*(\lambda - \lambda z)}{z - B^*(\lambda - \lambda z)} \tag{6.79}$$

$$\sum_{l=1}^{l_{max}} p_l \left[(F(z) - f_0) \left(1 - \left[\frac{B^*(\lambda - \lambda z)}{z} \right]^l \right) - R_l(1) + R_l\left(\frac{B^*(\lambda - \lambda z)}{z} \right) \right]$$

where

$$R_l(z) = \begin{cases} 0, & l = 1 \\ \sum_{k=1}^{l-1} \pi_0^{(k)} z^{l-k}, & l = 2, 3, ..., l_{max} \end{cases}$$

What is noteworthy in $F(z)$ and $\Pi(z)$ is the presence of l_{max} unknown boundary probabilities f_0, and $\pi_0^{(1)}, \pi_0^{(2)}, ..., \pi_0^{(l_{max}-1)}$ which can be found by using Rouche's theorem and the normalization condition. From $F(z)$ and $\Pi(z)$, the derivation of the mean response time in terms of the l_{max} boundary probabilities is straightforward.

Since Burke's Theorem (see Theorem 2.2) holds, the queue length distribution immediately before the arrival instants is the same as the distribution immediately after the service completion instants. Furthermore, due to PASTA (see Theorem 2.3), the steady-state mean queue length (at an arbitrary point in time) is equal to the steady-state mean queue length at the service completion instants. Hence, the mean queue length is given by the normalized z-transform derivative,

$$E[N] = \frac{1}{\Pi(z)} \cdot \frac{d}{dz}\Pi(z)\Big|_{z=1} . \tag{6.80}$$

Using Little's law (see Theorem 2.1), the mean waiting (i.e., queueing) time can be found

$$E[W] = E[N]/\lambda - b . \tag{6.81}$$

Results reported in [106] show that there is a close agreement between simulative estimates of $E[W]$ and analytical results for utilization factor values up to 0.7. These results are obtained by estimating the $v(t)$ and p_l via a simulative analysis.

LaMaire extended this work to the case of an $M/G/1/K$ queueing system with the exhaustive limited with limit variation service discipline [107]. The queue length distribution and the Laplace-Stieltjes transforms of the waiting time, busy period and cycle time distributions are derived in [107]. In addition, an expression for the mean response time is given.

6.3.2 Worst Case Model for Synchronous Traffic (Model 6)

In ([41], [45]) the FDDI MAC protocol is analyzed to evaluate whether or not the quality of service guaranteed by a highly congested FDDI network is adequate to support real time applications. The FDDI model used to perform such an analysis is based on the assumption that the interactions among stations generate a "worst" case (see Section 3.4) for the performance indices of the station (i.e., the tagged station) supporting real-time applications.

According to ([41], [45]), the tagged station has an infinite buffer size,

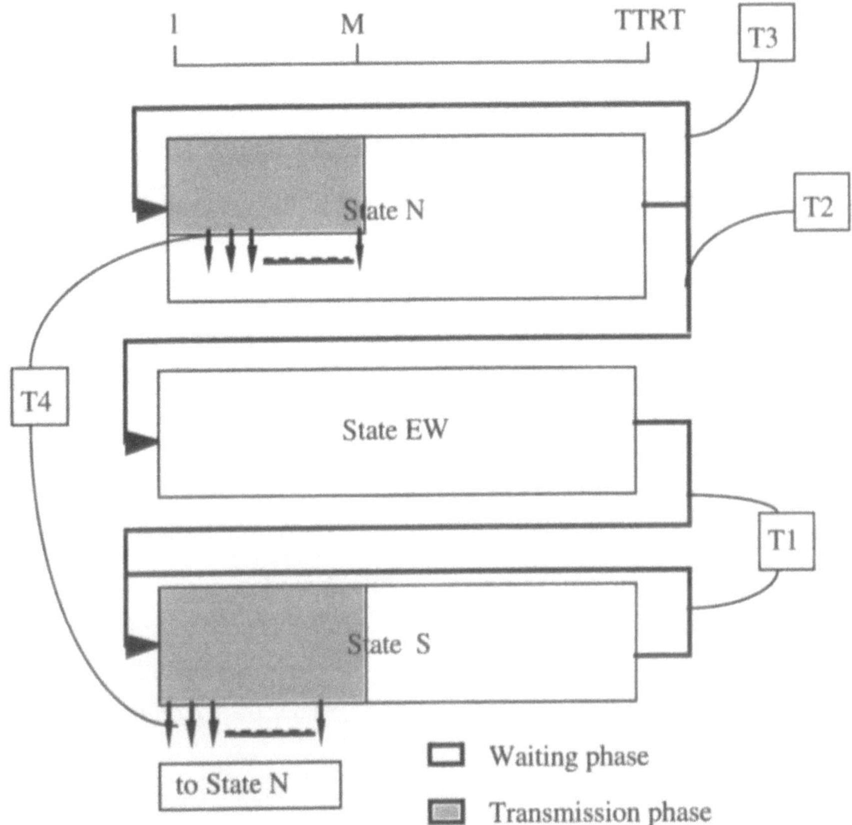

Figure 6.6: Model behavior

and supports both synchronous and asynchronous traffic. The arrival process which generates synchronous frames of fixed length for the tagged station is a superposition of discrete time Markov processes. Therefore, the maximum fraction of synchronous traffic that the tagged station is allowed to transmit in a service period can be translated into an integer number of synchronous frames, which will be denoted henceforth by M.[1]

In this model the tagged station can be in one of the following states: *Normal* (N), *Exceptional-waiting* (EW), and *Stabilization* (S).

The cycle length, as seen from the tagged station, can be either $TTRT$ or $2 \times TTRT$, depending upon the state of the tagged station and according

---

1. As explained in Section 3.4 the asynchronous traffic queue is never empty and hence the station always transmits M packets in a service period.

to the following rules:

1. in the Normal state the length of the cycle is $TTRT$, and up to M packets can be transmitted;

2. at the end of a cycle, if the state is Normal, it changes to Exceptional-waiting with probability P_{late}, or remains Normal with probability $(1 - P_{late})$;

3. in the Exceptional-waiting state the length of the cycle is $TTRT$ and the tagged station cannot transmit any packets;

4. at the end of a cycle in which the tagged station is in the Exceptional-waiting state, its state switches to the Stabilization state, with probability one;

5. in the Stabilization state the tagged station behaves as in the Normal state, it observes a cycle with length $TTRT$, and it transmits up to M packets. Nevertheless, at the end of the cycle the tagged station state does not change unless it becomes empty. When this occurs, the state of the tagged station returns to Normal.

Clearly, by means of probability P_{late}, rule 2 models early and late token arrivals. Furthermore, when the token is late rule 3 is used to model the maximum cycle length, i.e., $2 \times TTRT$. Since the cycle can be either $TTRT$ or $2 \times TTRT$ the average length of the cycle is greater than $TTRT$.

REMARK. While in a real FDDI the memory of a late token is lost when the accumulated delay has been "absorbed" during subsequent token rotations, rule 5 specifies that the worst-case model loses its memory of a late token arrival when the tagged station becomes empty. Hence the memory of a late token is lost when the effect of the exceptional waiting state on the queue length is lost.

$\Diamond$

The evaluation of the "distance" of the worst case model that has just been introduced with respect to a real system has already been carried via simulation. Specifically, Figure 6.7 compares results obtained via a simulation of the tagged station with those obtained via a numerical solution of the worst-case model. As Figure 6.7 shows, the worst-case model always overestimates the distribution of the tagged station buffer size. Furthermore, by focusing on percentiles in the range [90-th,99-th] the worst case model gives estimations that are very close to those obtained by simulation.

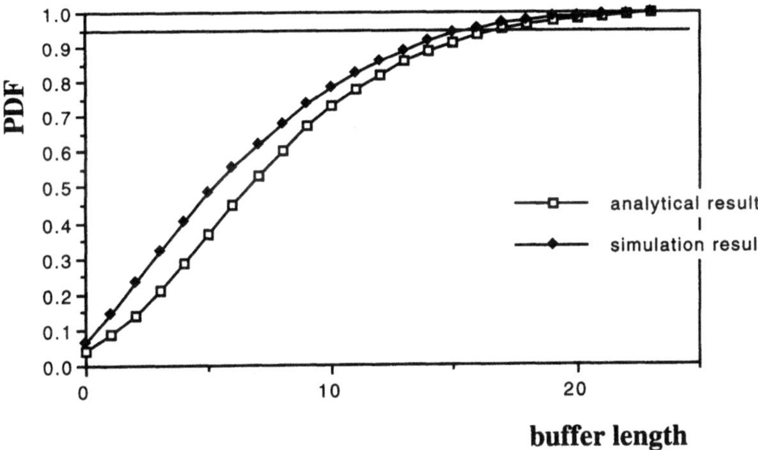

Figure 6.7: Comparison between analytic and simulative PDFs

The resulting stochastic model is an extension of the single server queueing system with vacation and E-limited service discipline with limit M (see Section 3.2.1). Since the tagged station can transmit at most M frames per cycle, the model described above represents a worst case model since in a real FDDI the average cycle length is less than or equal to $TTRT$.

Following the approach proposed in Section 3.2.1 this model can be solved with the embedded Markov chain technique. Two groups of embedding points are defined: one related to the Normal state and the other related to the Stabilization state. In the Normal state the embedding points are the service completion instants, while in the Stabilization state (to simplify the description of the dynamics of the system) the vacation termination instants are also taken into consideration.

The state of the system at an embedding point during the i-th cycle $(i = 1, 2, ...)$ is described by a tuple $\{\xi, N, \Sigma\}_{(i)}$. $\xi = 0$ indicates the vacation termination instant, while $\xi = m$, $m = 1, 2, ..., M$ denotes the end of the m-th frame transmission within a service period. $\Sigma = 0$ represents the Normal state, while $\Sigma = 1$ identifies the Stabilization state. As usual N is the number of frames in the station at an embedding point. If $0 < P_{late} < 1$, the resulting embedded Markov chain is irreducible and aperiodic.

For system stability the average number of arrivals in a $TTRT$ interval should be less than M. This stability condition can be understood by noting that if a delayed token is received, which results in a cycle length of $2 \times TTRT$, it is not possible to have another delayed cycle until the tagged station becomes empty.

Assuming that the system is stable, the following steady-state joint probabilities can be defined

$$\pi_k^{(m)} = \lim_{i \to \infty} P\{\xi = m, N = k, \Sigma = 0\}_{(i)}, \quad k = 0, 1, \ldots; m = 1, \ldots, M \quad,$$

$$s_k^{(m)} = \lim_{i \to \infty} P\{\xi = m, N = k, \Sigma = 1\}_{(i)}, \quad k = 0, 1, \ldots; m = 0, 1, \ldots, M \quad,$$

where $\pi_k^{(m)}$ ($s_k^{(m)}$) is the steady-state probability that the system is in the Normal (Stabilization) state immediately after the m-th frame transmission and that there are k frames in the buffer.

Before writing the steady-state equations for $\pi_k^{(m)}$ and $s_k^{(m)}$, the following random variables and probabilities need to be introduced:

- V: number of frame arrivals during the server vacation;
- $A^{(i)}$: number of frame arrivals during the i-th service time;
- $F^{(i)}$: total number of frame arrivals during the i-th, $(i+1)$th,...,M-th service times;
- C: number of frame arrivals during a cycle;
- $a_k^{(i)}$: probability that k cells arrive at the tagged station during the i-th service time, i.e., $a_k^{(i)} = P\{A^{(i)} = k\}$;
- c_k: probability that k cells arrive at the tagged station during a cycle of length $TTRT$, i.e., $c_k = P\{C = k\}$;
- $\langle X_1 + X_2 + \ldots + X_n \rangle_j$: probability that the sum of the r.v. in brackets is equal to j.

With the above definitions, the steady-state probability $\pi_k^{(1)}$ is given by

$$\pi_k^{(1)} = (1 - P_{late}) \left\{ \sum_{i=1}^{M} \pi_0^{(i)} \left[\sum_{j=1}^{k+1} \langle F^{(i+1)} + V \rangle_j \cdot a_{k-j+1}^{(1)} + \right.\right. \tag{6.82}$$

$$\left.\left. \beta_0 \sum_{n=1}^{\infty} (\alpha_0)^{n-1} \sum_{m=1}^{k+1} c_m a_{k-m+1}^{(1)} \right] + \sum_{j=1}^{k+1} \pi_j^{(M)} \langle A^{(1)} + V \rangle_{k-j+1} \right\} +$$

$$I_{\{k=0\}} s_1^{(0)} a_0^{(1)} \quad,$$

where

- $\beta_0 = [\langle F^{(i+1)} + V \rangle_0 (1 - P_{late}) + \langle F^{(i+1)} + V + C \rangle_0 P_{late}]$;

- $\alpha_0 = [c_0 (1 - P_{late}) + (c_0)^2 P_{late}]$; and
- $I_{\{.\}}$ is the indicator function.

Equation (6.82) gives the probability of having k frames in the system at the first service completion instant in the Normal state. The various terms in the right-hand side of (6.82) can be justified as follows.

1. Starting from an empty system at the i-th embedding point (i.e., $\xi = i$)
   - the first term in square brackets is the transition probability associated to the events in which before the end of the cycle at least one frame arrives, the token is early and the remaining frames arrive during the first service time;
   - the second term in square brackets is the transition probability associated to the events in which no frames arrive before the end of the cycle (β_0) and the system remains empty for $(n-1)$ cycles $(\alpha_0)^{n-1}$. In the n-th cycle at least one frame arrives, the token is early and the remaining frames arrive during the first service time.

2. The last summation in the braces contains the transition probabilities related to a non empty system at the embedding point M, the token is early and the remaining frames arrive before the end of the first service.

3. The last term is the transition probability given that at the vacation termination instant the tagged station was in the Stabilization state and becomes empty at the first embedding point within the service period.

Following the same reasoning, it is quite simple to derive the steady-state probability that an embedded point is the m-th service completion instant in the Normal state and that there are k frames in the system

$$\pi_k^{(m)} = \sum_{j=1}^{k+1} \pi_j^{(m-1)} a_{k-j+1}^{(m)} + I_{\{k=0\}} s_1^{(m-1)} a_0^{(m)}, \quad m = 2, ..., M, \ k > 0 . \quad (6.83)$$

Furthermore, equation (6.84) defines the steady-state probability that there are k frames in the system at the vacation termination instant

$$s_k^{(0)} = P_{late} \left\{ \sum_{i=1}^{M} \pi_0^{(i)} \left[\sum_{j=1}^{k+1} \langle F^{(i+1)} + V + C \rangle_k \right. \right.$$

$$\left. \left. + \beta_0 \sum_{n=1}^{\infty} (\alpha_0)^{n-1} \langle C + C \rangle_k \right] + \sum_{j=1}^{k+1} \pi_j^{(M)} \langle V + C \rangle_{k-j} \right\} + \sum_{j=1}^{k} s_j^{(M)} \langle V + C \rangle_{k-j} \qquad (6.84)$$

while equation (6.85) gives the steady-state probability that an embedded point is the m-th service completion instant in the Stabilization state and there are k cells in the system

$$s_k^{(m)} = \sum_{j=1}^{k+1} s_j^{(m-1)} a_{k-j+1}^{(m)}, \quad m = 1, 2, 3, ..., M, \ k > 0 \quad . \tag{6.85}$$

By defining $\Pi_m(z) = \sum_{k=0}^{\infty} \pi_k^{(m)} z^k$ and $S_m(z) = \sum_{k=0}^{\infty} s_k^{(m)} z^k$, after some lengthy algebraic manipulations, closed formulas for these PGFs are obtained as a function of $2M$ unknown boundary probabilities $\{ \pi_0^{(1)}, \pi_0^{(2)}, ..., \pi_0^{(M)}, s_1^{(0)}, s_1^{(1)}, ..., s_1^{(M-1)} \}$, see [41] and [45]. Specifically, the closed formula for $S_0(z)$, has the following structure

$$S_0(z) = \frac{\sum_{i=1}^{M} \pi_0^{(i)} \psi(z) + s_1^{(0)} \phi(z) + \sum_{i=1}^{M} s_1^{(i)} \vartheta(z)}{[z^M - C(z)] \, [z^M - C(z)(1 - P_{late})]} \quad ,$$

where

- $C(z)$ is the PGF of the number of arrivals in a cycle of length $TTRT$,
- $\psi(z)$, $\phi(z)$ and $\vartheta(z)$ are functions of z which contain known parameters.

Hence, $\{ \pi_0^{(1)}, \pi_0^{(2)}, ..., \pi_0^{(M)}, s_1^{(0)}, s_1^{(1)}, ..., s_1^{(M-1)} \}$ can be found by using Rouche's theorem and the normalization condition $\Pi(1) + S(1) = 1$ where $\Pi(z) = \sum_{m=1}^{M} \Pi_m(z)$ and $S(z) = \sum_{m=0}^{M} S_m(z)$.

For the special case $P_{late} = 1$ the average and the standard deviation of the distribution of the number of frames in the tagged station are directly computed from the closed formula of $S(z)$. Percentiles of the above distribution are overestimated by applying Chebyshev's inequality [85]. This inequality was obtained by using a two-moment approximation (see Section 4.2.1) to obtain an approximate distribution of the number of frames in the tagged station. In [46] the author shows, for this type of model, how to derive the delay statistics from the PGF.

6.3.3 M/G/1 with Vacation and Vacation-dependent, Time-limited Service Discipline (Model 7)

In [112] two vacation models for stations connected to a timed-token networks are analyzed. The first model employs the *constant time-limited (CTL) service discipline*, while the second model uses the *vacation-*

dependent, time-limited (VDTL) service discipline. The *CTL* service discipline can be used to model the transmission of synchronous traffic in the timed-token networks. Under the *VDTL* service discipline, the token holding time depends on the length of the last vacation. This modeling feature is introduced to capture the dependence of the *Token Holding Timer* (see Section 5.3.4) and the previous vacation. Hence *VDTL* can be used to model the transmission of asynchronous traffic in the timed-token networks.

Both models have an infinite buffer size. Packets arrive according to a Poisson process and are served on a FIFO basis. The packet service time and the duration of a vacation are sampled from general distributions. The inter-arrival, service and vacation times are mutually independent.

In the vacation model with the *CTL* service discipline, at the beginning of a service period a timer is started with a fixed value, referred to as *visit-time limit.* The service period ends when either all the packets in the system have been served or the timer expires, whichever occurs first. In the latter case, if a packet is in service when the timer expires, the packet service is completed before the server starts the new vacation.

In the model with the *VDTL* service discipline, the visit-time limit for a service period depends on the previous vacation in the following manner. A fixed target cycle time *(TCT)* is specified. The time limit for the next service period is equal to the difference between *TCT* and the length of the last vacation. If the vacation lasts longer than the *TCT*, the queue is not served and the server immediately takes a new vacation.

For analytical tractability, the visit-time limit and the *TCT* are approximated by an Erlangian distribution with a suitable number of time stages [99].

VACATION MODEL WITH CONSTANT TIME-LIMITED SERVICE *(CTL)*. In this vacation model, the service period is controlled by a constant time limit, T_S, which is approximated by a fixed number, J, of time stages. Each stage corresponds to a time interval exponentially distributed with rate $\alpha = J/T_S$. The server keeps track of the number of time stages that have elapsed since the beginning of the service period. At the completion of each packet service, the server starts to serve a new packet only if the number of time stages that have elapsed is less than J. If J or more stages have elapsed, the service period is terminated and the server takes a vacation. If the queue becomes empty before the number of stages elapsed reaches J, the server takes a vacation immediately after the service completion of the last packet.

It is well-known that, as the number of time stages increases to infinity the Erlangian distribution converges to a deterministic distribution [99]. Thus, $J = \infty$ can be used to represent the service period controlled by a constant timer, as happens in the timed-token networks. In [112] the authors show that this convergence occurs rapidly for a moderate value of J. For this reason, throughout this section, the service discipline based on the time stages is still denoted as CTL policy.

To analyze the CTL model, the system is observed at the epochs of packet departures. The system state immediately after the departure epoch of the n-th packet is characterized by $\{N, S\}_n$, where N is the queue length and S is the number of time stages that have elapsed since the beginning of the service period in which the n-th packet is served. Since the arrival process is Poisson and stages are exponential, it follows that $\{\{N, S\}_n, n \in \mathbb{N}\}$ is a Markov chain with the following transition probability matrix P

$$
P = \begin{bmatrix}
B_0 & B_1 & B_2 & B_3 & \dots \\
C_0 & A_1 & A_2 & A_3 & \dots \\
0 & A_0 & A_1 & A_2 & \dots \\
0 & 0 & A_0 & A_1 & \dots \\
\dots & \dots & \dots & \dots & \dots
\end{bmatrix} ,
\tag{6.86}
$$

where the elements of P are matrices with the following size: $B_0 \in \mathbb{R}^{1 \times 1}$, $B_i \in \mathbb{R}^{1 \times \infty}$, $C_0 \in \mathbb{R}^{\infty \times 1}$, $A_j \in \mathbb{R}^{\infty \times \infty}$.

Before describing the internal structure of matrices A_i and B_i the following probabilities are introduced

1. $a_{i,j}$ is the probabilily that

   - i packets arrived, and
   - j time stages elapsed during a packet service time;

2. $d_{i,j}$ is the probability that

   - i packets arrived during a vacation and the following packet service time, and
   - j time stages elapsed at the packet service time;

3. $c_{i,j}$ is the joint probability that

- i packets arrived during a vacation and the following packet service time, and
- j time stages elapsed at the packet service time, given that at least one packet arrived during the vacation.

Given the distributions of service time and vacation time, the probabilities $\{a_{i,j}\}$, $\{c_{i,j}\}$, and $\{d_{i,j}\}$ can be calculated by inverting the related PGFs [112]. In general, this approach may require a significant amount of computation. However, the computation can be simplified considerably when service time and vacation time distributions are of phase-type [125].

By exploiting the above definitions it can be verified that

$$B_0 = \sum_{j=0}^{\infty} c_{1,j} \ , \tag{6.87}$$

$$B_i = \left[c_{i+1,0}, \ c_{i+1,1}, \ c_{i+1,2}, \ \ldots \right], i \geq 1 \ , \tag{6.88}$$

$$C_0 = A_0 \times \mathbf{e} \ , \tag{6.89}$$

$$A_i =
\begin{array}{r}
(S=0) \\
(S=1) \\
\ldots \\
\ldots \\
\ldots \\
(S=J-1) \\
(S=J) \\
\ldots \\
\ldots \\
\ldots
\end{array}
\left[
\begin{array}{cccccccc}
a_{i,0} & a_{i,1} & a_{i,2} & \ldots & a_{i,J-1} & a_{i,J} & a_{i,J+1} & \ldots \\
0 & a_{i,0} & a_{i,1} & \ldots & a_{i,J-2} & a_{i,J-1} & a_{i,J} & \ldots \\
0 & 0 & a_{i,0} & \ldots & a_{i,J-3} & a_{i,J-2} & a_{i,J-1} & \ldots \\
\ldots & \ldots & \ldots & & \ldots & \ldots & \ldots & \ldots \\
\ldots & \ldots & \ldots & & \ldots & \ldots & \ldots & \ldots \\
0 & 0 & 0 & \ldots & a_{i,0} & a_{i,1} & a_{i,2} & \ldots \\
d_{i,0} & d_{i,1} & d_{i,2} & \ldots & d_{i,J-1} & d_{i,J} & d_{i,J+1} & \ldots \\
d_{i,0} & d_{i,1} & d_{i,2} & \ldots & d_{i,J-1} & d_{i,J} & d_{i,J+1} & \ldots \\
d_{i,0} & d_{i,1} & d_{i,2} & \ldots & d_{i,J-1} & d_{i,J} & d_{i,J+1} & \ldots \\
\ldots & \ldots & \ldots & & \ldots & \ldots & \ldots & \ldots
\end{array}
\right], i \geq 0 \ . \tag{6.90}$$

To understand the structure of the A_i and B_i matrices it is necessary to distinguish between the following classes of events

1. $\{N_n > 1 \ and \ S_n < J\}$. In this case the $(n+1)$th packet is served immediately after the previous packet's departure. Assuming that i packets arrived and j time stages elapsed during the $(n+1)$th packet service time, the probability that $\{N_{n+1}, S_{n+1}\} = \{N_n + i - 1, S_n - j\}$ is $a_{i,j}$.

2. $\{N_n > 1 \ and \ S_n \geq J\}$. In this case, the maximum number of time stages for the service period has been reached and the server has to take a vacation before serving the $(n+1)$th packet. Hence, the probability that $\{N_{n+1}, S_{n+1}\} = \{N_n + i - 1, j\}$ is $d_{i,j}$.[1]

3. $\{N_n = 1\}$ and no packet arrives during the $(n+1)$th packet service time. In this case, the queue length is zero at the $(n+1)$th packet departure epoch. As the system becomes empty at that time, the number of time stages that have elapsed during the last service time is irrelevant to the future evolution of the system and the probability of such an occurrence is thus C_0.

4. $\{N_n = 0\}$. In this case, the server takes vacations until the $(n+1)$th packet has arrived. If this packet leaves behind $i > 0$ packets in the system upon departure, there must be $i+1$ arrivals, including the $(n+1)$th packet, during the last vacation plus the $(n+1)$th packet service time. Thus, by definition, the probability of having $\{N_{n+1}, S_{n+1}\} = \{i, j\}$ is $c_{i+1, j}$. which is given by the probability vector B_i for $i \geq 1$. In case the $(n+1)$th packet leaves behind an empty system, the number of time stages elapsed in the last service time also becomes irrelevant. Thus, the probability for this case is B_0.

The structure of the matrix P in (6.86), after the sub-matrices have been appropriately truncated, at say J_{max} (see [112]), is of *M/G/1 -type*.

Using the standard notations already introduced for *M/G/1* system with vacation (see Section 2.1), in [112] the authors show that inequality $\lambda b (T_S + E[V]) < T_S$ gives a sufficient condition for stability for the *CTL* service. Hereafter, $\mathbf{x} = [\mathbf{x}_0, \mathbf{x}_1, ...]$ denotes the steady-state probabilities for the Markov chain, where

- $\mathbf{x}_0$ is the state probability of an empty system; and
- $\mathbf{x}_k = (x_{k,j})$, $k \geq 1$ and $j \geq 0$, is the probability of k packets remaining in the system and that j time stages elapsed at an arbitrary departure epoch.

---

1. Combining the above two cases with $N_n > 1$, the state transition is governed by matrices A_i with $i \geq 0$.

For a stable system, the steady state probability vector $\mathbf{x} = [\mathbf{x}_0, \mathbf{x}_1, \ldots]$ can be calculated by employing the matrix analytic method (see Section 3.3). The queue length distribution of packets, $\{\pi_k, k \geq 0\}$, at the departure epochs can be calculated by observing that $\pi_k = \mathbf{x}_k \times \mathbf{e}$.

Since packets are served one at a time and the arrival process is Poisson, by applying the Burke (see Theorem 2.2) and PASTA (see Theorem 2.3) theorems it follows that $\pi_k = \mathbf{x}_k \times \mathbf{e}$ is the distribution that there are $k = 0, 1, \ldots$ packets in the system at the arrival and arbitrary point in time epochs.

VACATION MODEL WITH VACATION-DEPENDENT TIME-LIMITED SERVICE. This service policy controls the cycle time (i.e., a vacation and the following service period) to be less than a *target cycle time*, T_c. As with the *CTL* model, the target cycle time is approximated by J exponential stages with rate α equal to J/T_c. Each time the server takes a vacation, the server keeps track of the number J_v of time stages that have elapsed since the start of the vacation. If $J_v \geq J$ at the end of the vacation, the server does not serve the queue and immediately takes a new vacation. Otherwise, the visit-time limit (measured in the number of time stages) for the following service period is $J - J_v$. The service period continues until this stage limit is reached or the queue becomes empty, whichever occurs first.

Note that the *VDTL* discipline only captures the interdependence between a vacation and the visit-time limit for the following service period, but it neglects the accumulated lateness due to all previous cycles, as occurs in the timed-token protocols. Nevertheless, as shown in [112] results for the *VDTL* service discipline reveal that the dependence of the visit-time limit and the accumulated lateness of previous cycles may be significant.

As with the *CTL* model, the system is observed at the packet-departure epochs, and the system state immediately after the departure epoch of the n-th packet is described by $\{N, S\}_n$, where N is the queue length and S is the number of time stages that have elapsed since the beginning of the last vacation until the service completion of the n-th packet. Since the arrival process is Poisson and stages are exponential, it follows that $\{\{N, S\}_n, n \in \mathbb{N}\}$ is a Markov chain with the following transition probability matrix P

$$P = \begin{bmatrix} B_0 & B_1 & B_2 & B_3 & \cdots \\ C_0 & A_1 & A_2 & A_3 & \cdots \\ 0 & A_0 & A_1 & A_2 & \cdots \\ 0 & 0 & A_0 & A_1 & \cdots \\ \cdots & \cdots & \cdots & \cdots & \cdots \end{bmatrix} . \tag{6.91}$$

The submatrices of P are defined as follows

$$B_0 = \sum_{j=0}^{\infty} g_{1,j} , \tag{6.92}$$

$$B_i = \left[g_{i+1,0}, \ g_{i+1,1}, \ g_{i+1,2}, \ \cdots \right], i \geq 1 . \tag{6.93}$$

$$A_i = \begin{matrix} (S=0) \\ (S=1) \\ \cdots \\ \cdots \\ \cdots \\ (S=J-1) \\ (S=J) \\ \cdots \\ \cdots \\ \cdots \end{matrix} \begin{bmatrix} a_{i,0} & a_{i,1} & a_{i,2} & \cdots & a_{i,J-1} & a_{i,J} & a_{i,J+1} & \cdots \\ 0 & a_{i,0} & a_{i,1} & \cdots & a_{i,J-2} & a_{i,J-1} & a_{i,J} & \cdots \\ 0 & 0 & a_{i,0} & \cdots & a_{i,J-3} & a_{i,J-2} & a_{i,J-1} & \cdots \\ \cdots & \cdots & \cdots & & \cdots & \cdots & \cdots & \cdots \\ \cdots & \cdots & \cdots & & \cdots & \cdots & \cdots & \cdots \\ 0 & 0 & 0 & 0 & a_{i,0} & a_{i,1} & a_{i,2} & \cdots \\ f_{i,0} & f_{i,1} & f_{i,2} & \cdots & f_{i,J-1} & f_{i,J} & f_{i,J+1} & \cdots \\ f_{i,0} & f_{i,1} & f_{i,2} & \cdots & f_{i,J-1} & f_{i,J} & f_{i,J+1} & \cdots \\ f_{i,0} & f_{i,1} & f_{i,2} & \cdots & f_{i,J-1} & f_{i,J} & f_{i,J+1} & \cdots \\ \cdots & \cdots & \cdots & & \cdots & \cdots & \cdots & \cdots \end{bmatrix}, i \geq 0 , \tag{6.94}$$

$$C_0 = A_0 \times \mathbf{e} , \tag{6.95}$$

where:

1. $f_{i,j}$ is the joint probability that
   - i packets arrived during a vacation and the following packet service time, and
   - j time stages have elapsed during the last vacation and the packet service time;

given that the system is non-empty prior to the vacations;

2. $g_{i,j}$ is the joint probability that

- i packets arrived during a vacation and the following packet service time, and
- j time stages have elapsed during the last vacation and the packet service time,

given that the system is empty prior to the vacations.

Probabilities $\{a_{i,j}\}$ can be calculated by the same method adopted for CTL service. The computation of $\{f_{i,j}\}$ and $\{g_{i,j}\}$ requires the inversion of discrete Fourier transforms [112].

The system operations for the $VDTL$ service discipline are identical to those of the CTL service discipline, except that the counting of the number of time stages that have elapsed starts from the beginning of each vacation under the $VDTL$ service policy. This fact has been accounted for by the probabilities, $\{f_{i,j}\}$ and $\{g_{i,j}\}$, according to their definitions. Since all other operations remain unchanged, replacing the $c_{i,j}$'s and the $d_{i,j}$'s in the transition matrices in (6.87), (6.88), and (6.90) by the $g_{i,j}$'s and the $f_{i,j}$'s, respectively, yields the corresponding vectors and matrices for the $VDTL$ service discipline.

By the similar reasons for the stability condition for the CTL service discipline, a sufficient condition for stability for the $VDTL$ service is given by $\lambda b T_c < T_c - E[V]$.

As with the CTL service discipline, the structure of the matrix P in (6.91), after the matrices have been appropriately truncated, at say J_{max} (see [112]), is of $M/G/1$-type. Hence, the steady-state probabilities $\mathbf{x} = [\mathbf{x}_0, \mathbf{x}_1, \ldots]$ for a stable system can be computed by applying the method described in Section 3.3.

Once $\mathbf{x}$ is known, the queue length distribution and moments of packet response time for the $VDTL$ service discipline can be computed in the same way as for the CTL service discipline. From vector $\mathbf{x} = [\mathbf{x}_0, \mathbf{x}_1, \ldots]$ the distribution of the time the server spends at the queue in each cycle servicing packets (service period) can also be calculated, along with the number of packets at the beginning of the server vacation.

6.3.4 M/G/1 with Vacations and Time-controlled Service Discipline with and without Accumulated Delay (Model 8)

Chiarawongse et al. [35] model timed-token MAC protocols (i.e., FDDI, token bus, etc.) by *M/G/*1 queueing systems with vacation and timer-controlled service disciplines. For this class of models, when the server arrives at the queue, the timer is set to a certain value depending on the policy employed. The server serves the waiting packets until either the system becomes empty, or the timer expires, whichever occurs first. In the latter case, the server completes the service on the customer being served, then takes a vacation.

Three policies for setting the timer value are defined and analyzed in [35]: *time-limited (TL)*, *cycle-time-limited (CL)*, and *cycle-time-limited with accumulated delay (CLL)*. Each policy gives rise to a service discipline for the *M/G/*1 queueing system with vacation.

TL SERVICE DISCIPLINE. Every time the server visits the queue, the timer will be set to a constant value called the server holding time *(SHT)*. Hence, the service period is bounded by the sum of the SHT and the maximum possible service time of a packet. This service discipline can be used to approximate the behavior of synchronous traffic transmissions in FDDI. The approximation is due to the fact that, in FDDI, a packet transmission is not authorized if the SHT value is less than the packet transmission time.

CL SERVICE DISCIPLINE. The timer is set to the difference between a constant parameter called *target cycle time (TCT)* and the last cycle time *(CT)*. If the difference is non-positive, the server does not serve any packets during this visit and immediately begins another vacation. The timer therefore regulates the cycle time by allowing service only when the last cycle time is less than a prespecified value TCT. Asynchronous traffic transmission in the token bus can be modeled using this service discipline.

CLL SERVICE DISCIPLINE. CLL differs from the CL service discipline because it takes into account the *n*-th cycle *delay* $L^{(n)}$ with respect to the TCT value. The delay on the *n*-th cycle, $L^{(n)}$, is defined as $max\,(0, -T^{(n)})$, where $T^{(n)}$

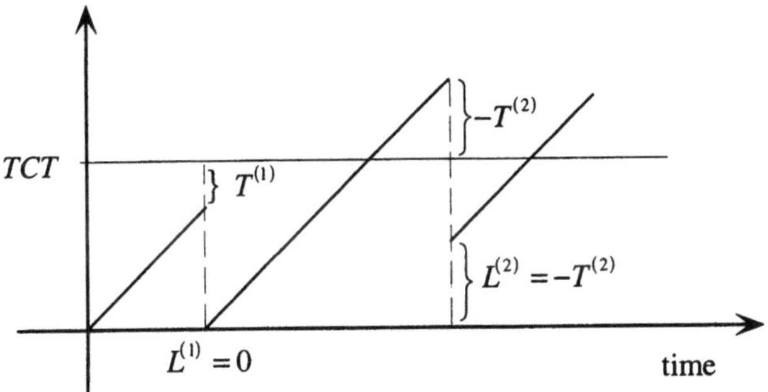

Figure 6.8: CLL cycles

denotes either the delay of server arrival when $T^{(n)} \leq 0$ or the anticipation $T^{(n)} \geq 0$ with respect to TCT (see Figure 6.8). Specifically, $T^{(n+1)} = TCT - CT - L^{(n)}$. If $T^{(n+1)}$ is non-positive, the server does not serve any packets during $(n+1)$th service period and immediately begins another vacation, otherwise $T^{(n+1)}$ is the time limit for this service period.

The CLL policy can be used to model asynchronous traffic transmission in FDDI. In fact, the TCT and delay correspond to the FDDI Target Token Rotation Time and accumulated delay, respectively.

In [35] the $M/G/1$ queueing systems with vacation and TL, CL, and CLL service disciplines are solved by the matrix-analytic technique developed in Section 3.3. Using the matrix-analytic approach, the probability distribution of the number of packets in the system at an arbitrary point in time, and the distribution of the time the server spends at the queue before it takes a vacation are computed.

Assuming that both packet service times $\{B_n, n \geq 1\}$ and the vacation times $\{V_n, n \geq 1\}$ are i.i.d. discrete and bounded random variables (i.e., $B_n \in \{1, 2, ..., b_{max}\}$, $V_n \in \{1, 2, ..., v_{max}\}$) the system behavior is described with the embedding point technique. Specifically, the embedding point is either the *departure instant* of a packet, or the *end of a vacation*. An *embedding interval* is defined as the time interval between two embedding points. Since service times and vacation times can assume only integer values an embedding interval only takes on integer values.

Let Z_i $(Z_i \in \mathbb{N})$ and X_i denote the i-th embedding instant and the corresponding state of the system, respectively. X_i is defined by a triple

$(N, T_1, T_2)_{(i)}$ where N indicates the number of packets in the system, T_1 represents either the elapsed time since the arrival of the server or, if the server arrives late (CL and CLL policies only), the amount of delay, and T_2 denotes the limit on the time the server is allowed to serve packets during the current visit period. For the TL policy, T_2 coincides with the server holding time, SHT. For the CL and CLL policies, T_2 is the difference between the target cycle time, TCT, and the most recently measured cycle time CT. However, for the CLL policy, CT includes the accumulated delay. If the server arrives late, i.e., if this difference is negative, then T_2 is set to zero.

With the above characterization of the status the embedded Markov chain has the following transition probabilities

$$P_{(n, t_1, t_2), (n', t'_1, t'_2)} = P\{X_{i+1} = (n', t'_1, t'_2) \mid X_i = (n, t_1, t_2)\} \quad . \tag{6.96}$$

To derive the transition probabilities it is useful to exploit the following relationship

$$P = \sum_{w=1}^{\infty} P(w) \quad , \tag{6.97}$$

where

$$[P(w)]_{(n, t_1, t_2), (n', t'_1, t'_2)} =$$
$$P\{X_{i+1} = (n', t'_1, t'_2), Z_{i+1} - Z_i = w \mid X_i = (n, t_1, t_2)\} \quad .$$

$P(w)$ can be represented by matrix (6.98), provided that the states of the system are lexicographically ordered; i.e., $(i, j, k) < (i', j', k')$, iff $i < i'$, or $i = i'$ and $j < j'$, or $i = i'$ and $j = j'$ and $k < k'$.

$$P(w) = \begin{bmatrix} B_0(w) & B_1(w) & B_2(w) & B_3(w) & \dots \\ A_0(w) & A_1(w) & A_2(w) & A_3(w) & \dots \\ 0 & A_0(w) & A_1(w) & A_2(w) & \dots \\ 0 & 0 & A_0(w) & A_1(w) & \dots \\ \dots & \dots & \dots & \dots & \dots \end{bmatrix} \quad . \tag{6.98}$$

By introducing the following definitions $\Phi^- = max(-\Phi, 0)$ and $\Phi^+ = max(\Phi, 0)$, each entry $B_n(w)$ is a matrix with the following elements

$$[B_n(w)]_{(t_1, t_2), (t'_1, t'_2)} = \begin{cases} v(w)a_n(w) & \text{if } (t'_1, t'_2) = \Gamma(t_1, w) \\ 0 & \text{otherwise} \end{cases} \quad , \quad (6.99)$$

where

$$a_n(w) = e^{-\lambda w}\frac{(\lambda w)^n}{n!}, \quad n \geq 0 \quad , \tag{6.100}$$

and

$$\Gamma(t_1, w) = \tag{6.101}$$

$$\begin{cases} (0, SHT) & \textit{for policy TL} \\ (0, (TCT - (t_1 + w)))^+) & \textit{for policy CL} \\ ((TCT - (t_1 + w))^-, (TCT - (t_1 + w))^+) & \textit{for policy CLL} \end{cases}$$

To justify equations (6.99) and (6.101) it is enough to remember that since there are no packets in the system at the beginning of the embedding interval, then the embedding interval is a vacation. Hence, $[B_n(w)]_{(t_1, t_2), (t'_1, t'_2)}$ in equation (6.99) gives the probability of a transition from the state $(0, t_1, t_2)$ (beginning of a vacation) to the state (n, t'_1, t'_2) (end of a vacation) when the embedding interval is a vacation of length w.

$\Gamma(t_1, w)$, specified by (6.101), identifies all the possible ranges of values for t'_1 and t'_2. Since the server begins a new cycle at the end of the vacation, for TL and CL policies the only possible value for t'_1 is zero. For the CLL policy the model keeps track of the delay, $L = TCT - (t_1 + w)$, as follows: if $L > 0$, t'_1 is set to L, otherwise t'_1 is set to zero.

For the TL policy, the following setting is always true: $t'_2 = SHT$. For both the CL and CLL policies two possible events may occur: if the server arrived late, t'_2 is set to zero; otherwise $t'_2 = TCT - (t_1 + w)$ and the server begins to serve packets.

By denoting with Θ the set of all possible timer values, it is useful to partition this set into the following two disjoint subsets

- $\Theta^{[B]} = \{ (t_1, t_2) \in \Theta | t_1 < t_2 \}$ the subset of timer values for which the server is allowed to serve a new packet, if any, and

- $\Theta^{[V]} = \{ (t_1, t_2) \in \Theta | t_1 \geq t_2 \}$ the subset of timer values for which the server has to depart on a vacation immediately.

Following the same line of reasoning already seen for $B_n(w)$, it can be verified that each entry $A_n(w)$ is a matrix with the following elements

$$[A_n(w)]_{(t_1, t_2), (t'_1, t'_2)} = \tag{6.102}$$

$$\begin{cases} b(w)a_n(w) & (t_1, t_2) \in \Theta^{[B]} \; (t'_1, t'_2) = (t_1 + w, t_2) \\ b(w)a_{n-1}(w) & (t_1, t_2) \in \Theta^{[V]} \; (t'_1, t'_2) = \Gamma(t_1, w) \\ 0 & \text{otherwise} \end{cases}$$

By defining

$$A_n = \sum_{w=1}^{\infty} A_n(w) \;, \tag{6.103}$$

$$B_n = \sum_{w=1}^{\infty} B_n(w) \;, \tag{6.104}$$

the matrix P of the embedded Markov chain is

$$P = \sum_{w=1}^{\infty} P(w) = \begin{bmatrix} B_0 & B_1 & B_2 & B_3 & \cdots \\ A_0 & A_1 & A_2 & A_3 & \cdots \\ 0 & A_0 & A_1 & A_2 & \cdots \\ 0 & 0 & A_0 & A_1 & \cdots \\ \cdots & \cdots & \cdots & \cdots & \cdots \end{bmatrix} \cdot \tag{6.105}$$

From the matrix-analytic theory (see Section 3.3.2) it is known that P is positive recurrent if and only if $\rho < 1$, and the matrix $\sum_{v=1}^{\infty} v B_v < 1$ is finite. In [35] the authors show that the former condition can be interpreted as follows: if the system is non-empty, the expected number of packets which arrive during an embedding interval (service or vacation) must be less than the expected number of service completions during an embedding interval. The condition $\sum_{v=1}^{\infty} v B_v < 1$ simply states that the average number of packet arrivals during a vacation period must be finite.

For P positive recurrent, the steady-state probability vector $\mathbf{x} = [\mathbf{x}_0, \mathbf{x}_1, \ldots]$ (where each element $\mathbf{x}_n$ is, in turn, a vector of $x_{(n, t_1, t_2)}$ for all admissible values of t_1 and t_2) exists and can be calculated by employing the matrix analytic method (see Section 3.3.2).

Once the vector $\mathbf{x}$ has been calculated, the distribution π_n, $n = 0, 1, \ldots$, of the number of packets that a departing packet leaves in the system can be calculated by introducing the following random variable

$$\xi = \begin{cases} 1 & \text{\textit{if the embedding point is a service completion}} \\ 0 & \text{\textit{if the embedding point is a vacation completion}} \end{cases}$$

Using ξ, π_n can be expressed as follows

$$\pi_n = \frac{P\{N = n, \xi = 1\}}{P\{\xi = 1\}} = \frac{\displaystyle\sum_{t_1, t_2 > 0} x_{(n, t_1, t_2)}}{\displaystyle\sum_{i=0}^{\infty} \sum_{t_1, t_2 > 0} x_{(i, t_1, t_2)}} . \tag{6.106}$$

The second equality follows from the fact that a service is allowed only when the value of T_2 is greater than zero, and the service completion instant, the elapsed time T_1 must be greater than zero, since $b(0) = 0$.

By applying the Burke (see Theorem 2.2) and PASTA (see Theorem 2.3) theorems, it follows that π_n is the steady-state probability that there are $n = 0, 1, \ldots$ packets in the system at a random point in time from which the mean response time of the system can be calculated by applying Little's theorem (see Theorem 2.1).

From vector $\mathbf{x} = [\mathbf{x}_0, \mathbf{x}_1, \ldots]$ the distribution of the time the server spends at the queue in each cycle servicing packets (service period) can also be calculated along with the number of packets at the beginning of the server vacation.

6.4 MODEL 4: DETAILS OF THE COMPUTATION

This section presents the details of the computation required to solve the model presented in Section 6.2.4. Specifically, the following subsections detail each step of the computation presented in Figure 6.5.

STEP 1: ANALYSIS OF $L_i^{(n)}$. When Q_i receives the token at the n-th scanning epoch it transfers the value of the $TRT(i)$ timer to the $THT(i)$ timer, and it continuously transmits asynchronous packets until $THT(i) < T_{Pri}(i)$. Hence, by denoting with $T_{THT(i,j)}^{(n)}$ the r.v. which represents the value of the $THT(i)$ timer after the transmission of the j-th packet at Q_i at the n-th scanning epoch, the following relationships hold

$$P\{L_i^{(n)} = h\} = \qquad\qquad (6.107)$$

$$\begin{cases} 1 - F^{(n)}{}_{THT(i,0)}(T_{Pri}(i)) & h = 0 \\ F^{(n)}{}_{THT(i,j-1)}(T_{Pri}(i)) - F^{(n)}{}_{THT(i,j)}(T_{Pri}(i)) & 0 < h < M_i \\ F^{(n)}{}_{THT(i,M_i-1)}(T_{Pri}(i)) & h = M_i \end{cases}$$

where $F^{(n)}{}_{THT(i,j)}(\ )$ is the distribution function of $T_{THT(i,j)}^{(n)}$.

To obtain an expression for $F^{(n)}{}_{THT(i,j)}(\ )$ it can be noted that

$$T_{THT(i,j)}^{(n)} = TRT(i) + \sum_{h=1}^{j} B_{i,h} \approx C_{i,n} + \sum_{h=1}^{j} B_{i,h} \ . \qquad (6.108)$$

From (6.108), since the service times are i.i.d. random variables which are also independent of $C_{i,n}$, by exploiting the properties of the LST

$$F^{(n)}{}^{*}_{THT(i,j)}(s) = C_{i,n}^{*}(s) \cdot [B_i^{*}(s)]^{j} \ , \qquad (6.109)$$

where $F^{(n)}{}^{*}_{THT(i,j)}(s)$, $C_{i,n}^{*}(s)$ are the LST of $T_{THT(i,j)}^{(n)}$ and $C_{i,n}$, respectively.

From (6.109) with routinary algebraic manipulations, the moments of the distribution of r.v. $T_{THT(i,j)}^{(n)}$ can be obtained, and its distribution is approximated with the two-moments approximation (see Section 4.2.1).

STEP 2: ANALYSIS OF $Tr_i^{(n)}$. The number of packets transmitted by Q_i at the n-th scanning epoch satisfies the following relationship

$$Tr_i^{(n)} = \min\{\hat{N}_i^{(n)}, L_i^{(n)}\} \ , \qquad (6.110)$$

hence its distribution function can easily be derived from the distributions of $\hat{N}_i^{(n)}$ and $L_i^{(n)}$ under the (approximate) assumption that $\hat{N}_i^{(n)}$ and $L_i^{(n)}$ are independent random variables.

STEP 3: ANALYSIS OF $Sp_i^{(n)}$. The service period at Q_i at the n-th scanning epoch $(Sp_i^{(n)})$ is made up of the sum of $Tr_i^{(n)}$ service times. Assuming that the packets' service times are independent of the number of packets to be transmitted,[1] the LST of the length of the n-th service period is

$$Sp_i^{(n)^*}(s) = [B_i^*(s)]^{Tr_i^{(n)}} \ . \tag{6.111}$$

STEP 4: ANALYSIS OF $C_{i,n+1}$. As shown in Figure 6.5, the length of the $(n+1)$th cycle observed by Q_i is

$$C_{i,n+1} = \sum_{j=1}^{i} E_j^{(n)} + \sum_{j=i+1}^{K} E_j^{(n+1)} \ , \tag{6.112}$$

where $E_j^{(n)} = Sp_i^{(n)} + S_i^{(n)}$ is the elapsed time between the n-th scanning epoch at Q_i and the n-th scanning epoch at Q_{i+1}.

A simple approximation of $C_{i,n+1}$ LST can be obtained by deriving the LST of $E_j^{(n)}$ $(E_i^{(n)^*}(s))$ from (6.111)

$$E_i^{(n)^*}(s) = Sp_i^{(n)^*}(s) \cdot S_i^*(s) \ , \tag{6.113}$$

and by assuming that the elapsed times at different stations are independent

$$C_{i,n+1}^*(s) \approx \prod_{j=i}^{K} E_j^{(n)^*}(s) \cdot \prod_{j=1}^{i-1} E_j^{(n+1)^*}(s) \ . \tag{6.114}$$

However, it is generally not easy to determine the distribution function from its LST. In Section 4.2.1 a fitting procedure based on the knowledge of the first two moments is presented to approximate the distribution of $C_{i,n+1}$. In principle, the first two moments of $C_{i,n+1}$ can be obtained from (6.114), but, in this case, the estimate of the second moment would be affected by the independence assumption used to derive (6.114). To improve the estimate of

---

1. Obviously, in the real case, the service times depends on $Tr_i^{(n)}$. In fact, $Tr_i^{(n)}$ has been derived by assuming that the sum of the transmission time of consecutive packets does not exceed the station token holding time.

$C_{i,n+1}$ distribution, in [152] the second moment of $C_{i,n+1}$ is computed by taking into consideration the dependencies between consecutive stations. To this end, the authors propose (when computing the variance of the cycle length) to take into consideration the covariance of the number of packets transmitted in a cycle by each couple of adjacent stations. These random variables would be expected to be negative correlated, as the number of packets transmitted by a station$\{i\}$ increases the cycle length observed by station$\{i+1\}$ and thus reduces the station$\{i+1\}$ token holding time.

Following these hypotheses (i.e., dependencies exist only between consecutive stations) the variance of the cycle length is

$$VAR\,[C_{i,n+1}] \approx \sum_{j=1}^{i-1} VAR\,[E_i^{(\bullet)}] + 2\sum_{j=1}^{K-1} COV\,[E_j^{(\bullet)} E_{j+1}^{(\bullet)}] \quad , \qquad (6.115)$$

where $E_j^{(\bullet)} = E_j^{(n+1)}$ if $j < i$, otherwise $E_j^{(\bullet)} = E_j^{(n)}$, and

$$COV\,[E_j^{(\bullet)} E_{j+1}^{(\bullet)}] = E\,[E_j^{(\bullet)} E_{j+1}^{(\bullet)}] - E\,[E_j^{(\bullet)}] \cdot E\,[E_{j+1}^{(\bullet)}] \quad . \qquad (6.116)$$

Under the additional assumption that only the number of packets transmitted at consecutive stations are correlated, (6.116) can be rewritten as

$$COV\,[E_j^{(\bullet)} E_{j+1}^{(\bullet)}] = E\,[B_j] \cdot E\,[B_{j+1}] \cdot COV\,[Tr_j^{(\bullet)} Tr_{j+1}^{(\bullet)}] \quad , \qquad (6.117)$$

and

$$COV\,[Tr_j^{(\bullet)} Tr_{j+1}^{(\bullet)}] = \sum_{n(j)=0}^{M_j} \sum_{n(j+1)=0}^{M_{j+1}} n(j)\,n(j+1) \cdot \qquad (6.118)$$

$$P\,\{Tr_j^{(\bullet)} = n(j), Tr_{j+1}^{(\bullet)} = n(j+1)\} - E\,[Tr_j^{(\bullet)}] \cdot E\,[Tr_{j+1}^{(\bullet)}] \quad .$$

Finally, to compute the joint probability of the couple of random variables $(Tr_j^{(\bullet)}, Tr_{j+1}^{(\bullet)})$, conditional probabilities are used

$$P\,\{Tr_j^{(\bullet)}, Tr_{j+1}^{(\bullet)}\} = \sum_{h=0}^{M_j} P\,\{Tr_{j+1}^{(\bullet)}|Tr_j^{(\bullet)} = h\} \cdot P\,\{Tr_j^{(\bullet)} = h\} \quad ,$$

and $P\,\{Tr_{j+1}^{(\bullet)}|Tr_j^{(\bullet)} = h\}$ is computed by assuming that in the last cycle station$\{j\}$ has transmitted exactly h packets (see the *conditional cycle* concept in Section 4.1.1)

$$C_{j+1,\bullet \mid (Tr_j^{(\bullet)} = h)} = \sum_{l \neq j} E_l^{(\bullet)} + \sum_{k=1}^{h} B_{j,k} \quad . \tag{6.119}$$

By exploiting (6.119) and by applying STEP 1 and STEP 2 to derive $P\{Tr_j^{(\bullet)} = h\}$ the joint probability $P\{Tr_j^{(\bullet)}, Tr_{j+1}^{(\bullet)}\}$ is finally derived.

STEP 5: ANALYSIS OF $R_i^{(n)}$. The number of packets queued at Q_i at the n-th scanning epoch and not transmitted in the n-th service period satisfies the following relationship

$$R_i^{(n)} = \max\{\hat{N}_i^{(n)} - L_i^{(n)}, 0\} \quad , \tag{6.120}$$

hence its distribution function can easily be derived from the distributions of $\hat{N}_i^{(n)}$ and $L_i^{(n)}$ under the (approximate) assumption that $\hat{N}_i^{(n)}$ and $L_i^{(n)}$ are independent random variables.

STEP 6: ANALYSIS OF $\hat{N}_i^{(n+1)}$. From the knowledge of the distributions of $C_{i,n+1}$ (see STEP 4) the distribution of the number of arrivals at Q_i during the $(n+1)$th cycle $(A_i(C_{i,n+1}))$ can easily be derived

$$P\{A_i(C_{i,n+1}) = h\} = \int_0^\infty \frac{e^{-\lambda_i t} \cdot (\lambda_i t)^h}{h!} P\{t \leq C_{i,n+1} < t + dt\} \, dt \quad . \tag{6.121}$$

Since $\hat{N}_i^{(n+1)} = R_i^{(n)} + A_i(C_{i,n+1})$, the distribution of the number of packets at Q_i at the $(n+1)$th scanning epoch can be obtained from the discrete convolution of (6.121) and the distribution of $R_i^{(n)}$ derived in the previous step.

7 Distributed Queue Dual Bus (DQDB)

The DQDB network has been standardized by IEEE as the main component
of a MAN (IEEE standard 802.6 [90]). A MAN is obtained by interconnecting several DQDB networks via bridges, routers or gateways (see
Figure 7.1).

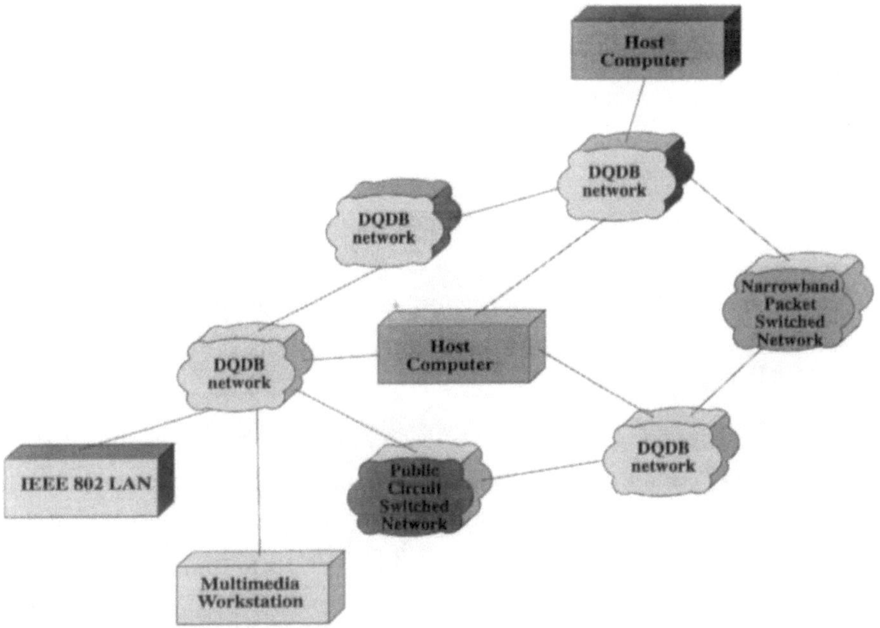

Figure 7.1: MAN configuration

DQDB supports integrated communications by providing   connectionless
data transfer, connection-oriented data transfer, and isochronous communications. These services are obtained by using a multi-access protocol
(DQDB MAC protocol) which manages the sharing of a common transmis-

sion media.

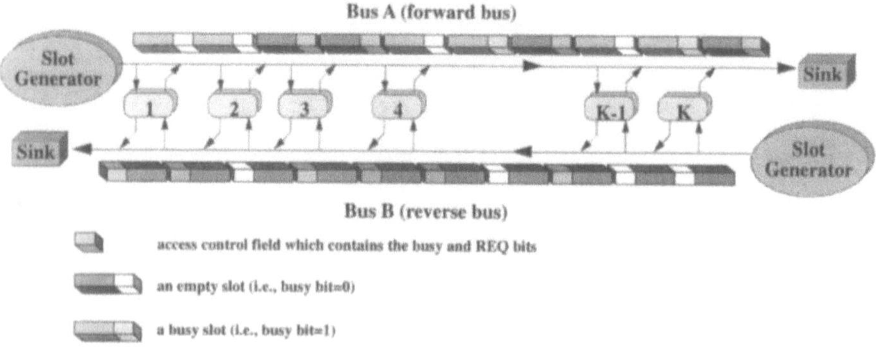

Figure 7.2: DQDB dual bus topology

The basic structure of a DQDB network is shown in Figure 7.2. The network consists of two high-speed, unidirectional buses (named Bus A and Bus B) carrying information in opposite directions, and of a multiplicity of intermediate nodes, addressed by an integer number (*node index*) ranging from 1 to K. The network nodes are distributed along the two buses and each node can transmit information to and receive information from both buses by means of one read and one write connection for each bus. The read connection is logically ahead of the write connection, and thus data are copied from the bus unaffected by the node's own writing. The writing of data is performed as a logical OR operation.

The node at the leading edge of each bus is designated as the head of that bus (*HOB*). Each HOB continuously generates either octets containing management information or fixed-length data slots which propagate along its respective bus. The data slots are used by DQDB to transfer user-generated traffic among the network nodes.

7.1 FUNCTIONAL ARCHITECTURE OF A NODE

As shown in Figure 7.3, the functional architecture of a DQDB node is subdivided into two layers: the DQDB Layer and the Physical Layer. The DQDB Layer can provide three types of service:

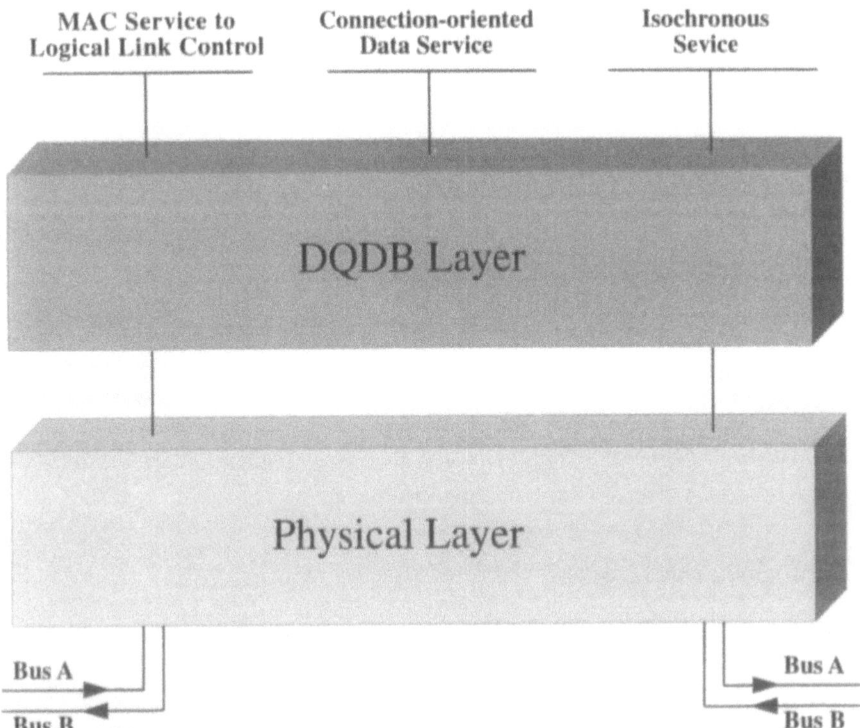

Figure 7.3: Functional architecture of a DQDB node

- the MAC service to the Logical Link Control (*LLC*) sublayer (i.e., connectionless data transfer);
- a connection-oriented data service;
- isochronous service.

To this end, the DQDB Layer enhances the services provided by the Physical Layer at the two Physical service access points. Each access point is associated to a duplex transmission link connecting the node$\{i\}$ to one of its adjacent nodes (i.e., node$\{i+1\}$ or node$\{i-1\}$).

The Physical Layer services enable the local DQDB Layer to transfer or receive an octet to or from a remote DQDB Layer. In addition to these services, the Physical Layer provides timing information and the state of the duplex transmission link associated to each service access point. The Physical Layer is subdivided into three functional blocks: a Transmission System, a Physical Layer Convergence Function and a Layer Management Entity.

The Transmission System defines the interface for accessing the transmission link connecting adjacent nodes, while the Physical Layer Convergence Function adapts the transmission system to meet the DQDB Physical Layer service requirements. Thus the DQDB Layer does not depend on the features of the transmission system. A Physical Layer Convergence Function is defined for standardized transmission systems.

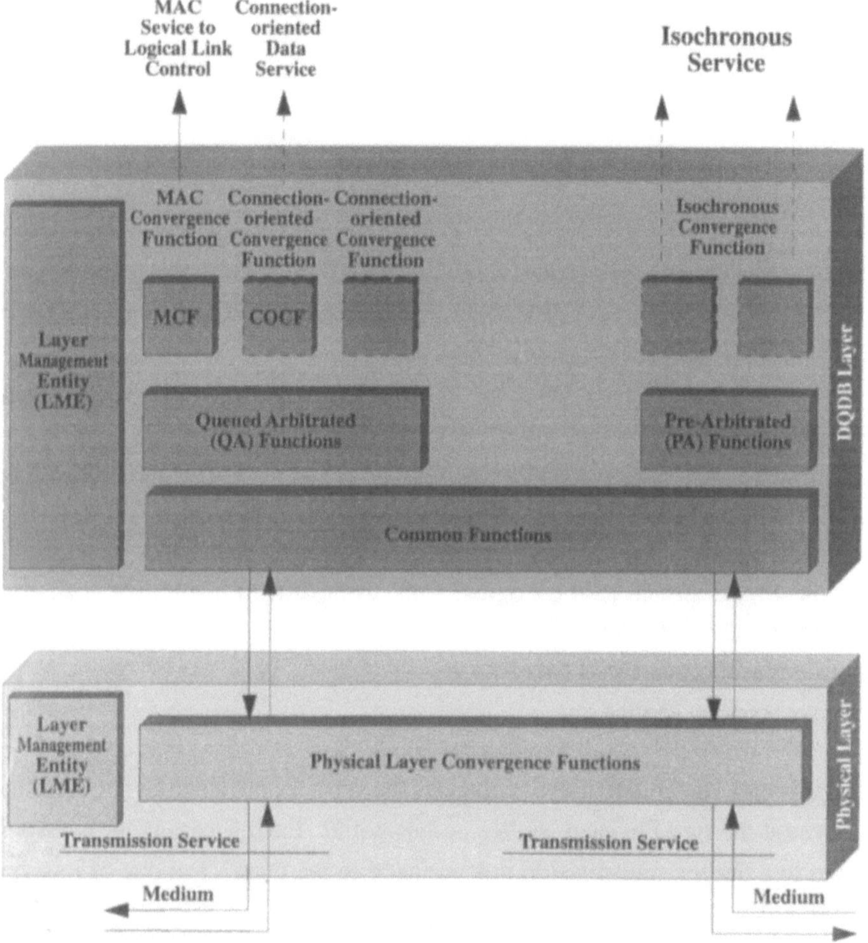

Figure 7.4: Functional architecture of the DQDB Layer

The functional architecture of the DQDB Layer is shown in Figure 7.4. It consists of four principal types of functions: the Common Functions, the Access Control Functions (Queued Arbitrated and Pre-Arbitrated), the Convergence Functions, and the management functions provided by LME.

The services provided by the DQDB Layer rely upon two access methods: Queue Arbitrated (*QA*) and Pre-arbitrated (*PA*). The former is used for the connectionless and connection-oriented data services while the latter supports isochronous service by transferring individual octets of data.

Fixed-length slots of 53 octets are the basic units of data transfer. The first byte in a slot constitutes the Access Control Field (ACF), which is utilized by the nodes in the network to coordinate their transmissions. The remaining 52 octets constitute the segment field. The format of the ACF is shown in Figure 7.5.

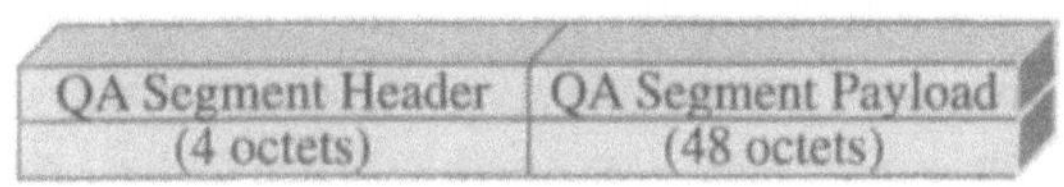

Figure 7.5: Access Control Field

The SL_TYPE bit in the ACF designates the access method for that slot: the slot is reserved for the QA access method if SLOT_TYPE=0, for the PA access method if SLOT_TYPE=1. The BUSY bit and the 3 REQUEST bits are used to implement the QA access method. The PSR bit is used by the slot reuse mechanism ([69], [132]). The two RESERVED bits are for further use.

The segment field (see Figure 7.6) is subdivided into two parts: the segment header (4 octets) and the segment payload (48 octets).

Figure 7.6: Segment field format

The segment header contains four fields (see Figure 7.7). The Virtual Channel Identifier (*VCI*) in the segment header is used to identify the virtual channel to which the segment belongs. The Header Check Sequence (HCS) provides error detection and single-bit error correction for the contents of the segment header. The Payload_Type and the Segment_Priority fields are for future use.

The QA functional entity provides an asynchronous data transfer service with a fixed-length payload (48 octets). The QA functional entity accepts the segment payloads from a convergence function, and generates the appro-

Figure 7.7: Segment header fields

priate segment header to create a QA segment. The segments generated by the MAC Convergence Function have, by default, a VCI field with all bits set to one. The QA segment is queued for access to the dual bus; access is controlled by the DQDB MAC protocol (see Section 7.2.2).

The QA access method only accepts fixed-length data units; since the LLC sublayer produces variable-length messages, a MAC convergence function is required. This MAC convergence function is based on a segmentation and reassembly protocol and manages the addressing of each segmentation unit.

The convergence function for the connection-oriented data service (the *COCF,* in Figure 7.4) is not yet standardized.

The PA function supports the isochronous service by receiving octets from the Isochronous Convergence Function (*ICF*) and transmitting them in the PA segment payloads. Before a PA data transfer can take place a connection must be established. As a result of the connection being established, the PA functional entity will be informed of the VCI value of the segments to be used for this connection, and the offsets of the octets it can use (within each slot with that VCI value). The information about octet offset is necessary because more than one node may share access to a PA slot. The PA segment payload consists of many octets, each of which may be used by a different node. The position of an octet in the segment field is identified by its offset, i.e., the distance measured (in octets) from the beginning of the segment payload field. The PA functional entity in each node maintains, for each VCI, a table indicating which octet offsets within the slot it should use for reading and which for writing. Hence, the PA entity accepts octets from the ICF and writes them into the pre-allocated positions within the payload of PA segments that bear the appropriate VCI value in their segment headers. To receive octets, the PA entity, upon receiving a PA segment with the correct VCI value, will copy octets from the pre-allocated positions within the segment payload.

VCI values are set by the Slot Marking Function in each HOB. PA segments are encapsulated in slots with both the BUSY bit and the SL_TYPE bit set to one (PA slots).

The ICF is not standardized, but its primary function is identified: to provide buffer space to allow for instantaneous rate differences between PA service and isochronous service.

The DQDB Layer management entity (*LME*) manages the local DQDB Layer. To fulfil its purpose it communicates with the DQDB LMEs at other nodes to provide distributed management of the DQDB Layer resources. Messages between DQDB LMEs are carried in the management octets generated by the Slot-marking Function.

The Common Functions block acts as a DQDB Layer relay for the transfer of slots and management information octets between the two service access points of the local Physical Layer. Thus, the Common Functions block allows the QA Functions block and PA Functions block to gain read and write access to the QA and PA slots, respectively. The Common Functions also support functions that operate at some or all of the nodes in the subnetwork. These functions include the head-of-bus function, the Configuration Control Function, and the Message Identifier (MID) Page Allocation Function.

The head-of-bus function is performed exclusively by the node located at the head of each bus. It includes the Slot Marking Function, which is the process of creating empty slots that are to be written onto the bus. The DQDB access mechanism operates by logical OR-writing. The node at the head of each bus must also write appropriate values into the octets reserved for management information.

The Configuration Control Function will be employed at subnetwork start-up to correctly configure the subnetwork. The Configuration Control and MID Page Allocation Functions manage DQDB Layer objects necessary for the communication between nodes on the subnetwork. Hence, these functions are part of the DQDB LME.

7.2 CONNECTIONLESS DATA SERVICE

In this book only the QA mode of operation is investigated. The PA part of the protocol is analyzed in [168], [169], [170].

7.2.1 MAC Convergence Function

The main purpose of the MAC Convergence Function (*MCF*) is twofold: it is responsible for the segmentation of the data received from the LLC sub-layer into fixed-length data units (*segmentation units*) which can be transmitted in the QA slots, as well as for the later reassembly of the data thus transmitted. The DQDB Layer receives the data to be transmitted, the source and destination addresses, and other control information from the LLC sub-layer. This information is collected by the MCF in a variable-length message (*Initial MAC Protocol Data Unit*), which is then subdivided into fixed-length *segmentation units* of 44 octets.[1] These segmentation units are transmitted individually and the message is reassembled at destination. To manage the transmission, the MCF builds a 48-octet Derived MAC Protocol Data Unit (*DMPDU*) by adding header and a trailer to each segmentation unit (see Figure 7.8). The DMPDU is then transmitted, according to the DQDB MAC

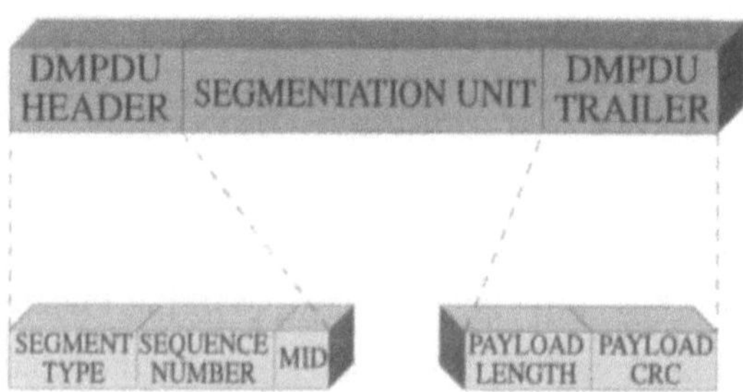

Figure 7.8: Format of a DMPDU

protocol, in the segment payload of a QA slot.

The DMPDU header contains three fields: *segment type*, *sequence number* and message identifier (*MID*). The segment type indicates whether the segment is the first, the last or an intermediate segment of a message.

---

1. Whenever necessary, the last segmentation unit of a message will be padded with trailing zero octets to make up the required length of 44 octets.

The sequence number indicates the position of a segment in a given message. Its value is increased by one after each segment transmission belonging to the same message. The MID field is used for addressing purposes. An MID value unambiguously identifies (for a DQDB network) all the segments belonging to the same message. The allocation of MID numbers to the DQDB nodes is controlled by a distributed algorithm, the *MID Page Allocation Scheme*. The trailer of a DMPDU contains a Cyclic Redundancy Check (*CRC*) field for error detection and a *Payload Length* field which indicates the length of the payload[1].

A DQDB node, upon receiving a DMPDU (from the bus) with the *segment type* field indicating the beginning of a message, will inspect (via its MCF) the destination address field. If the DMPDU is destined for this node, it will copy the segmentation unit of the DMPDU in the reassembling buffer and will also record (in a local table) the sequence number and the MID value from the header of the DMPDU. The next DMPDU belonging to the same message will bear the same MID value and a sequence number one higher than the previous one.

7.2.2 DQDB MAC Protocol

The DQDB Layer utilizes QA slots (i.e., slots with the SL_TYPE bit in the ACF equal to 0) to provide the connectionless service to the LLC Layer. Access to QA slots is controlled by the DQDB MAC protocol.
Two instances of the DQDB MAC protocol exist in each node, one for each bus. Without any loss of generality, bus A is named the *forward bus*, and bus B, the *reverse bus*. With reference to the forward bus and to a given node$\{i\}$, nodes with an index lower than i are called *upstream* nodes, while those with an index higher than i are called *downstream* nodes. Node$\{i\}$ transmits packets destined for its downstream nodes on the forward bus; the reverse bus is used when the destination is an upstream node.

In each node, arriving segments are put in the node's local queue (LQ) for bus A or that for bus B, depending on the destination address. There are two local node queues, one for each bus (see Figure 7.9). In the following discussion, only segments transmission on the forward bus is considered,

---

1. The Payload Length field contains 44 for all segments of a message except, sometimes, the last one.

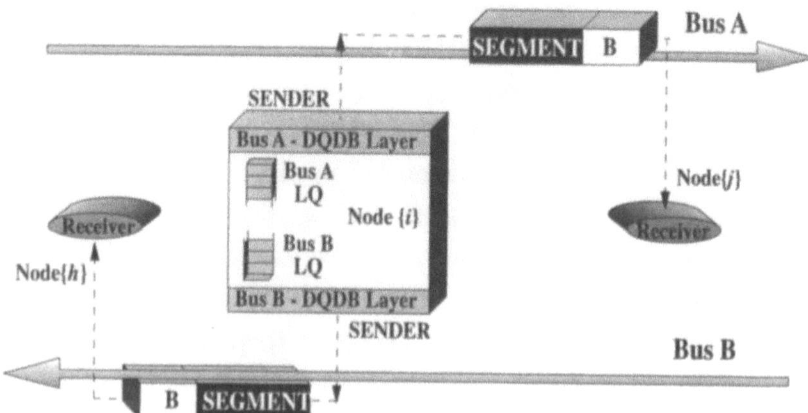

Figure 7.9: Use of the two buses

since the procedure for transmission on the reverse bus is the same.

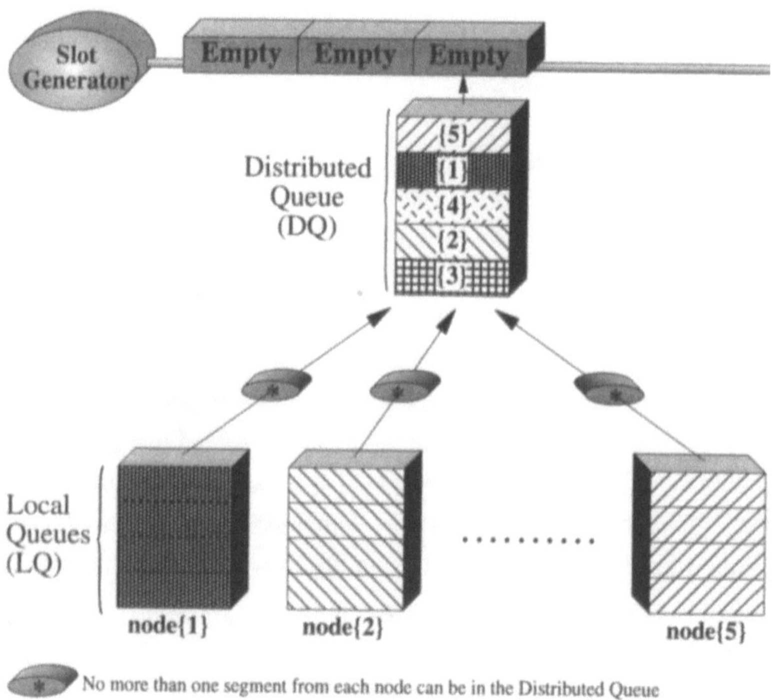

Figure 7.10: DQDB MAC protocol ideal model

The basic idea behind the DQDB MAC protocol is to control node access to a bus in a Round-Robin fashion [100], with a quantum of service equal to the slot time (Δ). Accordingly, the distributed queueing algorithm manages the access to the network by maintaining a FIFO queue, named the *Distributed Queue (DQ)*, in which, only one segment at a time from each node is allowed (see Figure 7.10). A segment arriving at a node which has no segment in the DQ immediately enters the DQ; otherwise, the segment is stored in the appropriate LQ. During each slot time, the segment at the top of the DQ is served[1] (i.e., the segment is transmitted). A node, after transmitting a segment, immediately transfers the segment (if any) at the top of its LQ into the DQ. The HOB generates *empty* slots, i.e., slots in which the BUSY bit is equal to 0. When a node uses an empty slot to transmit a segment, it makes that slot *busy* by setting the BUSY bit in the slot's ACF to 1.

The implementation of the DQ algorithm in a dual-bus topology must take node position into account. Due to their position with respect to the HOB (slot generator), nodes observe the shared resources (i.e., empty slots) in a sequential order.[2] When a node sets a slot's BUSY bit to 1, it logically removes that slot from the set of shared resources.[3]

A node transmits in an empty slot when its segment is at the top of the DQ. Hence, if a node$\{i\}$ observes an empty slot, no segment belonging to its upstream nodes is at the top of the DQ, since these nodes have already observed this slot and left it empty. Therefore, to determine whether it can use an empty slot, node$\{i\}$ only needs to take into account that portion of the distributed queue related to its downstream nodes (see Figure 7.11). In other words, to implement the DQ algorithm, a node$\{i\}$ needs to maintain up-to-date knowledge of how many segments (inserted by downstream nodes) are in the DQ ahead of and behind its own segment. In the DQDB MAC protocol, two counters are used in each node$\{i\}$ for this purpose: the CD counter and the RQ counter, which indicate the number of segments queued ahead and behind the node$\{i\}$ segment, respectively. When a node has no segment in the DQ, only the RQ counter is active, and its value corresponds to the number of segments in the DQ. The following example clarifies the relationship between the DQ as observed by a node and the values on that node's

---

1. The DQ has a deterministic service time equal to Δ.

2. Starting from node$\{1\}$ on bus A and starting from node$\{K\}$ on bus B

3. The Busy bit mechanism guarantees exclusive access to the node which sets the BUSY bit to 1.

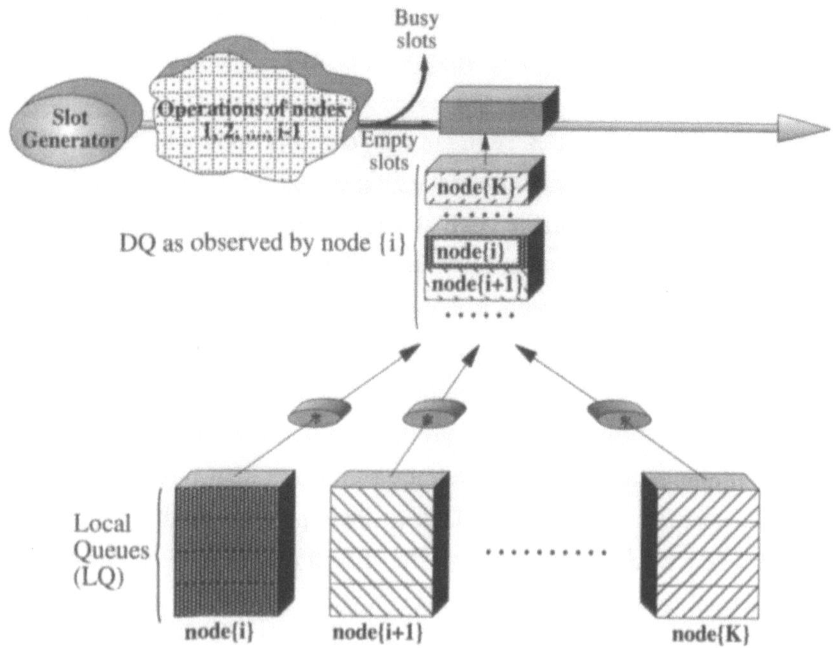

Figure 7.11: DQ as observed by node{i}

CD and RQ counters.

EXAMPLE 7.1 A network with five nodes is considered, and a *snapshot* of the DQ as observed by each of the first three nodes is shown. The DQ as observed by node{1} contains all the segments inserted in the distributed queue by the five nodes. The configuration of the DQ as seen by node{1} is shown in Figure 7.12.

The node{1} segment has one segment queued ahead of it (CD=1) and two segments queued after it (RQ=2). It is worth noting that these two counters indicates the number of segments queued ahead and behind the node{1} segment but they do not indicate which downstream node has inserted a given segment in the DQ (although in the figures this information is given, to assist the reader, by marking each segment with a node number and a pattern).

In the example, node{2} has no packet queued for transmission and, as shown in Figure 7.13, only its RQ counter is used to maintain the DQ status. Node{2} does not observe the node{1} segment in the distributed queue (since node{1} is upstream of node{2}), and thus the value of its RQ counter

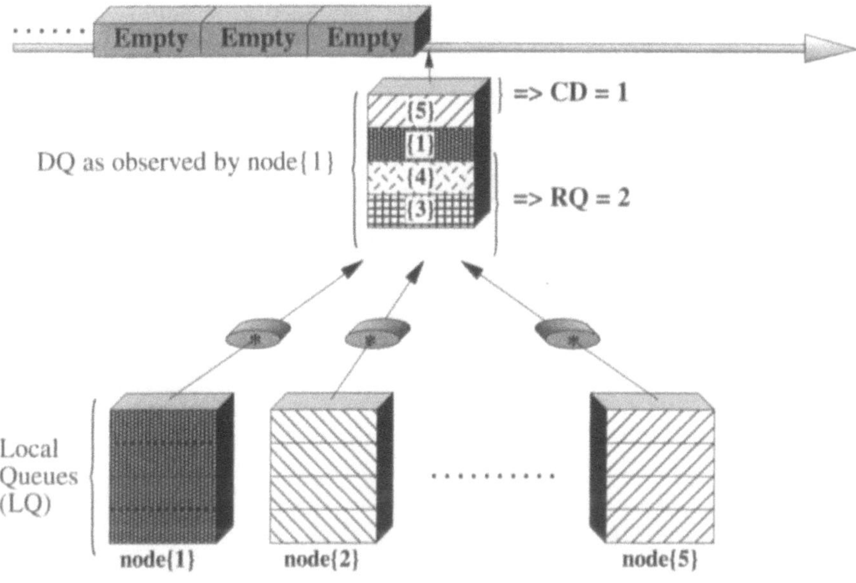

Figure 7.12: DQ as observed by node{1}

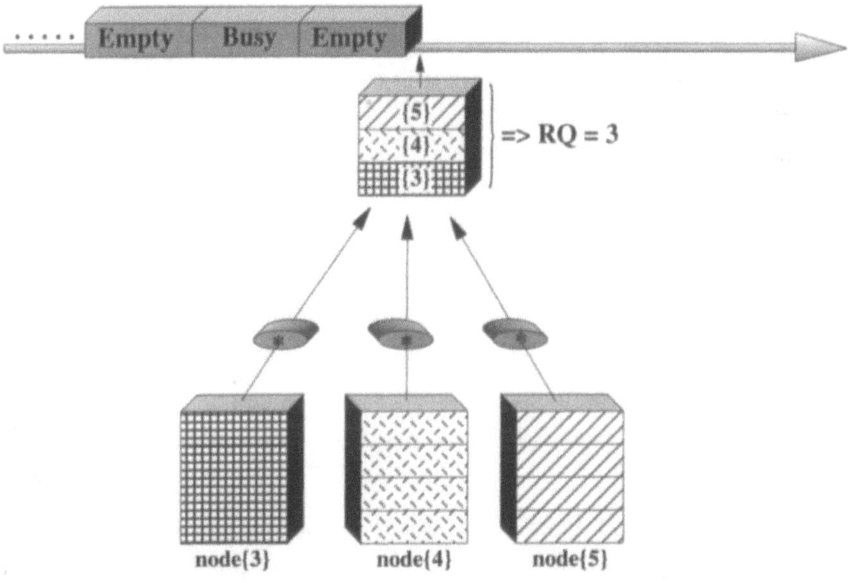

Figure 7.13: DQ as observed by node{2}

is equal to 3.

The last case of this example makes reference to node{3}. This node has a segment in last position in the distributed queue, and only observes the segments in the distributed queue belonging to node{5} and node{4}.

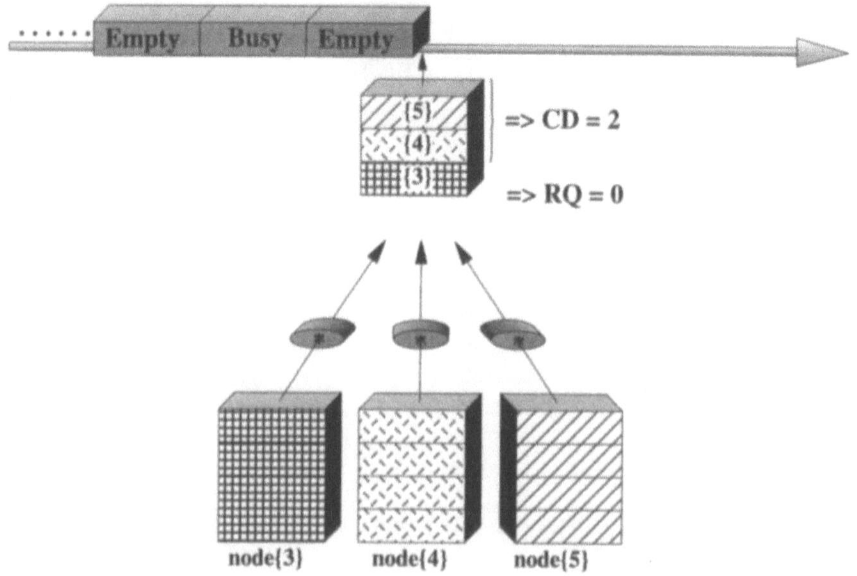

Figure 7.14: DQ as observed by node{3}

Hence, as shown in Figure 7.14, its RQ counter reads 0 and its CD counter reads 2.

$\Diamond$

Due to the relative position of the nodes, a node must inform those upstream of it when it inserts a segment in the distributed queue for a given bus. To this end, DQDB utilizes the three REQ bits contained in the ACF of the slots traveling on the other bus. Three REQ bits are provided to implement a three-level priority scheme to access QA slots. For ease of presentation, only one priority level is considered at first, and thus only one REQ bit is assumed.

When a node{i} inserts a segment in the bus A DQ, it notifies its upstream nodes (i.e., nodes 1,2,..., i-1) of this event by setting the REQ bit to one in a slot on bus B. When each upstream node observes a REQ=1 on the reverse bus, it increases the value on its RQ counters by one.

The procedure for segment transmission on the forward bus utilizes the BUSY bit in the ACF of the slots on the forward bus and one request bit (referred to as the REQ bit in the rest of this subsection) in the ACF of the slots on the reverse bus. The DQ algorithm in each node is implemented with a two-state protocol: the protocol is in the *idle* state when the node has no segment in the DQ; otherwise, it is in the *count_down* state. When it is in the *idle* state the protocol keeps count, via the RQ counter, of the number of outstanding segments (labelled REQ in Figures 7.15-7.18) in the DQ belonging to downstream nodes; the value on the RQ counter increases by

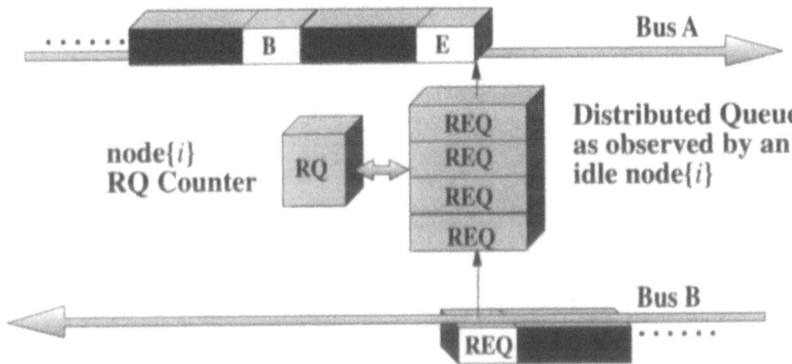

Figure 7.15: A node in the *idle* state

one for each REQ=1 observed on the reverse bus, and decreases by one (down to zero) for each empty slot observed on the forward bus.

When a node in the *idle* state receives a segment for transmission, it enters the *count_down* state and starts the transmission procedure by taking the following actions: *i)* the node transfers the value on the RQ counter to the CD counter, *ii)* it resets the RQ counter to zero, and *iii)* signals to its upstream nodes that it has inserted a segment in the DQ (see Figure 7.16). The operation at step *iii)* is implemented by generating a request which is inserted in a FIFO queue (named the *pending request queue*) while waiting for transmission. The request at the top of the pending request queue is transmitted on the reverse bus by setting REQ=1 in the first slot with REQ=0.

In the *count_down* state, the value on the RQ counter increases by one for each REQ=1 observed on the reverse bus, and the value on the CD counter is decreased by one for every empty slot observed on the forward bus,

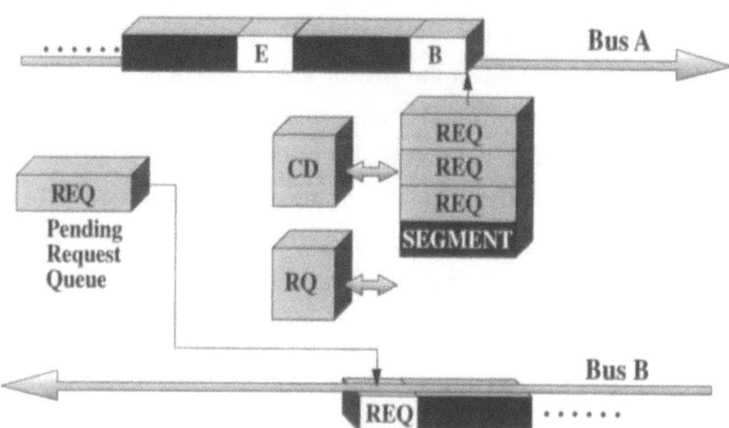

Figure 7.16: Node transition from *idle* to *count_down* state

until it reaches zero (see Figure 7.17). Immediately thereafter, the node

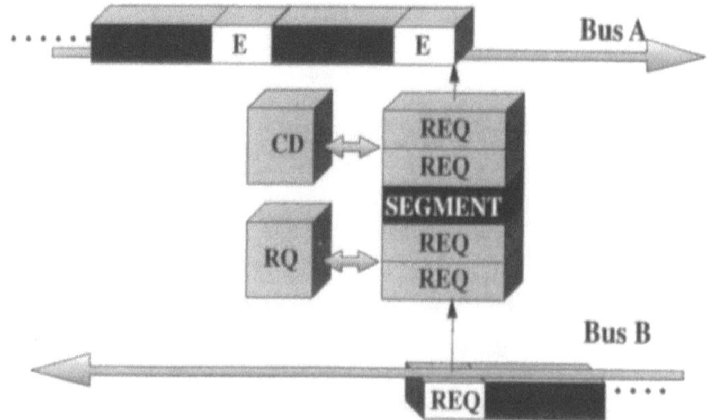

Figure 7.17: Node behavior in the *count_down* state

transmits its segment in the first empty slot on the forward bus (see Figure 7.18).

After transmitting the segment, if its LQ is empty the node returns to the *idle* state; if not the transmission procedure is repeated.

7.2.3 DQDB MAC Protocol with Priorities

In DQDB three priority levels are defined. The priority mechanism is designed to guarantee (in principle) that a segment will always be transmit-

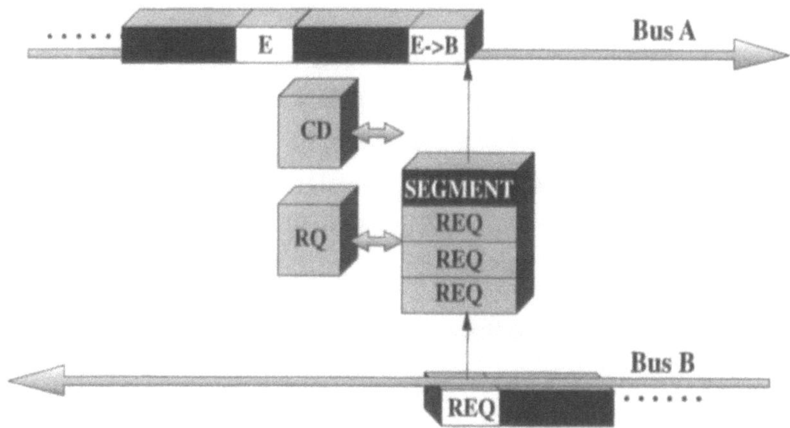

Figure 7.18: Transmission of a segment

ted before all lower priority segments.

To implement the priority mechanism, in each node, a Distributed Queue is maintained for each priority level I (I=0,1,2). To manage the three Distributed Queues in each node, three couples of two counters (RQ_I and CD_I) are utilized. In addition to these counters two other mechanisms are used. The first one controls priorities inside a node, the second one manages priorities between nodes. The former is implemented using a local message named SELFREQ_I. Every time a segment of priority I is inserted in the corresponding Distributed Queue a SELFREQ_I is issued to the other DQs in the same node. This message (SELFREQ_I) increases the value on the counters by one for priority levels J (J<I). More precisely, when a the level J (J<I) state machine is in the count_down state, the SELFREQ_I will increase the value on the corresponding CD_J by one; otherwise the value on the corresponding RQ_J is increased by one.

Priorities between nodes are managed via the REQ_I mechanism. Every time a segment of priority I is inserted in the level I Distributed Queue, a REQ_I bit is set to 1 in a slot on the reverse bus. A REQ_I bit set to 1 will cause an increases of the values on the RQ_J counters by one for priority levels J (J≤I) in the upstream nodes. The three REQ bits in the ACF are used for sending REQ_I, one bit for each priority level. For further details see [90] and Section 7.2.3.

7.3 DQDB PERFORMANCE AND FAIRNESS

As it was pointed out in the previous section, the DQDB MAC protocol was designed to implement a Round Robin policy in a distributed environment. A true round robin policy is well-known to be fair (see Section 1.3), and is achieved when the following times are negligible:

1. the waiting time for setting the REQ bit on the reverse bus;

2. the transfer time of the REQ bit to the upstream nodes; and

3. the transfer time of the empty slot from the head node to the node which issued the reservation.

Unfortunately, the DQDB MAC protocol only approximates a round robin policy because these times in a metropolitan environment are not always negligible. The degree to which delays 1-3 cause a deviation from the round robin scheduling discipline depends on several parameters (e.g., the physical size of the network, the position of the nodes along the buses, the network load, etc.).

In addition, the priority mechanism could significantly deviate from ideal behavior due to propagation delays. The main factor in this deviation is the time to spread reservations (REQ_I) to upstream nodes. As a result, priorities are only guaranteed to be respected between the segments within the same node.

Several metrics have already been devised to measure the degree of fairness of a real network ([37], [39], [72], [164]) depending on the load conditions. In this chapter the DQDB behavior is analyzed in normal conditions (i.e., when the offered load is lower [*underload*] or slightly higher [*overload*] than the medium capacity), and in *asymptotic* conditions (i.e., when each DQDB MAN node is trying to seize all the medium capacity). The following three indices of fairness are usually selected for the DQDB fairness analysis:[1]

1. the *average node access delay* (defined as the weighted sum of the access delays of bus A and bus B) in underload conditions;[2]

---

1. In [37] the MAC delay index was also used to investigate the DQDB fairness.

2. It should be noted that the average node access delay can not be used to measure fairness in overload as it is highly influenced by the buffer size of those nodes which are congested, but barely affected by the buffer size of the other nodes.

2. the *node throughput* in overload conditions;[1] and

3. the *bus throughput* in asymptotic overload conditions.

These fairness indices will be evaluated by assuming that each node generates the same average traffic and hence, in an ideal fair system, each node should have the same *average node access delay, node throughput* and *bus throughput*.

Further insight into DQDB behavior is gained from the analysis of the following performance indices

(i)   the *average bus access delay*, defined as the average time spent by a segment in the LQ plus that in the DQ (i.e., it includes the segment transmission time Δ); and

(ii)   the *percentage of packet loss* on a single bus, defined as $(OL - Throughput)/OL$, where OL is the *Offered Load*; OL and *Throughput* refer to the couple {node index, bus}.

These DQDB fairness and performance indices can be investigated with either simulative or analytical tools, depending on the index of interest. The main results obtained with simulative studies are discussed in this chapter; analytical results on DQDB will be deferred to Chapter 8.

Unless stated otherwise, the DQDB network model for which the simulation results were obtained was characterized by the parameter values reported in Table 7.1, and only one priority level was used. Results concerning several priority levels will be presented in Section 7.4.2.

Table 7.1 Network parameter values

Nodes were spaced equally along the two buses
Capacity of each bus = 150 Mbps
Length of each bus = 100 km
Bus latency (τ) = 400 microseconds
Dual-bus latency (2τ) = 800 microseconds
Total number of nodes (K) = 50

As the network load determines which of the performance and fairness indices are of interest results related to the DQDB analysis are separated here

---

1. This index is useful only in overload conditions; in underload, each node has a throughput equal to its input traffic (and thus DQDB is fair according to this metric).

into two categories: Underload and Asymptotic.

7.3.1 Underload Analysis

This section presents a simulative analysis of DQDB performance and fairness in uncongested situations, i.e., when the aggregate offered load to the network does not exceed the network capacity. The simulative analysis is

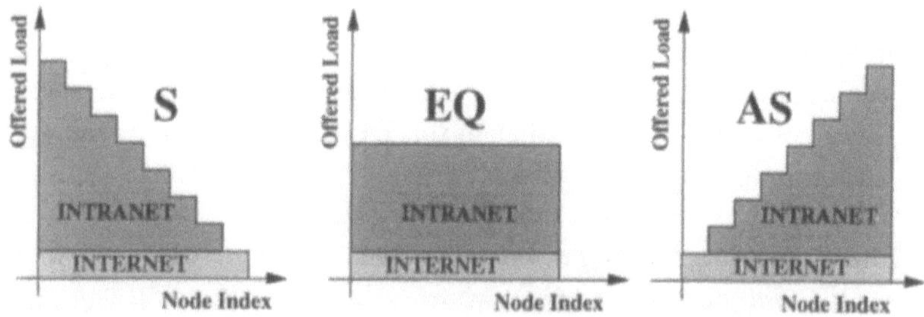

Figure 7.19: Workload types

carried out with several workloads characterized by the parameters *offered load, workload type, message interarrival time,* and *message length.*

The *offered load* (OL) designates the user-generated traffic normalized to the medium capacity. In simulation experiments, no overhead coming from the control part of a slot is taken into consideration. In the following sections, any given value of OL refers to both buses.

The *workload type* defines how the nodes contribute to the offered load. The workload type for each node is split up into two components: one which is kept equal for all nodes, and one which is dependent upon the node index. The rationale behind this choice is that in an internetworking environment, the fixed component represents the internet traffic, while the other represents the intranet traffic (see [40]). The policies selected for the latter component in the simulation experiments are *symmetric (S), asymmetric (AS),* and *uniform (EQ).* In all policies, the traffic generated by each node is the same. The difference between the policies is in the way the destination addresses of segments are generated (see Figure 7.19). In workload type *S,* the destination address is uniformly distributed. In other words, the probability that a node transmits a segment to any other node is constant. As a consequence, the probability that a node transmits a segment on a given bus is proportional to

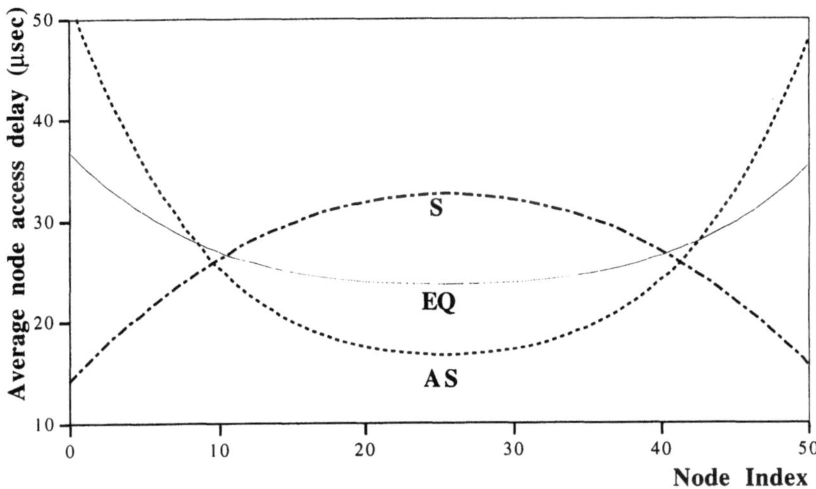

Figure 7.20:   Average node access delay (OL=0.80, C=1, MSG=1)

the number of downstream nodes on that bus. Workload type AS can be derived from workload type S by swapping the traffic on the two buses, i.e., the probability that a node transmits a segment on a given bus is then proportional to the number of upstream nodes on that bus. In workload type EQ, each node transmits the same average number of segments on each bus.

The *message interarrival time* is approximated by an exponential distribution (coefficient of variation, C, equal to 1) or a hyperexponential distribution (C=2 and C=4). The fourth parameter, *message length* (MSG), indicates the number of segments in a message.

Figure 7.20 plots the average node access delay obtained with symmetric, uniform, and asymmetric workload types. The figure indicates that when the network operates with offered load up to 80% of the network capacity, the average node access delay remains in the order of tens of microseconds. Figure 7.20 also indicates a slight degree of unfair behavior, as there are differences in the node access delay even though all nodes have the same load. DQDB unfairness in underload conditions is not very relevant, since all users obtain the transmission bandwidth they need and the differences in the delay experienced are slight.

This unfair behavior is also visible in Figure 7.21, which shows that in underload conditions, while the type of load has some influence on the aver-

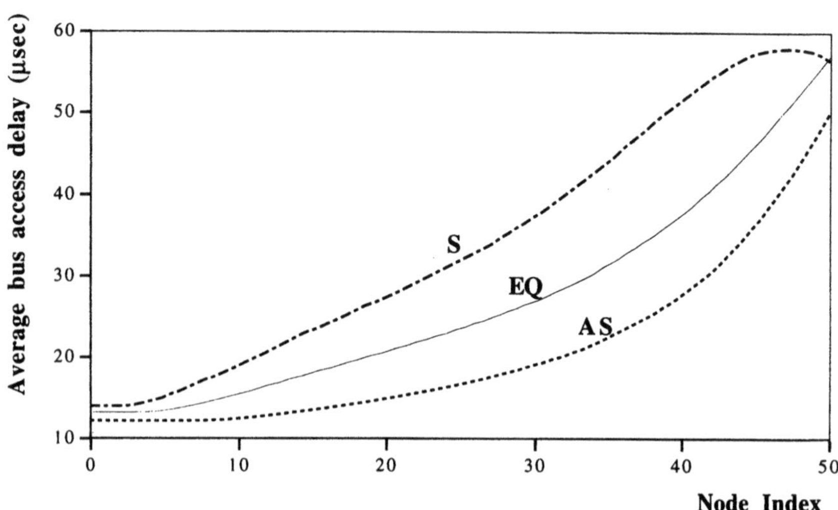

Figure 7.21: Average bus access delay: influence of the workload type (*OL*=0.80, *C*=1, *MSG*=1)

age bus access delay, the DQDB network always gives preferential treatment to nodes close to the head of the bus. These nodes, independently of their load conditions, always experience the lowest bus access delays. In

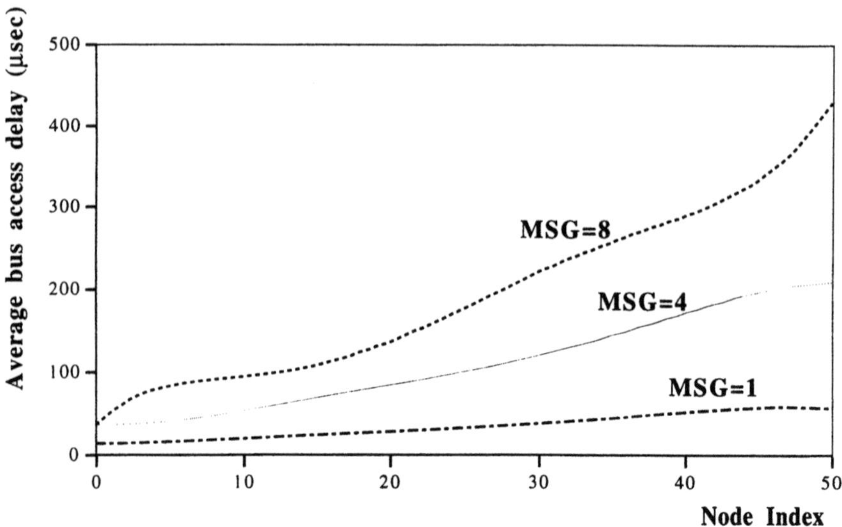

Figure 7.22: Average bus access delay: influence of the message length (workload type=*S*, *OL*=0.80, *C*=1)

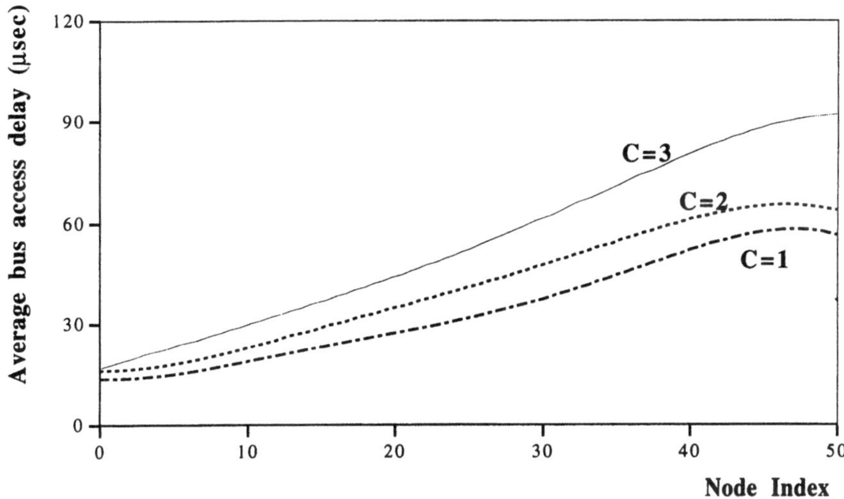

Figure 7.23: Average bus access delay: influence of burstiness (workload type=*S*, *OL*=0.80, *MSG*=1)

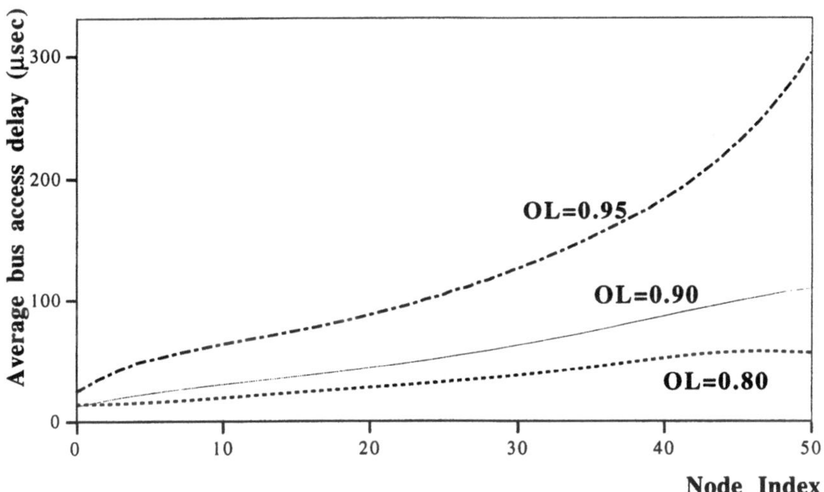

Figure 7.24: Average bus delay: influence of the offered load (workload type=*S*, *C*=1, *MSG*=1)

addition, Figure 7.21 shows that in the worst case, the value of the average bus access delay is about 60 μsec. This value is more than one order of magnitude lower than the access delay (under the same workload conditions and network parameters) in a network with a token passing protocol such as FDDI.

In Figures 7.22, 7.23, and 7.24 the influence of the other workload parameters on the average bus access delay is shown. Figure 7.22 shows the influence of the message length. It must be noted that while this variation in message length may not significantly affect the performance of a token-passing network such as FDDI, it causes a marked increase in the DQDB bus access delay (about one order of magnitude). The increase in the average bus access delay is due to the fact that DQDB is designed to manage a single segment at a time (i.e., DQDB allows each node to have only one packet at a time in the distributed queue). On the other hand, Figure 7.23 indicates that the DQDB average bus access delay does not significantly degrade when the burstiness of the input traffic increases.

As can be seen from Figure 7.24, when the OL increases from 0.80 to 0.95, the increase in the average bus access delay varies with the node index: it is almost negligible for the nodes close to the HOB, but more than fourfold as the node index approaches 50.

Last, it is interesting to show the worst case in underload found in the set of simulative experiments presented above: workload type=*AS*, *OL*=.95, *C*=3, *MSG*=8. The rationale behind this simulative experiment is to under-

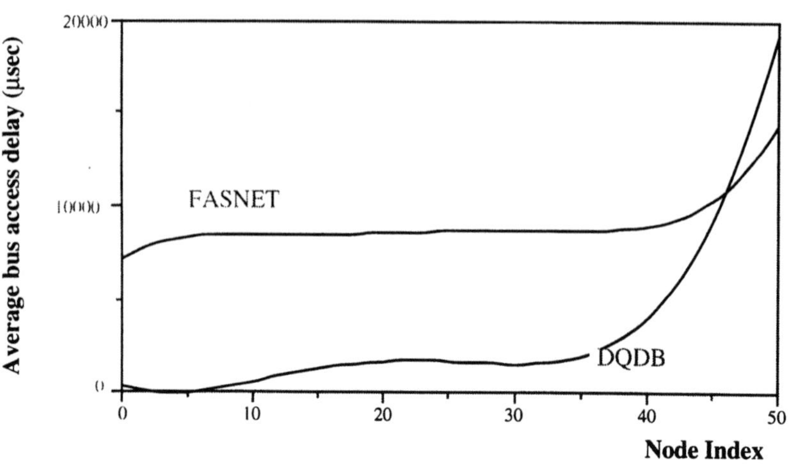

Figure 7.25: A worst case (workload type=*AS*, *OL*=0.95, *C*=3, *MSG*=8)

stand the compound effect of these worst case values on DQDB behavior. Figure 7.25 shows that a worst case scenario may cause a remarkable degradation in the bus access delay figures. For some nodes, access delays are about two orders of magnitude greater then those presented in Figure 7.21. The amount of this degradation is made clear by comparing the DQDB MAC protocol with a token passing MAC protocol under this workload situation and node buffer size. Specifically, this comparison is performed by considering the FASNET MAC protocol [113], as FASNET uses the same dual bus topology and, thus, identifying the same workload situation is straightforward.[1] From this comparison (see Figure 7.25), it becomes evident that even though DQDB is much better than FASNET for almost all nodes, the most unlucky nodes in DQDB experience an average bus access delay comparable to FASNET access delays.

7.3.2 Asymptotic Analysis

The unfair behavior of DQDB in underload conditions does not cause major problems to the network users but, as the load increases, the unfair behavior persists, eventually generating an unacceptable sharing of the network bandwidth.

The DQDB unfairness in bandwidth allocation was demonstrated in [165] by analyzing the nodes' throughput in a simple network configuration with only two active nodes, and the propagation delay between the nodes equal to h slot times (i.e., the propagation delay is $h\Delta$). It is assumed that initially (time $t \leq t_o$) only the upstream node on the forward bus (node$\{1\}$) sends data, and that this node has enough traffic to fill all the slots travelling on that bus. This is allowed by the DQDB protocol since the upstream node, while it is the only active node, does not see any REQ bit set to one on the reverse bus. The downstream node (node$\{2\}$) becomes active at time $t_0 + h\Delta$, i.e., at least h slot times after node$\{1\}$ starts transmitting. When node$\{2\}$ becomes active it immediately switches to the countdown state (with its CD counter reading zero), issues a request on the reverse bus, and

---

1. The only free parameter to set in the FASNET MAC protocol is the maximum number of segments that can be transmitted when the token is captured by a FASNET station P_{max}. For this comparison $P_{max} = 150$ was adopted [30].

then waits for an empty slot to appear on the forward bus. Figure 7.26 shows the network state when node{2} becomes active, with the propagation delay between the two nodes equal to three slot times. In this scenario, node{1}

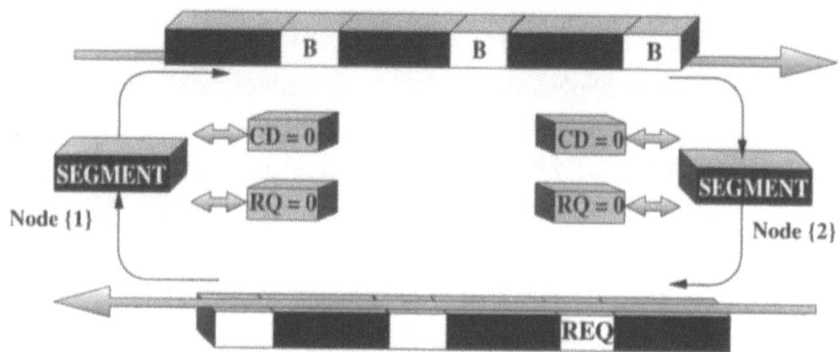

Figure 7.26: Network state at time $t_0 + h\Delta$

only leaves an empty slot in response to a node{2} request. Obviously, it takes h slot times before the upstream node can observe a request from node{2}. When the upstream node reads the node{2} request, it allows a slot to remain empty. After a propagation delay of $h\Delta$, the empty slot reaches node{2}, which transmits a segment. Now a new node{2} segment is authorized to enter the distributed queue, and a request is sent on the reverse bus. The network state at time $t_0 + 3h\Delta$ coincides with the network state at time $t_0 + h\Delta$; thus, the downstream node will only transmit one segment every $2h$ slots. For example, with a distance of 30 km between the two active nodes, $h = 50$ slots and, consequently, the downstream node can use about one slot out of a hundred.

Hence, in the scenario shown in Figure 7.26, the throughput achieved by the two nodes depends upon the time instants at which they become active, and the distance between the nodes. In the following discussion, results computed via simulation in more complex configurations with several nodes operating in asymptotic conditions are presented. Each throughput curve is associated with a different initial scenario: *forward*, *backward* or *random*. In the first scenario (forward), the nodes are sequentially activated from the head to the end node, while in the second (backward), nodes are activated in the reverse order. In the last one (random), nodes are activated randomly according to a uniform distribution. The time between the activa-

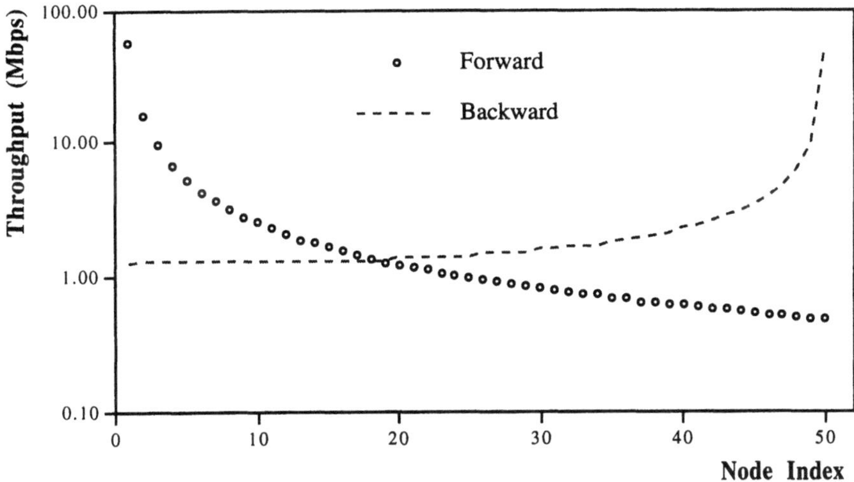

Figure 7.27: Node throughput analysis: forward and backward scenarios

tion of two consecutive nodes is equal to the dual bus latency $(2\tau$; see Table 7.1). Therefore, in all scenarios, all nodes are operational after an elapsed time equal to $(50 \cdot 2\tau)$ μsec., after which the collection of statistics begins. In the figures, only results related to steady-states are presented. Figure 7.27 shows that, in the forward and backward scenarios, the privileges (in terms of throughput) that a node acquires at the time of activation will not be completely lost in the steady state. It is clear from the curves that the later a node is activated, the lower its throughput will be in the steady-state. These curves show two opposite forms of unfair behaviors of DQDB. The causes of such unfairness have been studied in [156] and [37], and can be summarized as follows: a node, as soon as it becomes active, observes a stream of busy slots in the forward scenario, or a stream of REQ=1 in the backward scenario. In both cases, the mechanisms (REQ and Busy bit) used by a node to get permission to transmit can only partially reduce the initial, unbalanced situation.

To make the analysis more complete, the DQDB steady-state condition was also studied, starting from some randomly selected initial states. Figure 7.28 plots the throughput on the forward bus vs. node index in the steady-state condition for three different, random, initial states. It can easily be observed that there is unfair behavior in these cases, too, which is high-

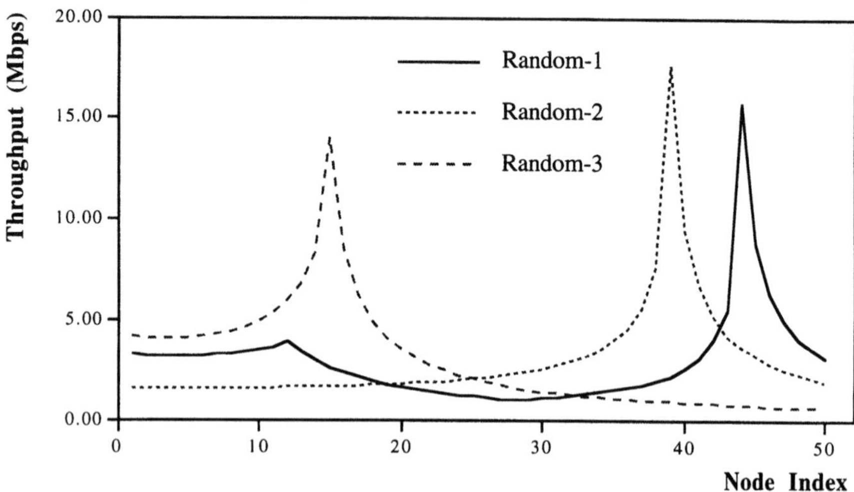

Figure 7.28: Node throughput analysis: random scenarios

lighted by the presence of peaks. Furthermore, in these experiments as well, those nodes activated first acquire privileges that will not be completely lost when the steady-state condition is reached. In particular, this tendency seems (at least in these experiments) to be more marked for the nodes located towards the end node.

Finally, it is interesting to investigate the influence of an increase in medium capacity[1] on DQDB fairness. Simulative results (obtained with the forward activation scenario) are reported in Figure 7.29.

Figure 7.29 reveals an amplification of the difference between the percentage of bandwidth obtained by the most and least favored nodes as the medium capacity shifts from 150 Mbps to 1.2 Gbps. This amplification was somewhat expected, as DQDB bandwidth-sharing is driven by the control information carried in the header of each slot. The number of slots on each bus at 1.2 Gbps is eight times higher than that at 150 Mbps, and this gives rise to an increase in information which has already been sent by a node but not yet received by all the others. Thus DQDB bandwidth-sharing, and thus DQDB fairness, depends on the ratio between the bus latency and the slot duration. In the literature, this ratio is often represented by a [104].

---

1. This is equivalent, to increasing the bus length while maintaining the same bus capacity.

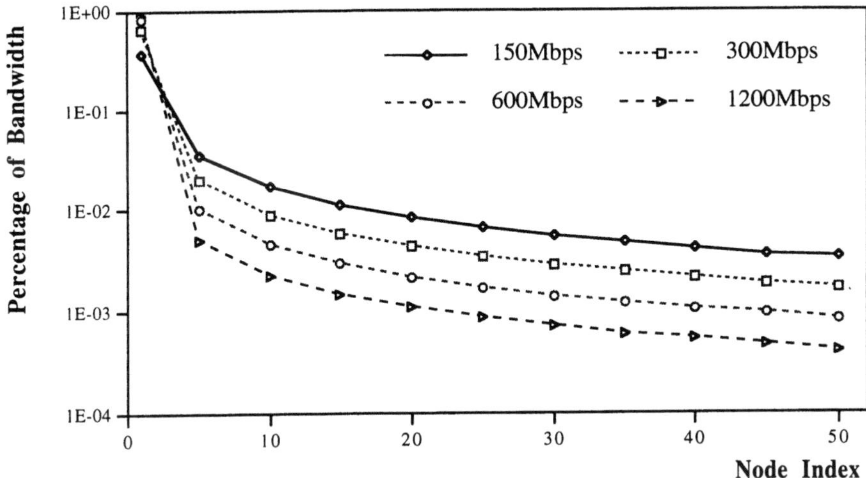

Figure 7.29: Node throughput analysis for several channel capacities

The results presented in this section indicate that two aspects of the quality of service provided by DQDB (i.e., access delay and bandwidth allocation) are strongly dependent on a node's position in the network. For this reason an additional mechanism, named the Bandwidth Balancing mechanism (*BWB*), has been added to the DQDB MAC protocol to improve its fairness.

7.4 THE BANDWIDTH BALANCING MECHANISM

DQDB unfairness in saturated conditions arises because the protocol allows a node to use every unused slot (i.e., an empty slot not reserved by a downstream node) on the bus. To overcome this problem, the *Bandwidth Balancing (BWB)* mechanism ([80]) has been added to the DQDB standard.

The BWB mechanism follows the basic DQDB protocol (see Section 7.2.2), except that a node can only take a fraction of the unused slots. Specifically, the BWB mechanism limits the throughput of each node to M times the unused bus capacity (U); this limit is defined as the *control rate*. Nodes with a demand lower than the control rate (*non-rate-controlled* nodes) get all the bandwidth they want. Therefore, if λ_i represents the node$\{i\}$ offered load, the node$\{i\}$ throughput, γ_i, satisfies the following relationship [80]

$$\gamma_i \ = \ \min\,[\,\lambda_i,\, M \cdot U\,] \ = \ \min\,[\,\lambda_i,\, M \cdot (1 - \textstyle\sum_{m=1}^{K_c} \gamma_m)\,] \ . \qquad (7.1)$$

This scheme is fair; in fact if there are K_c rate-controlled nodes, and S is the aggregate offered load of the non-rate-controlled nodes, all rate-controlled nodes get the same bandwidth, $\gamma \ = \ M \cdot (1 - S) / (1 + M \cdot K_c)$.

To implement the *BWB* mechanism, the DQDB protocol uses a counter, called *BWB_CTR*, to keep track of the number of transmitted segments, and a threshold parameter named *BWB_MOD*. Once the BWB_CTR reaches the BWB_MOD value, this counter is cleared and the value on the RQ counter is increased by one. The value of BWB_MOD can vary from 0 to 63. The value 0 means that the BWB mechanism is disabled, and hence the basic DQDB MAC protocol presented in the previous section is in operation. It must be noted that the BWB_MOD parameter value may vary from node to node.

A complete overview of the DQDB MAC protocol is provided by the protocol state machine shown in Figure 7.30. Here, events starting with "IN=" denote events which trigger transitions in the state machine (e.g., a REQ arrival, an empty slot, etc.). The action induced by an "IN=" event is written below the event itself. The event "IN=BWB_Signal" indicates that the BWB machine has reached the threshold value (i.e., BWB_MOD), while the action "BWB_UP Signal to BWBM" indicates that the value on the BWB counter has to be increased by one.

7.4.1 Performance of DQDB with one Level of Priority and the BWB Mechanism

The effectiveness of the BWB mechanism has been studied extensively via simulation ([39], [156], [37], and [13]). In the following discussion, selected results obtained with BWB_MOD=8 (unless stated otherwise) are presented.

Figure 7.31 shows that the BWB mechanism has almost no effect on the access delay in underload, as was to be expected.

In overload conditions, some nodes achieve a throughput lower than their offered load. Thus, for these nodes, the congestion of the local node queue increases greatly, and this generates two effects: first, some segments

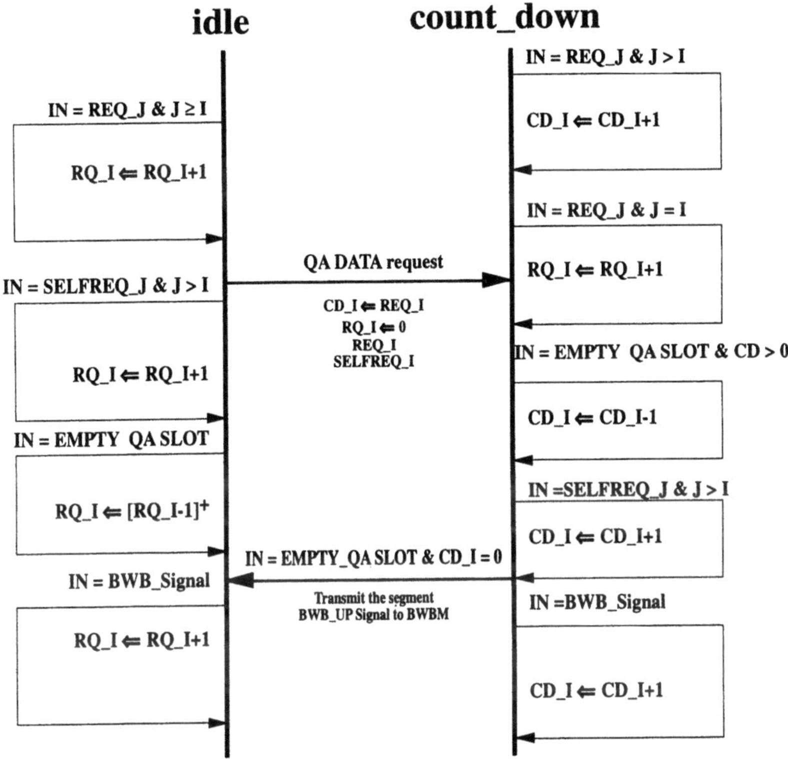

Figure 7.30: DQDB protocol state machine

are lost and second, the average bus access delay depends almost linearly on the LQ capacity.

Figures 7.32 and 7.33 show the average bus access delay in overload when the BWB mechanism is disabled and enabled, respectively. In these cases there are marked differences and this implies that there should be significant differences from the packet loss standpoint. In fact, by examining Figures 7.34 and 7.35 (which show the packet loss and the throughput vs. node index when the offered load is equal to 1.20), it can be seen that the BWB mechanism is really effective: nodes with an offered load lower than 1/50 of the channel capacity (i.e., 7057 packets/sec in this case) are protected against packet loss only if the BWB is enabled! Obviously, with BWB enabled, 1/50 of the channel capacity is a lower bound on the guaranteed throughput on a single bus. This bound is achieved when all nodes are rate-controlled. In the case shown in Figure 7.35, only the first 22 (approxi-

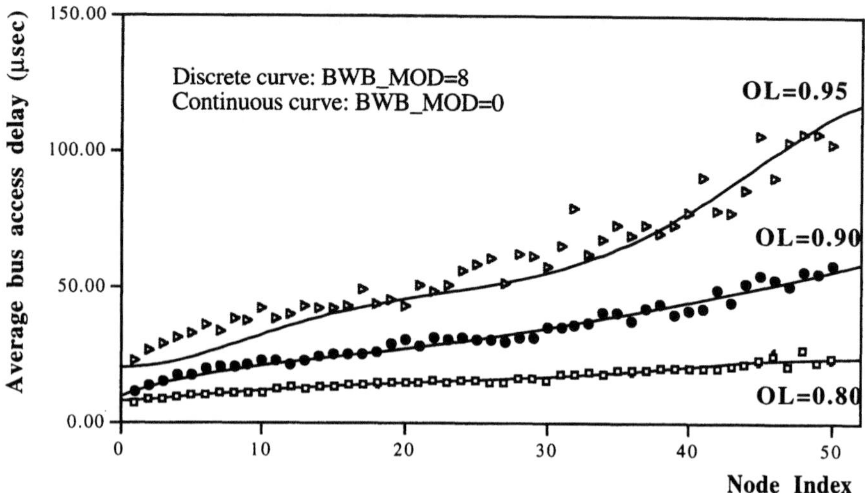

Figure 7.31: Average bus access delay in underload

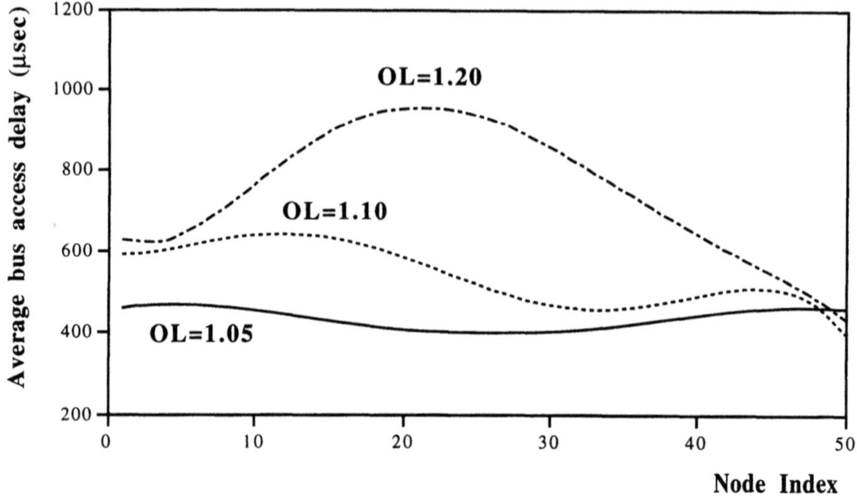

Figure 7.32: Average bus access delay in overload (BWB mechanism disabled)

mately) nodes are rate-controlled, and they achieve a throughput of about 9600 packets/sec (see equation (7.4) in Section 7.5).

Using the activation scenarios defined in Section 7.3.2, but with the BWB mechanism is enabled, one finds that DQDB always reaches a steady-

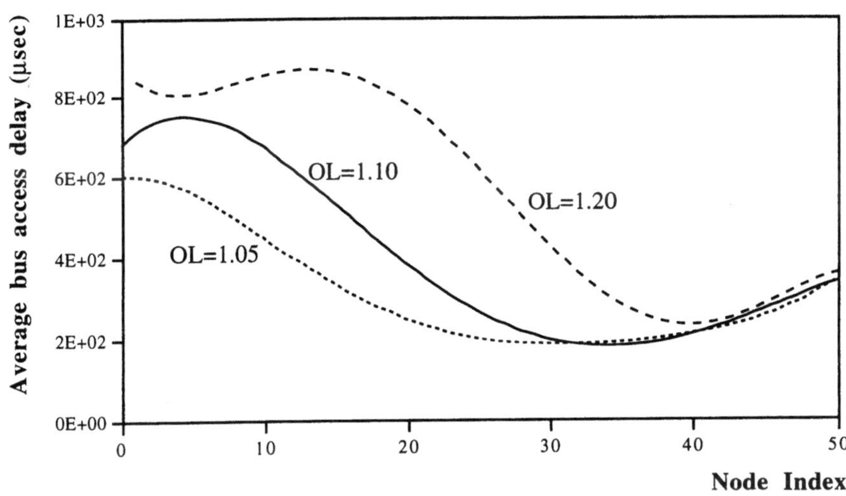

Figure 7.33: Average bus access delay in overload (BWB_MOD=8)

state condition where the bandwidth is equally shared among the nodes [37].

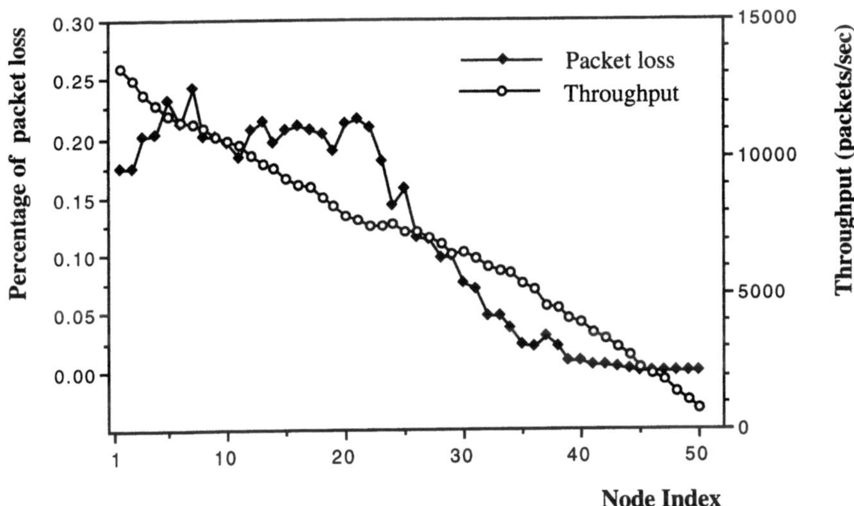

Figure 7.34: Packet loss vs. throughput (BWB_MOD=0)

This steady-state condition is achieved only after a significant amount of time (*transient time*). The duration of this transient time depends upon several factors: the initial state, BWB_MOD, and the number of slots travelling on the two buses [37]. Figure 7.36 shows, for the forward activation scenario

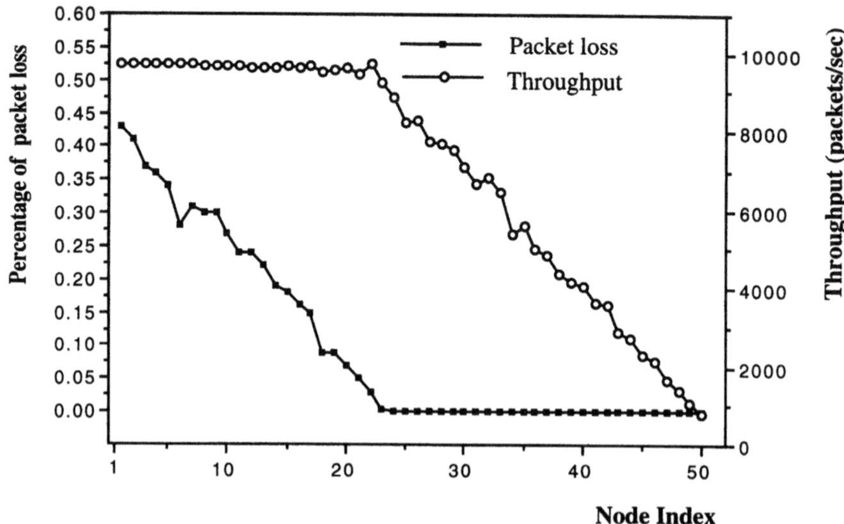

Figure 7.35: Packet loss vs. throughput (BWB_MOD=8)

and BWB_MOD=8, the evolution of the bandwidth sharing between nodes measured in different consecutive periods, each period being equal to $50 \cdot 2\tau$ μsec. It is evident from the figure that the steady state is entered in

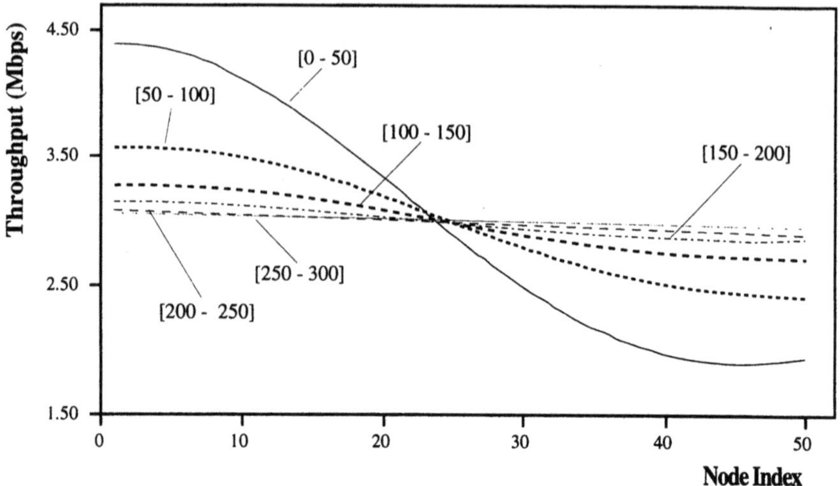

Figure 7.36: Transient behavior (forward activation, BWB_MOD=8)

the period $[250 - 300] \cdot 2\tau$ μsec, which corresponds to a transient time of at least 200msec. As shown in Figure 7.36, the transient time is not negligible, and thus it is interesting to analyze its dependency on the BWB_MOD value.

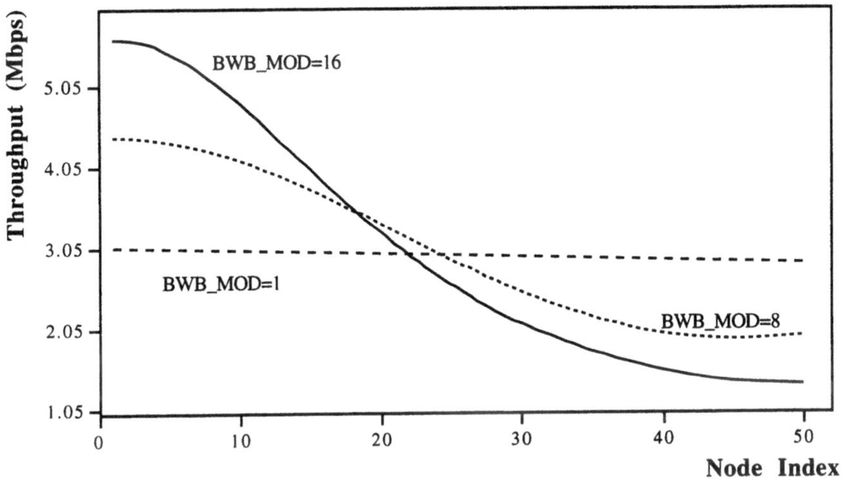

Figure 7.37:   Throughput in the period $[0 - 50] \cdot 2\tau$ for different BWB_MOD values (forward activation)

Figure 7.37 shows the throughput, estimated in the period $[0 - 50] \cdot 2\tau$ μsec, achieved by each node in a DQDB network for the forward activation scenario and for different values of the BWB_MOD parameter. As can be expected, the lower the BWB_MOD parameter, the lower the departure of the transient behavior from the steady-state behavior.

7.4.2 Performance of DQDB with Several Levels of Priority and the BWB Mechanism

In this section, the effectiveness of the DQDB priority mechanism in asymptotic conditions is investigated. The rationale behind this choice is that it highlights the major problems with the DQDB priority mechanism which still remain unsolved by the IEEE 802.6 standardization committee.

Since in asymptotic conditions the SELFREQ_I mechanism inside each node of the DQDB network is responsible for blocking all the lower-priority traffic produced by the node itself, only the highest-priority state machine can access the medium. Hence, each node only transmits segments of its highest-priority traffic.

The major problems with the DQDB priority mechanism can be observed by analyzing the simple network configuration shown in Figure 7.38. In the network there are only two active nodes, and these are transmitting traffic with different priorities. Table 7.2 shows the steady-state

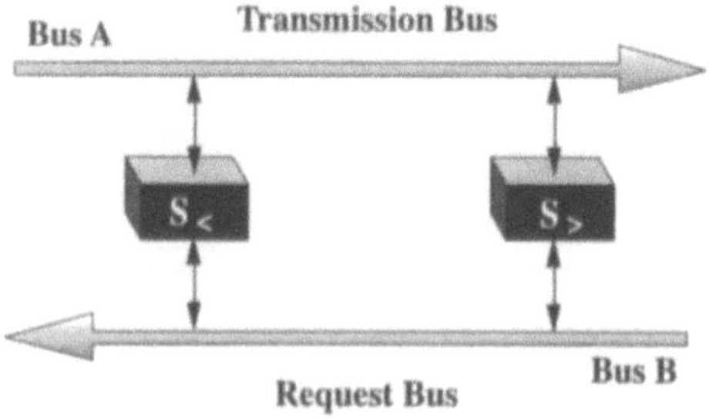

Figure 7.38: Two active nodes with different priorities

results, with forward and backward activation scenarios, for different values of the BWB_MOD parameter and a distance between the two nodes of about 90 km. $S_<$ and $S_>$ are, respectively, the nodes at lower and higher priority.

Table 7.2 Asymptotic bandwidth sharing (Mbps)

	BWB_MOD	$S_<$	$S_>$
Forward Activation	0	149	1
	1	50	50
	8	70	70
	16	72	72
Backward Activation	0	0	150
	1	50	50
	8	70	70
	16	72	72

The results obtained using BWB_MOD=0 are those with the BWB mechanism disabled, and they show the dependence between the throughput achieved at each priority level and the activation condition (see also [156]). Although an ideal result was obtained with backward activation, with forward activation the access mechanism is responsible for an inversion of the priority rights of the nodes. These results can be explained easily, considering that when the node transmitting first is the one at higher priority, the node at lower priority (but in a favorable position) observes a never-ending train of high-priority REQs on bus B, sent by the station $S_>$. As a result, once $S_<$ is activated, it never gains access to the transmission bus because its RQ counter indicates outstanding REQs of a higher priority than the segment $S_<$ has to send. The paradoxical steady-state configuration produced by forward activation is due to the propagation delay of the high-priority REQs on bus B. When the higher-priority node is activated, it experiences a train of busy slots used by the lower-priority node. In this situation, to obtain an empty slot, the node $S_>$ can send one and only one REQ, corresponding to the first segment in its local queue which has to be transmitted, to the upstream node. Therefore, in spite of the priority levels, the node at lower priority can access all the empty slots except that one which is reserved for the higher-priority node.

The results obtained using the BWB mechanism are those with BWB_MOD>0. In all the experiments, DQDB behavior seems to be absolutely independent of the priority of the two nodes. This is because the maximum throughput a node may achieve is BWB_MOD times the unused capacity. As shown in equation (7.5), increasing the BWB_MOD is the only way to increase a node's bandwidth.

7.5 DQDB MAC PROTOCOL CAPACITY

In this section the limitations on the aggregate throughput imposed by the DQDB MAC protocol are analyzed by investigating the MAC protocol capacity (see Section 1.1). Since the utilization factor may depend on the number of active nodes and on their contribution to the offered load, three indices are used: $\rho_{single}(i)$, ρ_{max}, and ρ.

The capacity indices for the DQDB MAC protocol when the BWB mechanism is disabled are equal to 1, as there is always at least one node with a non-empty LQ which will use all the non-reserved bandwidth. On the other hand, the BWB mechanism provides a fair bandwidth sharing by sparing a portion of the bus capacity, and the capacity indices depend on: the value of BWB_MOD in each rate-controlled node, number of rate-controlled nodes, and the aggregate offered load of non-rate-controlled nodes.

The relationship between the network parameters and the throughput achieved by each node when the BWB mechanism is enabled can be investigated by considering a network with channel capacity C, and K active stations. In addition, it is assumed that K_c $(K_c \leq N)$ stations are rate-controlled, while S is the aggregate offered load of the non-rate-controlled stations. Under these hypotheses, the throughput γ_i achieved by a rate-controlled station i is, according to (7.1),

$$\gamma_i = \left[C - \left(S + \sum_{j=1}^{K_c} \gamma_j \right) \right] \cdot \text{BWB\_MOD}_i \quad , \tag{7.2}$$

from which it results that

$$\sum_{j=1}^{K_c} \gamma_j = \frac{(C-S) \sum_{j=1}^{K_c} \text{BWB\_MOD}_j}{1 + \sum_{j=1}^{N_c} \text{BWB\_MOD}_j} \quad . \tag{7.3}$$

From (7.2) and (7.3), after routine manipulations, the throughput, γ_i, of each rate-controlled station is obtained:

$$\gamma_i = \frac{(C-S) \cdot \text{BWB\_MOD}_i}{1 + \sum_{j=1}^{K_c} \text{BWB\_MOD}_j} \quad . \tag{7.4}$$

Equation (7.3) shows that if a station i requires a bandwidth which is equal to k times the bandwidth of a station j, its BWB_MOD parameter should satisfy the following relationship:

$$\text{BWB\_MOD}_i = k \cdot \text{BWB\_MOD}_j \quad . \tag{7.5}$$

From (7.3), it results that ρ is

$$\rho = \frac{1}{C} \cdot \left[S + \sum_{j=1}^{K_c} \gamma_j \right] = \frac{C \cdot \sum_{j=1}^{K_c} \text{BWB\_MOD}_j + S}{C \cdot \left(1 + \sum_{j=1}^{K_c} \text{BWB\_MOD}_j \right)} \quad , \tag{7.6}$$

and thus the fraction of bandwidth wasted (B_w) in this case is

$$B_w = C \cdot (1 - \rho) = \frac{C - S}{1 + \sum_{j=1}^{K_c} \text{BWB\_MOD}_j} - S \quad . \tag{7.7}$$

ρ_{max} and $\rho_{single}(i)$ are immediately obtained from (7.6) by setting $(S = 0, K = K_c)$ and $(S = 0, K = 1)$, respectively:

$$\rho_{max} = \frac{\sum_{j=1}^{K} BWB\_MOD_j}{1 + \sum_{j=1}^{K} BWB\_MOD_j} \quad , \tag{7.8}$$

$$\rho_{single}(i) = \frac{BWB\_MOD_i}{1 + BWB\_MOD_i} \quad . \tag{7.9}$$

As can be expected, as shown in (7.7), the lower the BWB_MOD parameter, the higher the resulting waste of bandwidth. On the other hand, as shown in Section 7.4.1, the lower the BWB_MOD parameter, the lower the departure of the transient behavior from the steady-state behavior.

7.6 CURRENT USE OF DQDB

While FDDI is generally used as an high-speed LAN, DQDB is mainly used by telecommunication companies to offer data communications services in a metropolitan area. Furthermore, one of the use of the DQDB technology is to support the Switched Multimegabit Data Service (*SMDS*); i.e., a new data communications service that offers packet-switched data services at multi-megabit rates. At the moment, speeds are 1.2M to 34Mbps; higher rates will be introduced.

The SMDS interface protocol is a three-level connectionless protocol based on the DQDB protocol, for details on the SMDS service on DQDB see [116].

The Tuscany MAN gives an example of a distributed laboratory devoted to field trials of innovative end-user applications, employing the DQDB technology.

7.6.1 The Tuscany MAN Testbed

As shown in Figure 7.39, where the field trial layout is depicted, the topology of the Tuscany MAN encompasses several access nodes with a lot of users in the sub-networks of Florence, Pisa and Siena. Typical users are Hospitals, a Centre for People with Disabilities, Museums, Universities, Research Centres.

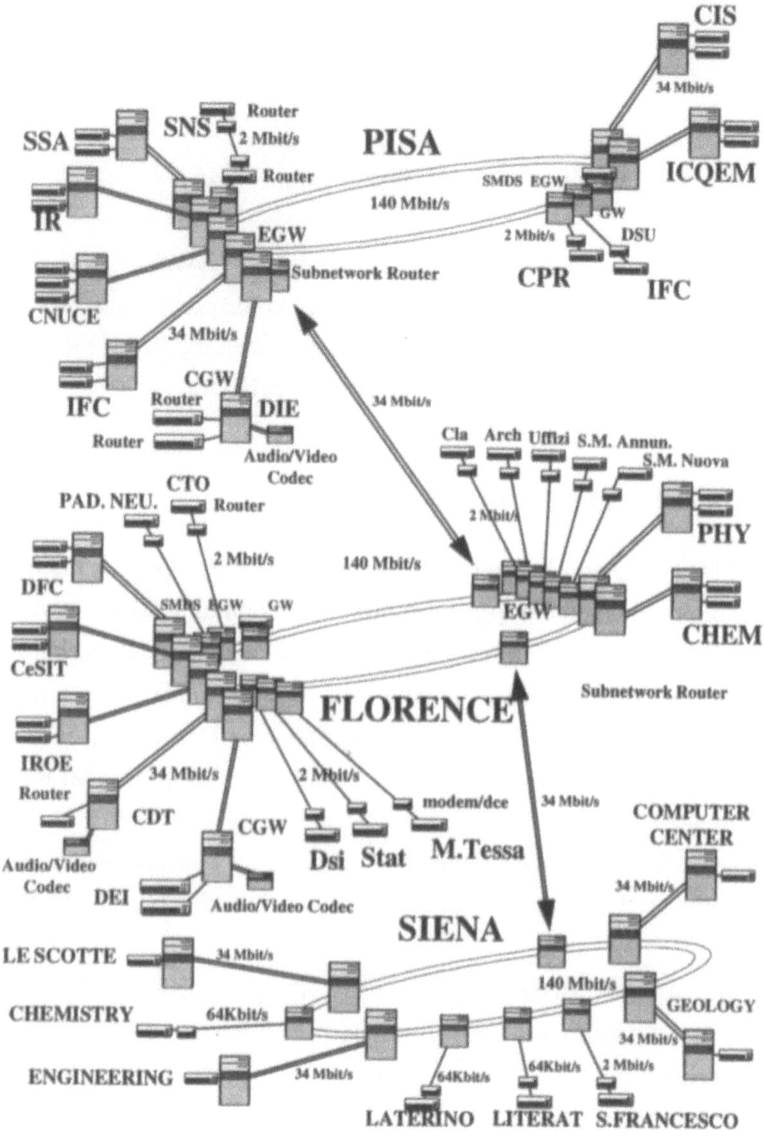

Figure 7.39: Tuscany MAN layout

In all cities a backbone operating at 140 Mbit/s interconnects the Edge Gateways (EGW), the Customer Network Interface Unit (CNIU) and the SMDS Access Unit (ACS). Each EGW has a 34 Mbit/s customer access network interconnecting a Customer Gateway (CGW). The CNIUs are interfaced with routers allowing, on the user side, access for LAN, while the ACS provide a 2 Mbit/s SMDS access.

The sub-networks are interconnected, via Sub-network Routers (SRs), by a link which operates at 34 Mbit/s.

The user access to the network, through the CGW, is made possible either by 10 Mbit/s Ethernet LAN interface, with bridge functionalities, or by 2 Mbit/s isochronous interface, based on Rec. CCITT G.703.

The Tuscany MAN is under the direct control of a Network Management Centre (NMC) connected to the EGW through HDLC links at the bit rate of 9.6 Kbit/s. Moreover to improve the operation, maintenance and management procedures, a further NMC has been set up in remote position.

APPLICATIONS RUNNING ON THE TUSCANY MAN. In the Tuscany MAN the main applications are about health care, as explained shortly thereafter. Environment protection and distributed computing are also themes of concern. It is worthwhile noting the connection of outstanding museums as the Uffizi. All applications run on terminals connected to MAN via LAN Ethernet solution.

HEALTH CARE. This is the most relevant class of applications which are routinely running on the Tuscany MAN. The most important application is Remote consulting and telediagnosis which allows to share digital images coming from Magnetic Resonance and Computerised Tomography devices. Voice communications are also supported by these systems to improve the effectiveness of the consulting sessions. Beyond remote consultancy other relevant applications include access to multimedia servers containing medical images and data and remote consulting and telediagnosis.

ENVIRONMENT CONDITIONS MONITORING AND PROTECTION. A prototype server located in Florence stores multimedia coming from radar, satellites and meteorological and hydrological sensor networks. Accesses to the information stored in this server are performed through multimedia workstations based on XWindow software.

DISTRIBUTED COMPUTING. The high speed connections offered by Tuscany MAN, between super-computers located in Florence and Pisa areas, make possible the joint utilisation of large computational resources. The experimental activities refer to distributed computing applications in the astrophysical, chemical and robotics fields.

ART. Multimedia kiosks allowing remote visit at "Museo della Scienza" in Florence have been developed. These kiosks let the users simulate experiments based on historical equipment and devices kept in the museum.

8 DQDB Models

DQDB has been the subject of considerable research related to performance modeling issues. Most of the existing results were obtained via simulation as it is extremely difficult, if not impossible, to analytically solve detailed models of this protocol. Due to the complexity of the protocol, models with analytical solutions have been developed to approximate protocol behavior under specific network configuration and workload conditions.

In this chapter a classification of the DQDB analytical models is proposed, and some relevant models are presented for each class.

8.1 INTRODUCTION

Due to the discrete time nature of the DQDB network, Markov chains are the most natural choice as the modeling tools to describe network behavior. The problem in using a Markov chain model is to define an appropriate state space. In [121] a discrete time Markov chain that exactly describes a DQDB network with K nodes, constant internode distance of d slots, and M buffers per node, is proposed and the size of its state space is investigated. The state of the Markov chain includes

- the values of the RQ_CTR and CD_CTR in each of the nodes (excluding the most downstream one, since it never receives any requests);
- the number of segments queued at each of the nodes;
- the number of requests queued per node to be transmitted on the reverse bus (excluding the most upstream, since it never sends requests);
- the value of the Busy bit for each slot in transit on Bus A;
- the value of the request bit for each slot in transit on Bus B.

However, not all the states can be reached, for example a state with

CD_CTR>0 and no segment in the node is clearly unreachable. In order to study the relationship between the number of possible states and the number of valid states (i.e., the reachable states), in [121] a network configuration with only two single buffer stations is analyzed.

The relationship between the number of possible states and the number of valid states for this configuration is shown in Table 8.1.

Table 8.1 DQDB modeling complexity

d	Total possible states	Number of valid states
1	216	61
2	1536	305
4	38400	5642
6	884736	92604

The table clearly shows that the state space explodes quite rapidly and analysis is only possible in a few simple cases. Thus a general solution for the DQDB network, i.e., one that encompasses any number of nodes and any internode distance, seems to be highly improbable. Simplifying assumptions therefore have to be made in order to obtain analytically-tractable solutions. Approximate solutions have been proposed for a general DQDB model, while exact solutions have been proposed for DQDB networks operating under specific conditions.

To provide a structured overview of DQDB analytical studies a model taxonomy is introduced (see Table 8.2). Two main classes of DQDB models

Table 8.2 Model taxonomy

	Model features		*Performance indices*
Network-wide models	node spaced models		throughput
	node concentrated models		average delay
Node-in-isolation models	L_NET models	1-st order	output process
		n-th order	output process
	Tagged node models	single buffer	average delay
		infinite buffer	average delay

are identified depending on whether the models consider explicitly all the network nodes or just focus on a tagged node: *Network-wide models* and *Node-in-isolation models* [120].

Network-wide models can be further subdivided into two classes: models which assume that the network nodes are spaced along the two buses (*Node-spaced models*) and models in which the nodes are concentrated in the same place (*Node-concentrated models*).

In Node-in-isolation models, the node under study is tagged and, with respect to the tagged node, the network is partitioned into L_NET and R_NET which represent the influence on the tagged node of the upstream and downstream nodes respectively. L_NET models characterize with a Markov chain the influence of the upstream nodes on the tagged node.

8.2 NETWORK-WIDE MODELS

8.2.1 Node-spaced Models

Models of a DQDB network which represent a few active stations spaced along the network buses are often used in literature to study the DQDB asymptotic behavior. These models are relevant as they can be used to evaluate, for example, bandwidth sharing among network nodes during simultaneous file transfer. Assuming that all active network nodes always have segments to transmit, the network behavior becomes deterministic, hence deterministic models can be utilized in this load condition. These models are used to obtain expressions for throughput achieved by every node as a function of the network span and the nodes activation scenarios (see Section 7.3.2). The complexity of the interdependencies among stations make the analysis possible for only a few active nodes. As noted in Section 7.3.2 a model of this type was applied to an earlier version of the DQDB MAC protocol to highlight its unfairness. In [80] the model proposed by Wong was extended to the standard version of DQDB (with the BWB mechanism disabled) by taking into consideration all the possible configurations of time instants at which nodes start to transmit. As in [165], in this model, a network configuration with two active nodes separated by h slots is assumed (see Figure 7.38). However, in the Wong model the node activation scenario is fixed, whereas in [80] the starting times of node{1} (t_1) and of node{2} (t_2) are assumed to be general and it is shown that the rate at which node{2} can generate its requests is a function of a quantity (hereafter X) which is the sum of the following components

(i)   the number of requests travelling on the reverse bus;

(ii)   the empty slots on the forward bus; and

(iii)   the requests queued in the node{1} (i.e., CD_CTR and RQ_CTR) at the time instant at which both nodes are active.

When the two nodes are active X becomes a constant and it determines the node{2} throughput, see (8.4). As shown below, the value of X is given by the following relationship

$$X = 1 + h - c(h) \quad , \tag{8.1}$$

where Δ is the segment transmission time (slot duration), and

$$c(h) = \begin{cases} (t_2 - t_1) & if \quad -h\Delta \le (t_2 - t_1) \le h\Delta \\ -h & if \quad -h\Delta > (t_2 - t_1) \\ h & if \quad h\Delta < (t_2 - t_1) \end{cases}$$

As in [80], Equation (8.1) is justified by observing that

- if node{2} becomes active at least $h\Delta$ time units after node{1}, i.e., $h\Delta < (t_2 - t_1)$, following the line of reasoning shown in Section 7.3.2, it is easily obtained that $X = 1$, and thus $c(h) = h$;

- at the other extreme, assume that node{1} becomes active at least $h\Delta$ time units after node{2}, i.e., $-h\Delta > (t_2 - t_1)$. When node{1} becomes active the request bus is already carrying h REQs and in the $h\Delta$ time units that it takes for node{1}'s first segment to reach node{2}, other h REQs will be issued by node{2}. Hence, in this case, $X \approx 2h$ and $c(h) \approx -h$.

Before showing the relationship between X and the node throughput, the following definitions must be introduced[1]

- $\gamma(1)$ and $\gamma(2)$ are the throughputs of node{1} and node{2}, respectively;

- $Q(1)$ is the average length of the distributed queue observed by node{1} just after it has inserted a segment in the distributed queue (i.e., one plus the value of the node{1} CD_CTR);

---

1. Throughput is measured in segments per slot time and the round-trip delay is measured in slot times.

- $Q(2)$ is the average length of the distributed queue observed at node$\{1\}$ immediately after a request from the downstream node has been inserted in the queue;
- T is the average delay between the time node$\{2\}$ issues a request and the time it receives the related empty slot.

The network behavior in steady state can be approximated by the following four equations [80]:

$$\gamma(1) + \gamma(2) = 1 \quad , \tag{8.2}$$

$$\gamma(1) = \frac{1}{Q(1)} \quad , \tag{8.3}$$

$$\gamma(2) = \frac{X}{T} \quad , \tag{8.4}$$

and

$$T = 2h + Q(2) \quad . \tag{8.5}$$

Finally, by solving the system of linear equations (8.2)-(8.5), with the approximation $T = 2h + Q(2) \approx 2h + Q(1)$, the nodes' throughputs are obtained

$$\gamma(1) \approx \frac{2}{2 - h - c(h) + \sqrt{(h - c(h) + 2)^2 + 4h \cdot c(h)}} \quad , \tag{8.6}$$

and

$$\gamma(2) = 1 - \gamma(1) \quad . \tag{8.7}$$

As the difference between $Q(2)$ and $Q(1)$ is most pronounced when $(h\Delta < (t_2 - t_1))$ and $h \gg 1$, it is worth discussing this approximation in this initial scenario. $Q(1) \approx 1$ since the queue length observed by node$\{1\}$ after inserting a data segment is usually one and occasionally two, the node$\{1\}$ data segment plus the REQ from node$\{2\}$. On the other hand $Q(2) = 2$ since the distributed queue at the REQ arrival instants contains a node$\{1\}$ data segment plus the REQ from node$\{2\}$. Even though $Q(2)$ and $Q(1)$ differ by a factor of two, h is large so the approximation for T is still justified. In [80] it is shown that the throughput predicted using equations (8.6) and

(8.7) matches the simulative results very well.

Note that for a short network $h \approx 0$, thus the nodes get equal through-put. In a large network $h \gg 0$, the minimum throughput of node$\{2\}$ is $\gamma(2) \approx 1/2h$ (this value is obtained when $c(h) \approx h$), while in node$\{1\}$'s most unfavourable scenario (i.e., $c(h) \approx -h$) this node is less penalized since $\gamma(1) \approx 1/\sqrt{2h}$.

BIBLIOGRAPHIC NOTES. Deterministic models to analyze the DQDB asymptotic behavior have also been presented in ([95], [57], [114]).

8.2.2 Node-concentrated Models

In [129] a DQDB network with K stations is analyzed. Messages arriving at a station for transmission on a bus are divided into fixed-length segments. Segments are then queued in the related Local Queue depending on their priority level; H priority levels are assumed for the transmission of the asynchronous traffic. To make the analysis possible below it is assumed that

- propagation and processing delays are zero;
- the request channel has an infinite capacity;
- the order in which segments are transmitted does not depend on the position of the station.

It can be observed that these hypotheses correspond to the ideal conditions according to which DQDB provides its nodes with a Round Robin (*RR*) service discipline (see Section 7.2.2). Therefore, to analyze the DQDB behavior under these ideal conditions a discrete time Multi-queue Processor Sharing (*MPS*) model which extends the classical RR model is proposed [100].

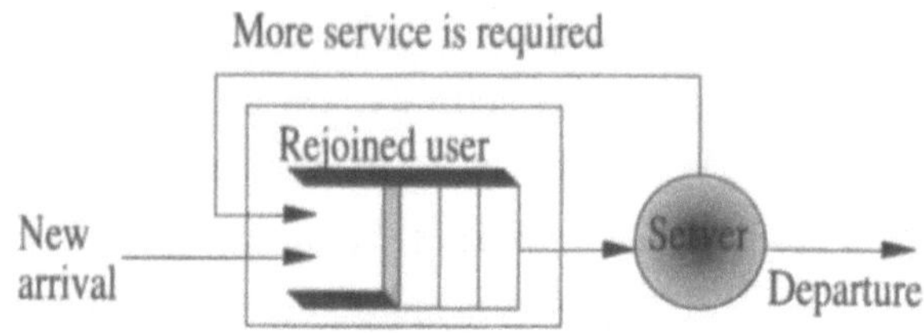

Figure 8.1: Classical Round Robin model

The classical RR model is based on a single server queue, in which newly arriving messages join the end of the queue. Messages are served on a FIFO basis. When a message is served, it receives a quantum of service (which corresponds to a time unit), and if it requires more service it rejoins the end of the queue. The MPS model extends the RR model by introducing multi-priority levels, LQs and discrete time services.

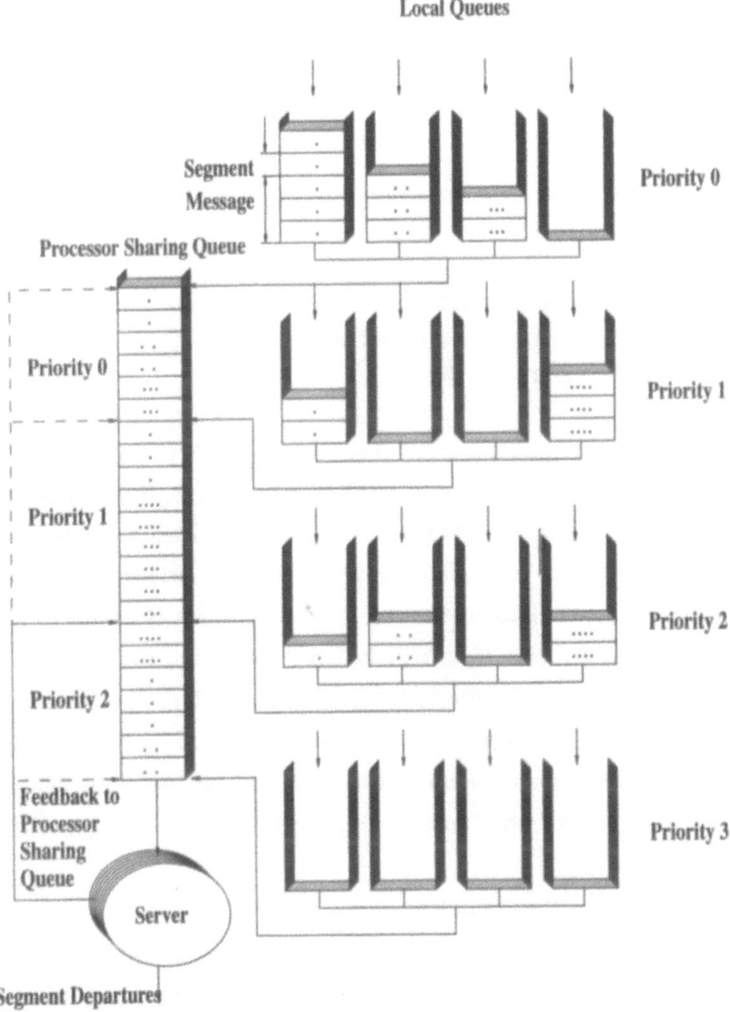

Figure 8.2: Four stations MPS model

In the MPS, an arriving message is queued in the LQ related to its priority level. There is a separate LQ for each priority in each station. One representative (if any) for each LQ is inserted into the Processor Sharing (*PS*) queue. In the PS queue, messages of different priority levels are served on a strict priority discipline (HoL [100]). Messages of the same priority level are served on an RR basis: a message in the PS queue recycles through the service facility, receiving (each time) a quantum of service equivalent to one segment transmission. The message leaves the PS queue after it has recycled enough times to service all the segments in the message.

A closed formula for the mean time $(D_p(n))$ that a priority-p message made up of n segments spends in the MPS system (i.e., the time from its arrival until it is transmitted) is derived under the following hypotheses on the arrival process

- the number of priority-p messages arriving in a slot time are independent and identically distributed (i.i.d.), and independent of the arrivals at the other LQs;
- the number of segments in the messages are discrete and i.i.d.[1]
- all message arrivals within any slot are assumed to arrive at a point in time just before the end of the slot.

$D_p(n)$ includes the time a priority-p message spends in the LQ, L_p, and the time it spends in the PS queue to serve its n segments. By defining $S_p(n)$ as the mean time a priority-p packet (consisting of at least n-segment), spends in the PS queue to complete n segments of service, it follows that $D_p(n) = L_p + S_p(n)$. Before starting the derivation it is useful to introduce some notations. $\bar{a}_p$ and $C^2_{a,p}$ will be used to denote the average and the squared coefficient of variation of the number of priority-p message arrivals in a slot, respectively; while b_p, $C^2_{b,p}$ and $F_{b,p}(\bullet)$ will be used to denote the average, the squared coefficient of variation, and the probability distribution function of the length (in segments) of priority-p messages, respectively.

$S_p(n)$ COMPUTATION. As in the RR model, in the MPS model it can be proved that $S_p(n)$ linearly increases with n. In fact, by indicating with $N_p(n)$ the mean number of priority-p messages in the PS queue which have already obtained exactly n quantum of service, i.e., n segments transmission,

$$S_p(n) = n \cdot (N_p(0)/\lambda_p) \quad , \tag{8.8}$$

---

1. The distributions may differ for different priority levels.

where λ_p is the total priority-p message arrival rate at the system in a slot time.

Formula (8.8) is derived by preliminarily observing that

$$N_p(n) = \lambda_p \cdot [1 - F_{b,p}(n)] \cdot [S_p(n + 1) - S_p(n)] \quad , \tag{8.9}$$

where

- $\lambda_p [1 - F_{b,p}(n)]$ is the average number of messages composed of more than n segments arriving within a slot and
- $[S_p(n + 1) - S_p(n)]$ is the mean time between the service completion of the n and $n+1$ segments, i.e., it is the average service time for the $(n+1)$th segment in a message.

Formula (8.9) is obtained by applying Little's theorem to a system which includes only those messages which have obtained exactly n quanta of service and have not yet completed the $(n+1)$th quantum of service. Specifically, $\lambda_p [1 - F_{b,p}(n)]$ is the arrival rate in this system and $[S_p(n + 1) - S_p(n)]$ is the system response time.

Let Γ_n be the probability that the service of the message is completed at the $(n +1)$th quantum given that it was not completed at the n-th. By definition for all $n \geq 0$

$$\Gamma_n = \frac{P_{b,p}(n + 1)}{1 - F_{b,p}(n)} \quad , \tag{8.10}$$

and also [129]

$$\Gamma_n = \frac{N_p(n) - N_p(n + 1)}{N_p(n)} \quad . \tag{8.11}$$

From equations (8.10) and (8.11) it results

$$\frac{N_p(n) - N_p(n + 1)}{N_p(n)} = \frac{P_{b,p}(n + 1)}{1 - F_{b,p}(n)} \quad , \tag{8.12}$$

and the solution for this set of equations is of the form (see (4.15 in [100])

$$N_p(n) = K [1 - F_{b,p}(n)] \quad . \tag{8.13}$$

The value of the constant K is obtained by substituting $n=0$, $K = N_p(0)$. Thus using equations (8.13) and (8.9) the following set of difference equations is obtained

$$N_p(0) = \lambda_p \cdot [S_p(n+1) - S_p(n)] \quad . \tag{8.14}$$

Equation (8.8) is the solution of (8.14).

$S_p(n)$ is derived by measuring the delay of a "long test message" as proposed by Kleinrock [100]. Specifically, the time spent by a long test message at priority-p made up of x segments in the PS queue tends towards the sum of

1. the service time x of the test message;

2. the service times required by all the priority-p messages which join the PS queue during the service of the test message (i.e., the priority-p message representatives of the other LQs), and

3. the service times of the messages with priority q higher than p which join the PS queue during the service of the test message.

The resulting expression for the $S_p(x)$ is

$$\lim_{x \to \infty} S_p(x) = x + S_p(x) \cdot \left[\frac{M_p - 1}{M_p} \cdot \rho_p + \sum_{q=p+1}^{H} \rho_q \right] , \tag{8.15}$$

where M_p is the number of priority-p LQs, and ρ_i is the priority-i bus utilization, $\rho_i = \lambda_p \cdot b_i$, $\quad i = 0, 1, .., H$.

After some algebraic manipulations, it can be shown that

$$S_p(n) = \frac{n}{1 - \sum\limits_{q=p+1}^{H} \rho_q - \frac{M_p - 1}{M_p} \cdot \rho_p} . \tag{8.16}$$

L_p COMPUTATION. To compute L_p, first the average delay experienced by the class-p segments, as a function of L_p, is derived ($E[D_{seg(p)}]$) , then by exploiting the equivalence in terms of the average segment delay between the MPS model and the $D[M/G/1]_{PR}$ model (see [133]) L_p is obtained. In fact, both in the discrete-time $M/G/1$ queueing system with preemptive resume priority service discipline, $D[M/G/1]_{PR}$, and in MPS, during any time slot, a segment of the highest priority that exists in the system is served and never returns. In both systems preemption is at the message level whereas the segment transmission cannot be interrupted. Hence, if the message arrival processes in the two systems are statistically identical also the distribution of the total number of segments in each priority level are identical.

To compute $E[D_{seg(p)}]$ it is useful to introduce $R_p(k)$, which is the probability that a randomly selected priority-p segment is the k in its own packet. It is shown below that

$$R_p(k) = (1 - F_{b,p}(k-1))/b_p \quad . \tag{8.17}$$

To derive equation (8.17) it is better to focus on the i.i.d. sequence of packet lengths $\{b_p(h), h \geq 1\}$ and to associate a renewal with the first segment of each packet. It is thus possible to define the discrete-time renewal point process $\{C_p(h), n \geq 1\}$ where $C_p(1) = 1$ and

$$C_p(h) = 1 + \sum_{j=1}^{n-1} b_p(j), \quad \text{for } n \geq 2 \quad .$$

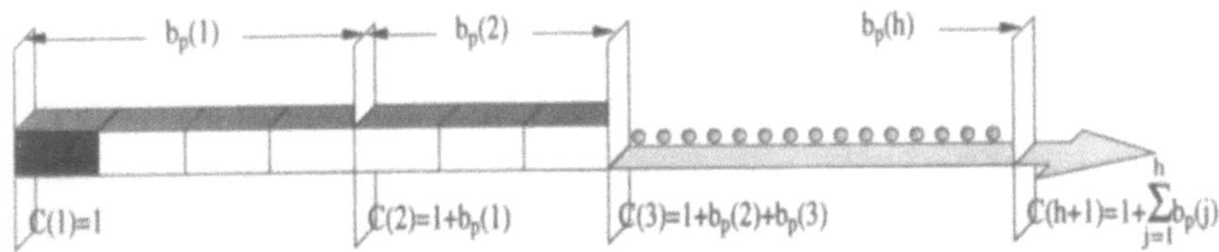

Figure 8.3: Renewal process

The $R_p(k)$ is computed by introducing the probability that the j-th segment in the $C_p(h)$ process is the k-th of its packet ($R_p(k, j)$) and noting that

$$R_p(k) = \lim_{j \to \infty} R_p(k, j) \quad . \tag{8.18}$$

Let $W_p(h)$ be the probability that the h-th priority p segment represents a renewal, thus

$$R_p(k, j) = W_p(h - (k-1)) \cdot [1 - F_{b,p}(k-1)] \quad . \tag{8.19}$$

Equation (8.17) holds since if the h-th segment is the k-th of its packet the $h - (k-1)$ segment is a renewal point and no renewals occur after the $h - (k-1)$ and before the h-th segment. Furthermore, since $\lim_{j \to \infty} W_p(j) = 1/b_p$ using (8.18) and (8.19), equation (8.17) is derived.

The average segment delay in the MPS queueing system can be expressed as

$$E[D_{seg(p)}] = \sum_{k=1}^{\infty} [L_p + S_p(k)] \cdot R_p(k) \tag{8.20}$$

$$= L_p + \frac{[b_p \cdot (1 + C_{b,p}^2) + 1]}{2\left(1 - \sum_{q=p+1}^{H} \rho_q - \frac{M_p - 1}{M_p} \cdot \rho_p\right)} \quad .$$

To compute L_p from (8.20) the equivalence is exploited, in terms of the average segment delay, between the MPS model and the discrete-time $M/G/1$ queueing system with preemptive resume priority service discipline [133], hereafter referred to as $D[M/G/1]_{PR}$.

Finally, by equating expression (8.20) with the expression of the average priority-p segment delay in the equivalent $D[M/G/1]_{PR}$ system a closed-form expression for L_p can be derived

$$L_p = \frac{b_p \cdot (C_{b,p}^2 + (C_{b,p}^2 \cdot \lambda_q)/M_q) + \sum_{q=p}^{H}\left(\frac{\rho_q \cdot v_q}{(1 - \sum_{q=p}^{H}\rho_q)}\right)}{2(1 - \sum_{q=p+1}^{H}\rho_q)} \tag{8.21}$$

$$- \frac{[b_p \cdot (1 + C_{b,p}^2) + 1]}{2\left(1 - \sum_{q=p+1}^{H}\rho_q - \frac{M_p - 1}{M_p} \cdot \rho_p\right)} + \frac{1}{2} \quad .$$

In [129] it is clearly shown, by using simulative results, that the performance indices obtained with the MPS provide an adequate estimation of the DQDB performance figures, given that the distance between stations is small.

8.3 NODE-IN-ISOLATION MODELS

According to this approach[12] the node under study is tagged and, with respect to the tagged node, the network is partitioned into (see Figure 8.4)

- L_NET (Left Network) which includes all the upstream nodes (from the tagged node);

- *tagged node* itself; and
- *R_NET* (Right Network) which includes all the downstream nodes (from the tagged node).

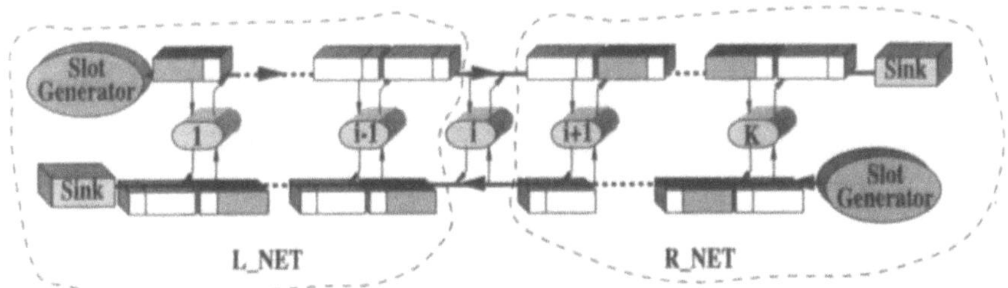

Figure 8.4: Complexity reduction

As shown in Figure 8.5, with respect to the tagged node, R_NET is a generator of requests on Bus B (R_NET process) while L_NET is a generator of Busy/Empty slots on Bus A (L_NET process). For this reason the L_NET process is also referred to as a *slot-occupancy-pattern* process.

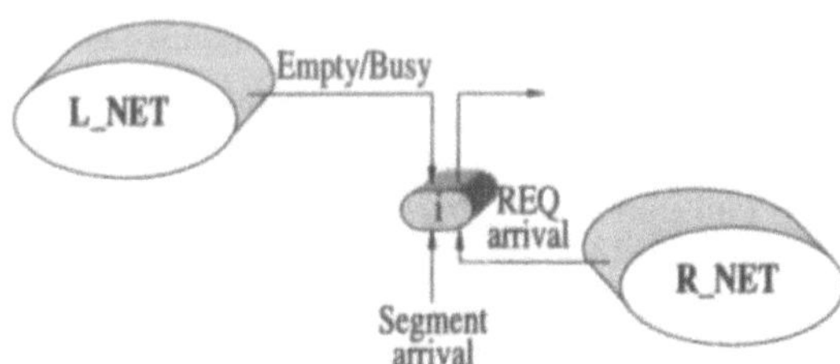

Figure 8.5: Tagged node model

Although characterizing L_NET and R_NET processes is very difficult, when DQDB operates in underload conditions the modeling complexity can nevertheless be reduced ([39], [42]) as the number of empty slots is greater than the number of segments to be transmitted. Since the time it takes a REQ sent by node$\{j\}$ to affect upstream nodes may have a duration of several slots in a MAN environment, it follows that a segment is often transmitted in a slot which is ahead of the one corresponding to its REQ when DQDB operates in underload conditions.

To clarify this better it is useful to analyze the reservation mechanism in detail. At time $t=0$ it is assumed that node$\{i\}$ inserts a segment in the DQ and transmits a REQ=1 on the reverse bus. After these operations the node will wait until it transmits its segment either in an unused slot (i.e., an empty slot not reserved by any downstream station) or in its reserved slot.

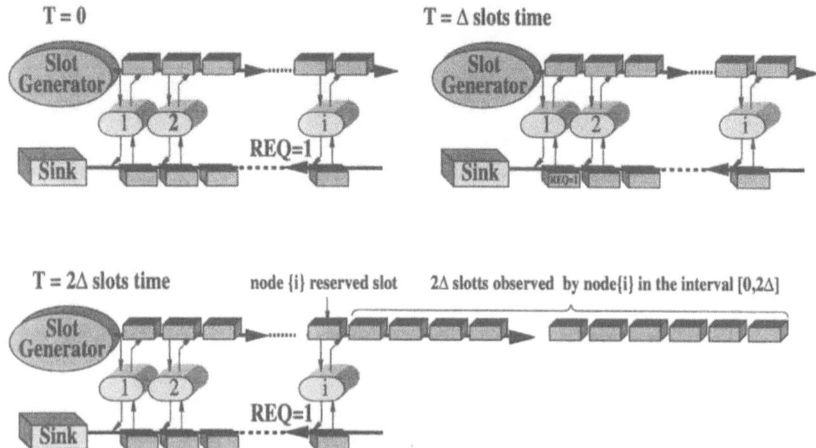

Figure 8.6: Relation between time and reservation

Obviously if the propagation delay between node$\{i\}$ and node$\{1\}$ is Δ slot times, the latter will transmit in its reserved slot only if there is no unused slot among the 2Δ consecutive slots it observes on the forward bus (see Figure 8.6).

An easy way to estimate the probability that node$\{i\}$ transmits in its reserved slot is based on the assumption that the status of the slots is modeled with a Bernoulli process with

$$P\{empty\ slot\} = 1 - \tilde{\rho}_i \ , \tag{8.22}$$

where $\tilde{\rho}_i = \sum_{j=1}^{i-1} \lambda_j \cdot b_j$.

Table 8.3 clearly shows that under medium/light conditions this probability is almost negligible for nodes with a distance of a few kilometers[1] from node$\{1\}$. Therefore, the correlation between the transmission of a segment and its REQ is almost negligible, and the only effect of node$\{i\}$ REQs is to widen the time interval between consecutive transmissions of its upstream nodes.

These observations indicate that in the analysis of a tagged node$\{i\}$,

---

Table 8.3 Probability that node{i} transmits in its reserved slot

Distance from Node{1}		$\tilde{p}_i$			
		0.60	0.70	0.80	0.90
$\Delta = 1$	0.564 km	0.36	0.49	0.64	0.81
$\Delta = 2$	1.128 km	0.13	0.24	0.41	0.65
$\Delta = 5$	2.882 km	<0.01	0.03	0.10	0.35
$\Delta = 10$	5.64 km	<1E-4	<1E-3	0.01	0.12
$\Delta = 20$	11.28 km	1E-8	1E-6	1E-5	0.01

R_NET can be simply modeled with a node{i} arrival process with a rate equal to the downstream REQs arrival rate. This significantly reduces the complexity as downstream nodes are aggregated in a very simple process. In all the existing analytical models R_NET generates requests according to a memoryless distribution (Poisson or Bernoulli process), while research on modeling DQDB nodes in isolation has concentrated on solving the following subproblems

- L_NET modeling;
- tagged node modeling.

8.3.1 L_NET Modeling

Results reported in ([37], [39], [13]) show that the number of consecutive busy slots observed by nodes close to the head node has a nearly geometric distribution, i.e. the state of the consecutive slots is independent. On the other hand, while moving towards the end node the correlation among the state of consecutive slots sharply increases. In the literature, L_NET is frequently modeled by a Bernoulli process ([12], [154]). Obviously, this model does not take into consideration the correlation between the state of consecutive slots. To overcome this the correlation between consecutive slots is considered in the models presented in ([43], [48]). In the previous section it was pointed out that a node (e.g. node{i}) may often transmit a segment in an empty slot positioned ahead of the slot forced empty by the node's own REQ. This event may occur when

(i)   no node upstream of the node{i} has a segment to transmit, or

(ii)   the empty slot (somewhere ahead of that requested by node$\{i\}$) has been forced by a REQ which has already been satisfied (hereafter, a worthless REQ).

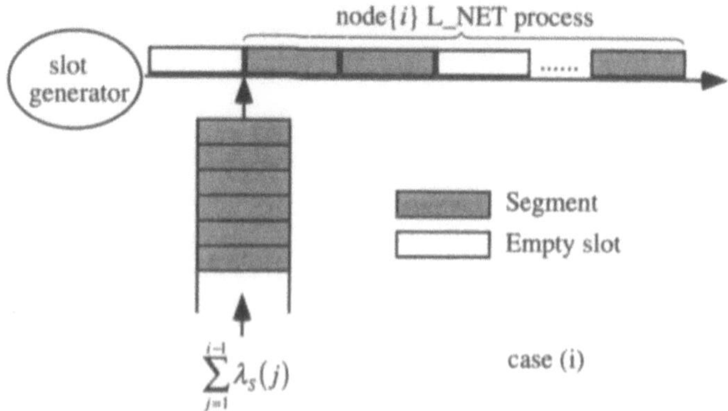

Figure 8.7: Case (i): *M/D/1* busy period

If the L_NET process observed by a node were only due to point (i), the L_NET process would correspond exactly to the busy period process of an *M/D/1* system [99] where the input traffic is obtained by the superposition of the segment-arrival processes in the upstream nodes (see Figure 8.7). However, in the light of point (ii), the busy periods of the *M/D/1* system are subdivided into sub busy periods separated by empty slots induced by the "service" of worthless REQs (see Figure 8.8). It thus follows that L_NET

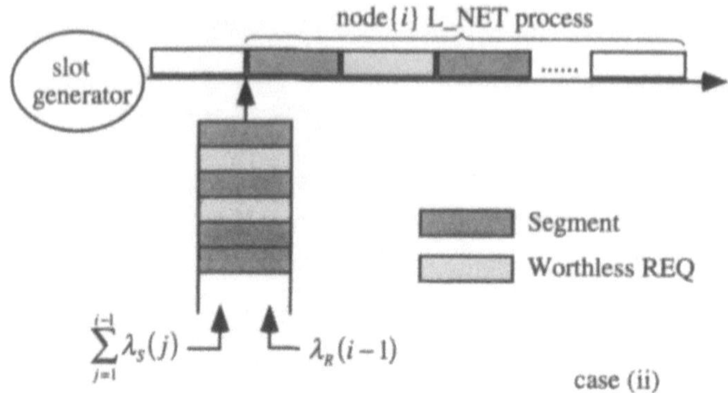

Figure 8.8: Case (ii): *M/D/1* busy period with worthless REQ

can be studied by referring to a *Simplified DQDB* model which is characterized by the following assumptions:

1. the reverse bus is modeled by introducing, for each node, a Poisson process to characterize the arrival of REQs generated by downstream nodes;

2. the forward bus is slotted and the status of the slots is modeled via the L_NET process;

3. the MAC protocol is modeled by a queue where REQs and segments are stored on a FIFO basis. Consequently, for each node$\{i\}$, where $1 \leq i \leq K$, the input traffic is made up of segments and REQs;

4. the arrival process is Poisson with $\lambda(i)$ parameters, where $\lambda(i) = \lambda_S(i) + \lambda_R(i)$; $\lambda_S(i)$ and $\lambda_R(i)$ are the segment and the REQ arrival rates, respectively. Obviously, $\lambda_S(i)$ depends on the workload characterization, whereas $\lambda_R(i) = \sum_{j=i+1}^{K} \lambda_S(j)$;

5. the transmission time of both segments and REQs is constant and equal to the slot duration, and both can be transmitted in the first empty slot seen by a node. An empty slot used for a transmission remains empty if there is a REQ at the head of the queue, but becomes busy if there is a segment at the head of the queue. The probabilities that a node$\{i\}$ queued packet is a REQ $(P_{REQ}(I))$ or a segment $(P_{SEG}(I))$ are

$$P_{REQ}(i) = \frac{\lambda_R(i)}{\lambda_S(i) + \lambda_R(i)} \ , \tag{8.23}$$

and

$$P_{SEG}(i) = 1 - P_{REQ}(i) \ . \tag{8.24}$$

ACCURACY OF THE SIMPLIFIED DQDB MODEL. The major difference between the simplified model and the DQDB is the scheduling algorithm for segment transmission. In the simplified model, segments and REQs are served following a FIFO scheduler, while in DQDB a more complex scheduler based on the count_down mechanism is used (see Section 7.2.2). Since these two schedulers differ only when there is more than one segment in the node queue, in [48] the accuracy of the simplified model is analyzed by computing the distribution of the number of segments, N_seg, in the node queue. Since differences between the two schedulers may occur only when at least one segment is ready for transmission, the comparison is performed by studying $P\{N\_seg = 1 | N\_seg > 0\}$.

Different workload characterizations and network configurations have been analyzed in [48]. Results are shown in Figure 8.9 and Figure 8.10 for a 10 and 50 node network configurations, respectively.

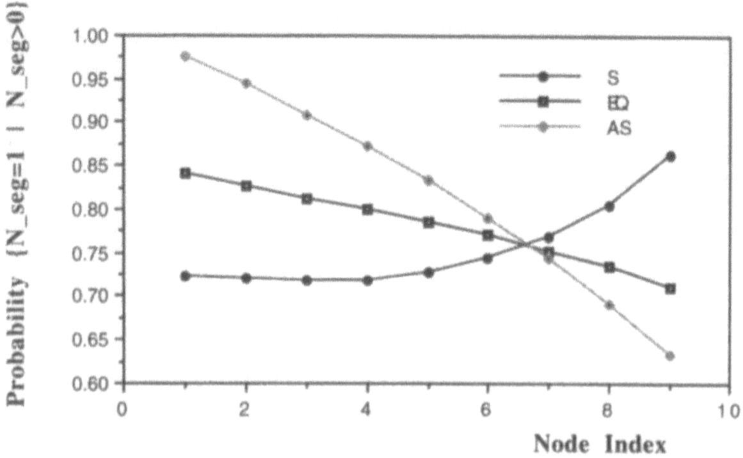

Figure 8.9: *OL*=0.80 and 10-node network

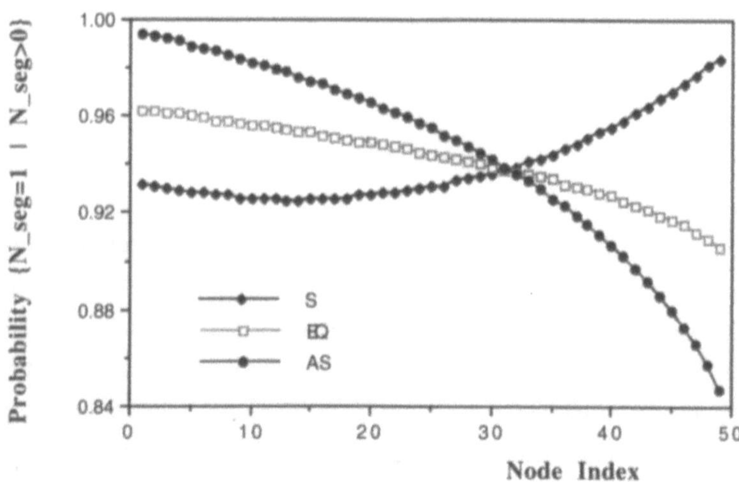

Figure 8.10: *OL*=0.80 and 50-node network

Figures 8.9 and 8.10 plot the graphs of $P\{N\_seg = 1|N\_seg > 0\}$ for $OL = 0.80$ and a network configuration with 10 and 50 nodes, respectively.

These figures show that for AS and S workload types the node segment arrival rate is the predominant factor in the $P\{N\_seg = 1|N\_seg > 0\}$ and that the probability is minimal in those nodes with the highest arrival rate. For the EQ workload type the position of the node is an important factor. Nodes close to the end of the bus experience the lowest $P\{N\_seg = 1 | N\_seg > 0\}$. This behavior occurs for both network configurations. However, in the 50 node configuration, since the offered load is split among a large number of nodes, the probability of seeing more than one segment in the node queue is almost negligible. Therefore the simplified model provides an accurate representation of the DQDB scheduling algorithm.

L_NET MODELING. In ([43], [48]) the L_NET process of the Simplified DQDB is characterized by observing, for each node$\{i\}$, the L_NET process immediately ahead of

$$S_{inp}(i) := \{ S_j^{(i)}; \ j \in \mathbb{N} \} \quad ,$$

and behind

$$S_{out}(i) := \{ (S_j^{(i)}, A_j^{(i)}); \ j \in \mathbb{N} \} \quad ,$$

node$\{i\}$. The random variable $S_j^{(i)}$ takes the value B if the slot is busy, and the value E if the slot is empty, while $A_j^{(i)}$ represents the action of node$\{i\}$ on the slot, $A_j^{(i)} \in \{0, 1\}$. Thus the random vector $(S_j^{(i)}, A_j^{(i)})$ takes the following values

(i) $B0$: if the slot is already busy in the input process;
(ii) $E1$: if the slot was empty and the node uses it for segment transmission;
(iii) $E0$: if the slot was empty, and the node queue is either empty or there is a REQ on top of it.

$S_{inp}(i)$ and $S_{out}(i)$ are referred to as *input* and *output processes*. respectively. The states of consecutive slots are not independent and thus processes $S_{inp}(i)$ and $S_{out}(i)$ do not satisfy the Markov property. However, the correlation between the status of two slots separated by n slots tends to weaken as n increases. In [48] $S_{inp}(i)$ and $S_{out}(i)$ are therefore approximated with discrete time Markov processes $S_{inp}^{(n)}(i)$ and $S_{out}^{(n)}(i)$, which only capture the dependencies between n consecutive slots (n^{th}-order discrete-time Markov process).

   The state space of $S_{inp}^{(n)}(i)$ is

$$\{s_1, s_2, \ldots, s_n \,|\, s_j \in \{B, E\}, 1 \le j \le n\} \quad , \tag{8.25}$$

and its transition probabilities are

$$P_i\{S_{n+1} = s_{n+1} \,|\, (S_1 = s_1, S_2 = s_2, \ldots, S_n = s_n)\} \quad . \tag{8.26}$$

Each n-tuple $(S_1 = s_1, S_2 = s_2, \ldots, S_n = s_n)$ describes the state of the last n consecutive slots observed on the forward bus by the tagged node. While the state space of $S_{out}^{(n)}(i)$ is

$$\{s_1 a_1, s_2 a_2, \ldots, s_n a_n \,|\, s_j a_j \in \{B0, E0, E1\}, 1 \le j \le n\} \quad , \tag{8.27}$$

and its transition probabilities are

$$Q_i\{S_{n+1}A_{n+1} = s_{n+1}a_{n+1} \,|\, (S_1 A_1 = s_1 a_1, S_2 A_2 = s_2 a_2, \ldots, S_n A_n = s_n a_n)\} \quad .$$

$S_{inp}^{(n)}(1)$ is known, since all slots observed by node$\{1\}$ are empty, and since $S_{inp}^{(n)}(i+1)$ can be constructed from $S_{out}^{(n)}(i)$. The L_NET study is thus performed by defining an algorithmic approach to compute the nth-order Markov model of the output process of node$\{i\}$ starting from the nth-order Markov model of the input process (see [48]). Specifically, first the joint probability distribution function of the status of $(n+1)$ consecutive slots in the output process $(J^{n+1} - pdf)$ is computed, and then the transition probabilities in the output process by using the definition of the conditional probability. Two algorithmic approaches are presented one for the 1st- order model and the other for the models which take into consideration higher orders of correlation.

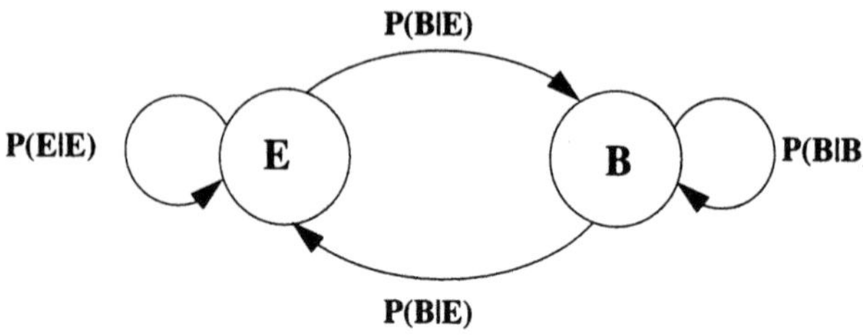

Figure 8.11: First-order Markov chain

1st-ORDER MARKOV MODEL. In this case $S_{inp}(i)$ and $S_{inp}(i+1)$ are approximated as a first-order Markov chain. Specifically, as shown in Figure 8.11, relations (8.25) and (8.26) reduce to

$$\{ s_n | s_n \in \{ B, E \} \} \quad , \tag{8.28}$$

$$P \{ S_{n+1} = s_{n+1} | S_n = s_n \} \quad . \tag{8.29}$$

The first-order approximation allows to simplify the general solution methodology. Specifically,

- the $S_{inp}^{(1)}(i + 1)$ can be derived directly from $S_{inp}^{(1)}(i)$, without passing through the $S_{out}^{(1)}(i)$ process, and
- the characterization of the $S_{inp}^{(1)}(i + 1)$ process is obtained by exploiting the Theorem 8.1.

THEOREM 8.1 *$S_{inp}^{(1)}(i)$ and $S_{inp}^{(1)}(i + 1)$ processes are regenerative with respect to the sequence $T = \{ T_j ; j \in \mathbb{N} \}$, where T is a renewal process defined by the successive instants (renewal times) at which the queue of node{i} in the Simplified DQDB becomes empty.*

The proof of this theorem is beyond the scope of this book, the interested reader can refer to [43].

By exploiting the regenerative property of the $S_{inp}^{(1)}(i)$ and $S_{inp}^{(1)}(i + 1)$

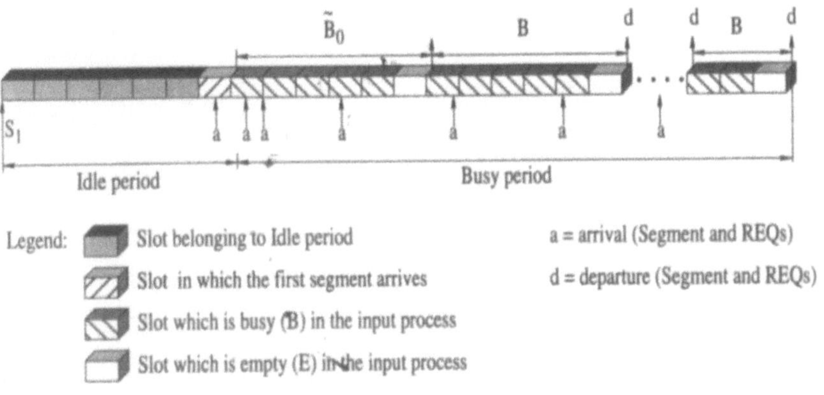

Figure 8.12: Renewal period structure

processes, the $S_{inp}^{(1)}(i + 1)$ process can be derived by observing node{i} in a renewal period, see Theorem 2.4. The structure of the T renewal period is shown in Figure 8.12. At the beginning of the renewal period the tagged node has an empty queue (*Idle period* in Figure 8.12). After the first arrival (segment or REQ) in a renewal period the tagged node uses all the empty slots up until its queue becomes empty again (*Busy period* in Figure 8.12).

To derive the transition probabilities of $S_{inp}^{(1)}(i+1)$, only the joint probability of two consecutive busy slots after node$\{i\}$, $P\{B,B\}$, need to be computed (see [43]).

$P\{B,B\}$ is calculated by applying Theorem 2.4:

$$P\{B,B\} = \frac{E[N_{BB}]}{E[Cycle]} \, , \tag{8.30}$$

where $E[N_{BB}]$ is the average number of (B,B) couples in a generic renewal period, and $E[Cycle]$ is the average length of a generic renewal period.

$E[Cycle]$ is computed by observing that the length of a renewal period has the same distribution as the delay period of an $M/G/1$ system with exceptional first service [99], [149]. In fact, users (segments and REQs) arriving during a busy period (i.e., when a queue was found not empty) experience i.i.d. service times $\tilde{B}$, while the user initiating a busy period (i.e., queue found empty) experiences an exceptional service time denoted by $\tilde{B}_0$. Thus $E[Cycle]$ is computed following the methods for busy period calculation in $M/G/1$ with exceptional first service ([99], [149], [160]).

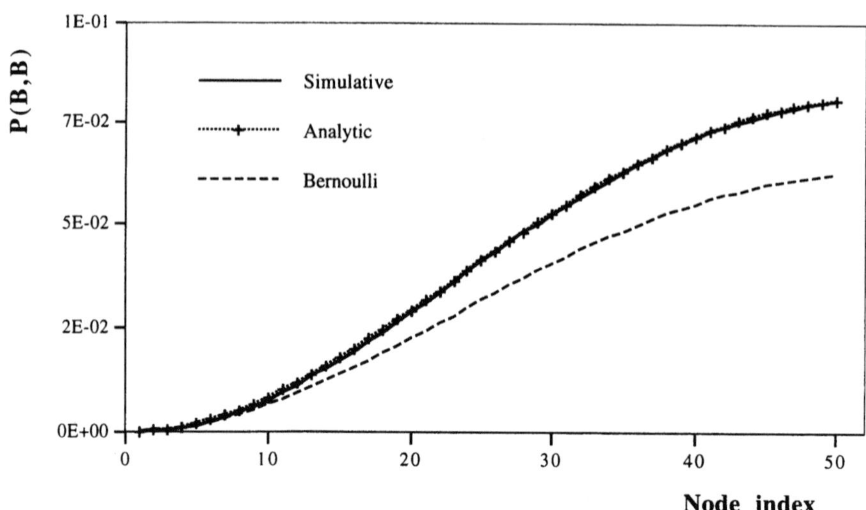

Figure 8.13: $P\{B,B\}$ comparison for OL=0.25

$E[N_{BB}]$ is computed by exploiting the following relation

$$E[N_{BB}] = E[N_{BB}^{Idle}] + E[N_{BB}^{Busy}] \quad , \qquad (8.31)$$

where $E[N_{BB}^{Idle}]$ and $E[N_{BB}^{Busy}]$ are the average number of (B,B) in the Idle period and Busy period, respectively.

$E[N_{BB}^{Idle}]$ is derived from the $S_{inp}^{(1)}(i)$ process only, while the derivation of $E[N_{BB}^{Busy}]$ is more complex since $E[N_{BB}^{Busy}]$ is obtained by taking into account both the services in the busy period and the $S_{inp}^{(1)}(i)$ process (see [43]).

The results presented in [43] (see Figure 8.13) show that even at light loads the interdependence between slots is significant, and that the Bernoulli hypothesis generally used for modeling the L_NET process diverges from the real behavior at almost any load condition. On the other hand, by using a simulative analysis it is shown that the first-order Markov model is able to capture the 1st-order dependencies in the output process of a DQDB network for a wide range of offered loads (up to OL=0.70) as shown in Figure 8.14. In [51] this model was extended to include the effect of the BWB mechanism on the L_NET process.

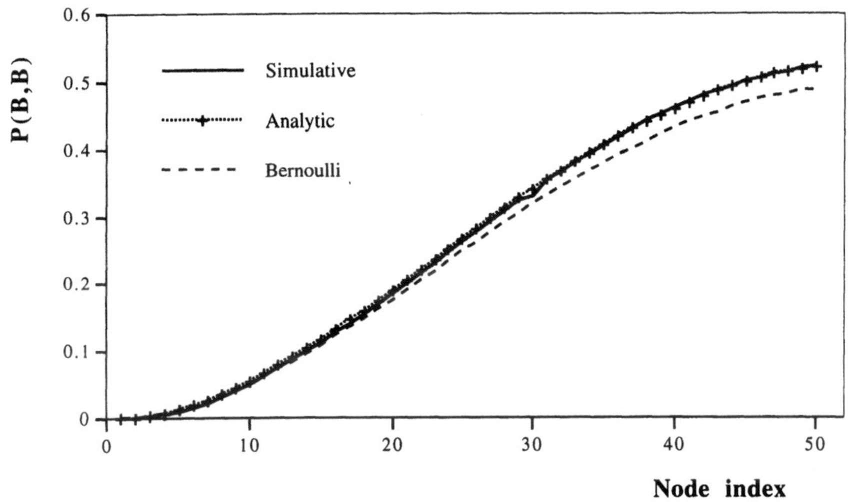

Figure 8.14: $P\{B,B\}$ comparison for OL=0.70

n^{th}-ORDER MARKOV MODELS. This section presents a method to develop models for the slot-occupancy-pattern process which captures the dependency between n consecutive slots, n^{th}-order correlation. The methodology developed in the previous section cannot be easily extended to approximate the slot-occupancy-pattern process via a nth-order. as it requires that the number of consecutive busy slots in the L_NET process be represented by i.i.d. random variables.

The approach presented here is based on the study of the following auxiliary discrete-time Markov process embedded at the end of the k-th slot

$$\{ (S_1 A_1, S_2 A_2, ..., S_n A_n, L_n)_k, k \in \mathbb{N} \}^{1} , \tag{8.32}$$

where $A_1, A_2, ..., A_n$ are random variables which describe the actions performed by node$\{i\}$ on the last n slots it observed, and $L_n, (L_n \in \mathbb{N})$ is the length of the node$\{i\}$ queue immediately after the sequence of actions $A_1, A_2, ..., A_n$.

Once the steady state probabilities of the auxiliary process are known the $(J^{n+1} - pdf)$ and $(J^n - pdf)$ of the output process are derived using the following relation

$$\begin{aligned}
P\{S_1 A_1, S_2 A_2, ..., &S_n A_n, S_{n+1} A_{n+1}\} \\
= &P\{S_{n+1} A_{n+1} | S_1 A_1, S_2 A_2, ..., S_n A_n\} \cdot P\{S_1 A_1, S_2 A_2, ..., S_n A_n\} \\
= &P\{S_{n+1} A_{n+1} | S_1 A_1, S_2 A_2, ..., S_n A_n, L_n > 0\} \\
&\cdot P\{S_1 A_1, S_2 A_2, ..., S_n A_n, L_n > 0\} \\
&+ P\{S_{n+1} A_{n+1} | S_1 A_1, S_2 A_2, ..., S_n A_n, L_n = 0\} \\
&\cdot P\{S_1 A_1, S_2 A_2, ..., S_n A_n, L_n = 0\} \quad .
\end{aligned} \tag{8.33}$$

In equation (8.33) $P\{S_1 A_1, S_2 A_2, ..., S_n A_n, L_n > 0\}$ and $P\{S_1 A_1, S_2 A_2, ..., S_n A_n, L_n = 0\}$ are the steady state probabilities of the auxiliary process (see (8.32)), while the other components are obtained from the following relations

$$\begin{aligned}
P\{S_{n+1} &= s_{n+1}, A_{n+1} = a_{n+1} | S_1 A_1, S_2 A_2, ..., S_n A_n, L_n = 0\} \tag{8.34} \\
&= P\{S_{n+1} | S_1, S_2, ..., S_n\} \cdot I_{\{a_{n+1} = 0\}} ,
\end{aligned}$$

---

1. The state variable can be simplified. Only the joint distribution of the number of users in the system and the status of the last n slots in the input process are strictly required. The state description used in this section was chosen to simplify the presentation.

$$P\{S_{n+1} = s_{n+1}, A_{n+1} = a_{n+1} | S_1 A_1, S_2 A_2, \ldots, S_n A_n, L_n > 0\} \qquad (8.35)$$

$$= P\{S_{n+1} | S_1, S_2, \ldots, S_n\} \cdot \left(I_{\{(s_{n+1} = B), (a_{n+1} = 0)\}} \right.$$

$$\left. + P_{SEG} \cdot I_{\{(s_{n+1} = E), (a_{n+1} = 1)\}} + P_{REQ} \cdot I_{\{(s_{n+1} = E), (a_{n+1} = 0)\}} \right) ,$$

where $P\{S_{n+1} | S_1, S_2, \ldots, S_n\}$ are the transition probabilities of the input process.

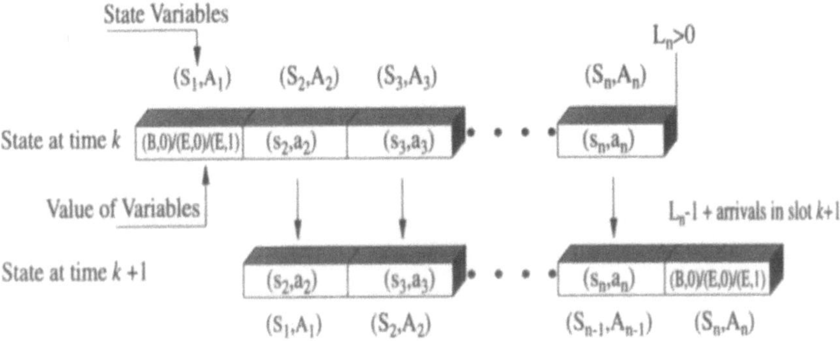

Figure 8.15: A one step transition from a generic state

Figure 8.15 shows a generic one step transition of the auxiliary process which describes the state of the system, see (8.32). As shown in the figure, by considering the transition from the k-th and $(k+1)$th embedding points, it follows that

1. the value of L_n can decrease at most by one unit. This happens when an empty slot is observed, $L_n > 0$, and no arrivals occur during the $(k+1)$th slot

2. in the state at time $(k+1)$, the value of the j-th pair $(S_j A_j)$ should be equal, for each $1 \le j < n$ to the value of the pair $(S_{j+1} A_{j+1})$ at time k.

3. in the state at time $(k+1)$, the value of the newest pair $(S_n A_n)$ takes one of the values $B0$, $E1$, $E0$. However, if the queue is empty the value $E1$ is not allowed.

Observations 1.-3. imply that the auxiliary process is a Markov chain of $M/G/1$-type in which the L_n value is the level, whilst the value of $(S_1 A_1, S_2 A_2, \ldots, S_n A_n)$ is the phase, see Section 3.3. Hence, the size of the square blocks A_j in the transition matrix is $3^{order\ of\ correlation}$

In [48] steady state probabilities are numerically computed by solving an *M/G/1*-type system. A comparison with the simulative analysis is reported in Table 8.4. Specifically, the table presents the joint probability of $n+1$ consecutive busy slots and the standard deviation of the number of consecutive busy slots computed with the nth-order analytical, simulative and Bernoulli models. The results show that the *n*th-order Markov models can capture almost all the significant dependencies in the output process of a DQDB network for a wide range of offered loads ($OL \leq 0.60$). Furthermore, the *n*-th order-Markov-model characterization always outperforms the Bernoulli characterization.

Table 8.4 Comparison between analytical, simulation and Bernoulli models in a 50 node DQDB network under *OL*=0.6

$n+1$	Joint			Standard deviation		
	Analysis	Simulation	Bernoulli	Analysis	Simulation	Bernoulli
4	0.201	(0.196-0.202)	0.124	2.53	(2.62-2.75)	1.92
5	0.151	(0.145-0.152)	0.074	2.60	(2.62-2.75)	1.92
6	0.115	(0.110-0.160)	0.044	2.64	(2.62-2.75)	1.92

In [47] it is shown how to reduce the space complexity of the *M/G/1*-type Markov chain (8.32). This complexity reduction makes possible to perform a more accurate analysis of L_NET process by capturing in the model orders of correlation higher than the 5-th.

8.3.2 Tagged Node Models

The exact queueing model of the DQDB MAC protocol in the tagged node is a discrete time single server queue with two classes of packets: the segments generated by the tagged node and the reservations from downstream nodes. The two classes are served according to a two-state discipline which reflects the different behaviors of DQDB depending on whether or not the segment queue is empty. Specifically, when the segment queue is empty (*idle-discipline*) the server attends the reservation queue continuously. When a segment arrives the *busy-discipline* is applied. According to the latter discipline the server alternates between the queues until the segment queue becomes empty. The service disciplines are: gated for the reservation queue and one limited for the segment queue.

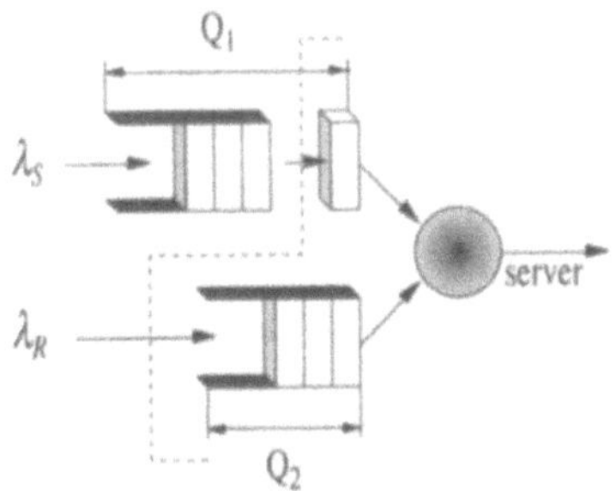

Figure 8.16: The quasi-gated queueing system

This type of discipline in [14] is referred to as a *quasi-gated discipline*, while, in [108], its discrete time version, is referred to as consistent *gated/ limited priority policy with head of line service* (*c-G/L/Hol*). Figure 8.16 describes a quasi-gated queueing model. REQs and segments arrive with rates λ_R and λ_S, and are stored in Q_2 and Q_1, respectively. Q_1 is made up of two parts. The first one contains the buffer space for one segment only, i.e., the segment at the top of the queue. All other segments are stored in the second part of Q_1. As soon as the server has finished serving the head of Q_1 a new segment can enter the first part of Q_1.

Several papers in literature have analyzed the tagged node in isolation with the quasi-gated discipline. Exact solutions have been derived only when the tagged node has a single buffer for queueing a packet [12] or a message ([93], [94]). A general solution has not yet been derived in the case of a tagged node with an infinite buffer. Three different approximate solution methods have been proposed for the infinite buffer model

- methods which provide bounds on the average performance figures ([14], [108]);
- methods which provide approximations for the tagged node performance figures ([154], [42]); and
- methods based on the *M/G/1*-type theory [115], [142].

SINGLE BUFFER MODEL. In [12] the model reported in Figure 8.16 is simplified by assuming that Q_1 can store at most one packet. Furthermore, it is assumed that the traffic generated by L_NET and R_NET are Bernoulli processes with parameters α, and β, where α is the probability that a slot is busy and β is the probability that a slot on the reverse bus contains a REQ. Under these simplifying assumptions an exact analysis of the node model is performed. Specifically, the model solution provides the generating function of the access delay, $G_D(z)$.

The access delay of a packet (D) is equal to the waiting time in the buffer (W) plus a slot time (which corresponds in this analysis to the time unit) to transmit the packet itself, $D=W+1$. Hence

$$G_D(z) = zG_w(z), \qquad |z| < 1 \quad , \tag{8.36}$$

where $G_W(z)$ is the generating function of the waiting time.

$G_W(z)$ is obtained by conditioning on the number of outstanding requests (F) when the segment is stored in Q_1:

$$G_W(z) = \sum_{i=0}^{\infty} G_w(z;i) \cdot P\{F = i\} \quad , \tag{8.37}$$

where $(G_w(z;i))$ is the waiting time generating function conditioned on the event $\{F = i\}$.

To derive (8.37) the following recursion is exploited

$$G_w(z;i) = \begin{cases} 1, & (i = 0) \\ (1 - \alpha)\, zG_w(z;i-1) + \alpha z \cdot G_w(z;i), & (i \geq 1) \end{cases} \tag{8.38}$$

The solution of recursion (8.38) is

$$\left(G_w(z;i) = \left(\frac{(1-\alpha)\cdot z}{1-\alpha z} \right)^i \right), |z| < 1, i = 0, 1, 2, \ldots \quad , \tag{8.39}$$

and thus

$$G_w(z) = \sum_{i=0}^{\infty} G_w(z;i) \cdot P\{F = i\} = G_F((1-\alpha)\cdot z/(1-\alpha z)), |z| < 1 \tag{8.40}$$

where $G_F(z) = \sum_{i=0}^{\infty} z^i \cdot P\{F = i\}$.

Finally, by substituting (8.40) in (8.36) it follows that

$$G_D(z) = zG_w(z) = z \cdot G_F((1-\alpha)\cdot z/(1-\alpha z)), |z| < 1 \quad , \tag{8.41}$$

from where the average access delay is obtained

$$E[D] \;=\; \frac{1-\alpha}{1-\alpha-\beta}\cdot \tag{8.42}$$

$$\left[\beta+\frac{e^{-\lambda}}{1-e^{-\lambda}}\cdot\{\alpha\cdot\beta-(1-\alpha)(1-\beta)(1-\theta)\}+1-\theta(1-\alpha)\right],$$

where

- $1-e^{-\lambda}$ represents the probability of a segment generation in a slot by the tagged node in the hypothesis of exponential segment interarrival times, and
- $\theta\cdot(1-\alpha)=P\{F=0\}$ represents the probability that a segment finds no outstanding requests ahead of it when it is generated by the tagged node.

Therefore the computation of the average delay is reduced to the derivation of θ. To this end, it is convenient to introduce the Markov chain $\{F_n,\,n\geq 1\}$, where F_n is the number of outstanding requests at the arrival of the n-th packet. This Markov chain is homogeneous, irreducible, and aperiodic, and it is ergodic if $\beta<1-\alpha$ (i.e., the request arrival rate is less than the rate of the empty slots observed by the tagged node).

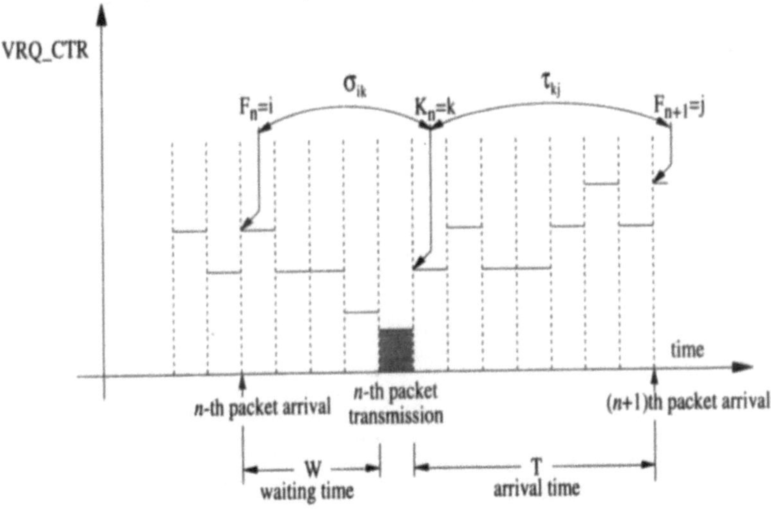

Figure 8.17: Generic transition of the embedded Markov chain

Figure 8.17 describes a transition probability $\psi_{i,j}=P\{F_{n+1}=j|F_n=i\}$

of the Markov chain. By introducing the following definition.

$$
\text{VRQ\_CTR} = \begin{cases} \text{CD\_CTR} + 1 & \text{buffer not empty} \\ \text{RQ\_CTR} & \text{buffer empty} \end{cases}
\tag{8.43}
$$

Figure 8.17 shows that the transition probability $\psi_{i,j}$ can be decomposed into two components,

$$
\psi_{i,j} = \sum_{k=0}^{\infty} \sigma_{i,k} \cdot \tau_{k,j} \quad ,
\tag{8.44}
$$

where

- $\sigma_{i,k} = P\{K_n = k \,|\, F_n = i\}$ is the conditional probability that there are $K_n = k$ outstanding requests after the transmission of a packet, given that there were $F_n = i$ outstanding requests when the packet arrived, and
- $\tau_{k,j}$ is the conditional probability that a packet finds j outstanding requests at its arrival, given that there were k outstanding requests just after the transmission of the previous packet.

$\sigma_{i,k}$ corresponds to the probability that k REQs arrive in the time required to observe $i+1$ empty slots where the distance between two consecutive empty slots has a geometric distribution with parameter α.

In [12], by defining $G_Y(z;i) = \sum_{k=1}^{\infty} \sigma_{i,k} \cdot z^k$, after several algebraic manipulations, it is shown that

$$
\begin{aligned}
G_Y(z;i) &= \sum_{t=1}^{\infty} (1 - \beta + \beta z)^{t+1} \binom{t-1}{i-1} (1 - \alpha)^i \cdot \alpha^{t-i} \\
&= (1 - \beta + \beta z) \cdot \left(\frac{(1-\alpha) \cdot (1-\beta+\beta z)}{1 - \alpha \cdot (1-\beta+\beta z)} \right)^i, \ |z| < 1 \ .
\end{aligned}
\tag{8.45}
$$

Hence, the $\sigma_{i,k}$ can be obtained numerically by inverting (8.45).

The computation of $\tau_{k,j}$ is less straightforward since it entails tracking the VRQ_CTR value when the node buffer is empty, given that its initial value is k. This can be performed through a random walk analysis with a barrier corresponding to VRQ_CTR=0.

Once the transition probabilities have been derived, the steady-state probabilities of the Markov chain $\{F_n, n \geq 1\}$ can be numerically computed and the unknown $\theta \cdot (1 - \alpha) = P\{F = 0\}$ in (8.42) is derived.

The model in [12] has been extended in [93] to the case where each node can queue one message of fixed length (l) segments. The results obtained from this model show that for a sufficiently large l the message delay behaves like a linear function of the message length that was also observed in the simulative study reported in [13].

BOUNDS ON THE PERFORMANCE FIGURES FOR INFINITE BUFFER MODELS. When the buffer is infinite, the quasi-gated queueing system presented in Figure 8.16, has not yet been analyzed exactly. In this section, bounds on the average delay figures are derived both in the continuous and in the discrete time domain.

Bounds for a model defined in the continuous time domain have been derived in [14]. Specifically, in this work the segment and REQ arrival processes are assumed to be Poisson processes with rates λ_S and λ_R, respectively, and, for each class of traffic, the service times are general independent and distributed as B_S and B_R, respectively.

The analysis is based on the observation that the quasi-gated queueing discipline is work conserving (see Section 2.4). Hence, the following relationship holds (see Theorem 2.7)

$$\rho_s \cdot E[D_S] + \rho_r \cdot E[D_R] = \frac{\rho \cdot E[\tilde{B}_+]}{(1-\rho)} + \rho_s \cdot E[B_S] + \rho_r \cdot E[B_R] \; , \; (8.46)$$

where
- $E[\tilde{B}_+]$, the residual service time of the message in service, $E[\tilde{B}_+] = 1/2 \cdot [\lambda_R \cdot E[B_R^2] + \lambda_S \cdot E[B_S^2]]$;
- $\rho_S = \lambda_S \cdot E[B_S]$, $\rho_R = \lambda_R \cdot E[B_R]$; and
- ρ is the total offered load $\rho = \rho_R + \rho_S$.

From (8.46), once $E[D_R]$ is known, $E[D_S]$ is easily obtained.

To derive $E[D_R]$ it is useful to distinguish three cases depending on the state of the system when the REQ arrives

(i)   the system is empty,
(ii)   a segment is in service,
(iii) a REQ is in service.

Event (i) occurs with probability $(1-\rho)$. The average delay conditioned on the event (i) $E[D_R^{(i)}]$ is $E[D_R^{(i)}] = E[B_R]$.

Event (ii) occurs with probability ρ_s. The average delay conditioned on event (ii) $E[D_R^{(ii)}]$ is

$$E[D_R^{(ii)}] = E[B_R] + E[B_{R+}] + E[B_R] \cdot E[N_{R|\text{event (ii)}}] \quad , \qquad (8.47)$$

where $E[B_{R+}]$ is the average residual segment service time and $E[N_{R|\text{event (ii)}}]$ is the conditional mean number of REQs in the system, given that the server serves a segment.

Event (iii) occurs with probability ρ_R. In this case the conditional average delay is

$$E[D_R^{(iii)}] = E[B_R] + E[B_{R+}] + E[B_R] \cdot E[N_{R|\text{event (iii)}}] \qquad (8.48)$$
$$+ P\{N_s > 0 | \text{event (iii)}\} \cdot E[B_S] \quad ,$$

where $E[N_{R|\text{event (iii)}}]$ is the conditional mean number of REQs in the system, given that the server serves a REQ; and $P\{N_s > 0 | \text{event (iii)}\}$ denotes the probability that the segment queue is not empty, given that the server is serving the REQ queue.

Combining the conditional delays of the three cases (i)-(iii) and using Little's law ($E[N_R] = \lambda_R \cdot E[D_R]$), it follows that

$$E[D_R] = \frac{E[\tilde{B}_+] + \xi \cdot E[B_S]}{(1 - \rho_S)} + E[B_R] \quad , \qquad (8.49)$$

where $\xi = P\{N_s > 0, \text{ event (iii)}\}$, i.e., the steady state probability the segment queue is not empty and the server is serving a REQ.

Now substituting (8.49) in (8.46) the average delay of a segment is obtained

$$E[D_S] = \frac{E[\tilde{B}_+] - (1 - \rho) \cdot \dfrac{\rho_R}{\rho_S} \cdot \xi \cdot E[B_S]}{(1 - \rho_R) \cdot (1 - \rho)} + E[B_S] \quad . \qquad (8.50)$$

The only unknown in equation (8.50) is ξ. The exact determination of ξ is very difficult but upper and lower bounds for it, are provided by the following lemma.

LEMMA 8.1   *The probability $\xi = P\{N_s > 0,\ \text{event (iii)}\}$ is bounded by*
$$0 \leq \eta \leq \xi \leq \theta \leq 1,$$
where

- $\eta = (1 - (1 - \rho_S)/B_S^*(\lambda_S))$;
- $B_S^*(s)$ *is the Laplace Stieltjes transform of the segment service time; and*
- $\theta = min((E[\tilde{B}_+]\rho_S)/(E[B_S]\cdot(1-\rho)),\rho_R)$.

The proof of this lemma is omitted here, the interested reader can find this proof in the Appendix of [14].

A simulative analysis in [14] shows that these bounds are tight, especially in light or heavy traffic conditions.

The discrete time version of the above model has been studied in [108]. In this model the time is assumed to be slotted and the service time of all classes is deterministic and equal to one slot. Three classes of traffic are considered: the busy slots traveling on the forward bus, the segments queued in the tagged node, and the requests traveling on the reverse bus. The segment and request arrival processes are assumed to be Bernoulli, while the busy slot arrival process is modeled by a 1st-order Markov model with at most one packet per slot [43]. The service time is deterministic and equal to one slot.

The upstream traffic queue has the highest priority and is served in accordance with the HoL discipline [100], while the segment and request queue are served following the consistent gated/limited discipline (c-G/L) (see Figure 8.16).

The analysis is based on a renewal/regenerative theory, a work conservation law and the theory for approximating the solution of infinite systems of equations [96]. This analysis provides upper and lower bounds on the average access delay in the tagged DQDB node. The basics steps of the solution method are outlined below.

The objective of the delay analysis is the computation of the average delay using Theorem 2.4 which leads to the following relationship

$$E[D_i] = \frac{E[C_i]}{\lambda_{C_i}\cdot E[X]}\ , \tag{8.51}$$

where

- $E[C_i]$ is the expected value of the cumulative delay of the i-th priority packets that arrived during a renewal cycle;
- λ_{C_i} is the arrival rate of the i-th class; and
- $E[X]$ is the average length of a renewal cycle.

In this model the renewal cycle is defined by the sequence of time instants in which the system becomes empty, $\{S_n, n \in \mathbb{N}\}$. By denoting with $C_{i,n}$ the cumulative delay of the i-th priority packets that arrived during the n-th cycle, it results that $\{C_{i,n}, n \in \mathbb{N}\}$ is a regenerative process with respect to the renewal process $\{S_n, n \in \mathbb{N}\}$. The length of the n-th renewal cycle, X_n, is $X_n = S_n - S_{n-1}$. and hence by applying classical renewal/regenerative arguments (see Theorem 2.4) the mean delay of an i-th priority packet can be computed from (8.51).

By analyzing the system in a renewal cycle a set of linear equations among average cumulative delays is obtained. The structure of the generic equation is

$$E[C^x(i, j)] = a^x(i, j) + \sum_{l=0}^{\infty} \sum_{m'=0}^{\infty} b(i, j, l, m) \cdot E[C^x(l, m)] \quad , \quad (8.52)$$

where

- $x \in \{REQ, SEG\}$;
- $a^x(i, j)$ and $b(i, j, l, m)$ are constants;
- $E[C^x(i, j)]$ is the average cumulative delay of all the x packets which arrived (and were served) over the time it takes the system to move from the state i, j (at time t_n) to empty;
- j is the amount of time that has elapsed since the gate was closed in a high priority queue for the n-th time;
- i is such that $i + j$ describes the time distance from t_n and the arrival of the packet at the head of the low priority queue.

In principle, the set of equations (8.52) is infinite, but by applying the theory of infinite dimensional linear equations [96], for each traffic class, say i, a lower bound of the average access delay $(E[D_i^{lo}])$ can be computed by solving a finite set of equations.

An upper bound on the average delay for each traffic class is obtained by observing that the $(c\text{-}G/L)$ service discipline is a work-conserving system. Hence a work conservation law can be used to define a relationship between the average access delay of the same system with a FIFO service discipline ($E[D_{FIFO}]$) and the weighted sum of the average access delay of each class in the *c-G/L/HoL* system

$$\lambda \cdot E[D_{FIFO}] = \sum_i \lambda_{C_i} \cdot E[D_i] \quad , \qquad (8.53)$$

where $\lambda = \lambda_{SEG} + \lambda_{REQ} + \lambda_{BUSY}$.

Considering that the average access delay for the busy-slot class of traffic is constant and equal to one, (8.53) provides a relationship between the segment and REQ access delays. Hence an upper bound on the segment average access delay is obtained by substituting in (8.53) the lower bound on the REQ average access delay

$$E[D_{SEG}^{up}] = \frac{1}{\lambda_{SEG}} \cdot \{\lambda \cdot E[D_{FIFO}] - (\lambda_{REQ} \cdot E[D_{REQ}^{lo}] + \lambda_{BUSY})\} \quad . \quad (8.54)$$

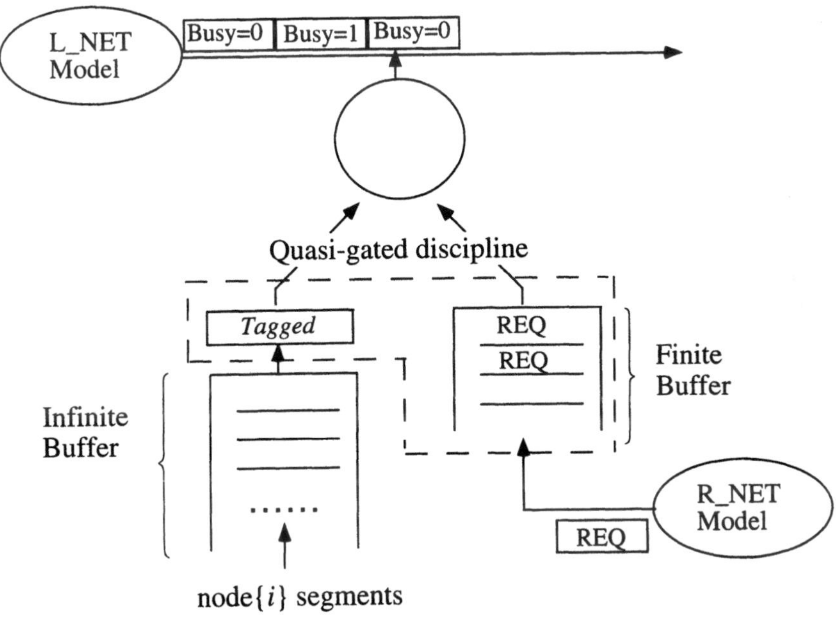

Figure 8.18: Tagged node model

FINITE REQ BUFFER MODELS. In this class of models the main simplifying assumption is that the REQ queue in the tagged node (Q_2 in Figure 8.16) has a finite size M. In DQDB this assumption implies that the sum of the

CD_CTR and RQ_CTR counters in the tagged node is limited (M is the maximum value of the sum). In addition, the L_NET process is assumed to be an n-th order discrete time Markov process (see Section 8.3.1) and therefore the tagged node model (see Figure 8.16) becomes the model reported in Figure 8.18. Under the above assumptions the tagged node model can be represented via an $M/G/1$-type Markov chain.

In the most general case, the system is observed at the beginning of each slot and its state at the h-th slot is described by $\{Q_h, \phi_h\}$, where Q_h is the number of segments waiting for transmission in the tagged node and $\phi_n = (CD\_CTR, RQ\_CTR, S_1, S_2, ..., S_n)$. It can be verified that the Markov chain that represents the system is $M/G/1$-type in which the levels correspond to the number of packets in the system Q_h and the phase corresponds to ϕ_n. In fact $Q_h \in \mathbb{N}$, and whenever $Q_h > 0$, after a transition its value can decrease by at most one. Furthermore, for $Q_h > 1$ the transition matrix is spatially homogeneous. As far as the number of state in each level is concerned, it is also possible to note that the level with $Q_h = 0$ contains less states than the other levels. This occurs because, when $Q_h = 0$, the only meaningful value for the r.v. CD_CTR is zero. Hence,

(i)   when $Q_h = 0$ the number of phases is $m_0 = 2^n \cdot (M + 1) \cdot 1$, where
- 2^n is the number of possible values for vector $(S_1, S_2, ..., S_n)$;
- $(M + 1)$ is the number of possible RQ_CTR values; and
- 1 is the number of possible CD_CTR values, i.e., CD_CTR=0.

(ii)   when $Q_h > 0$ the number of phases is $m_i = (2^n \cdot (M + 1) \cdot (M + 2))/2$, where
- 2^n is the number of possible values for vector $(S_1, S_2, ..., S_n)$, and
- $(M + 1) \cdot (M + 2)/2$ is the number of possible values for the couple (RQ_CTR, CD_CTR).

The different number of phases between $Q_h = 0$ and $Q_h > 0$ implies that the transition matrix has the following structure

$$
P = \begin{bmatrix}
B_0 & B_1 & B_2 & B_3 & \cdots \\
C_0 & A_1 & A_2 & A_3 & \cdots \\
0 & A_0 & A_1 & A_2 & \cdots \\
\cdots & 0 & A_0 & A_1 & \cdots \\
\cdots & \cdots & \cdots & \cdots & \cdots
\end{bmatrix} ,
$$

where

- the transition submatrix from level 0 to level 0, B_0, is a square matrix of order m_0;
- the transition submatrices B_i from level 0 to level i, $i>0$, have m_0 rows and m_i columns;
- the transition submatrix from level 1 to level 0, C_0, has m_i rows and m_0 columns;
- the transition submatrices A_i from a level j to level $j-i+1$ *are* square matrices of order m_i.

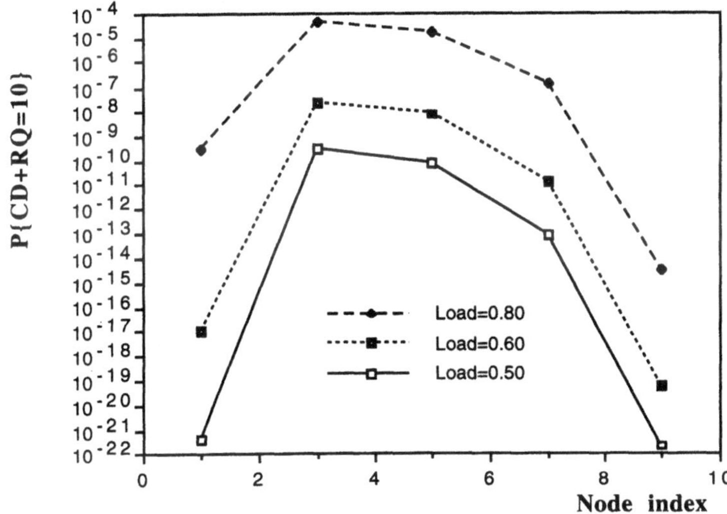

Figure 8.19: REQ loss probability (R_{loss})

As was pointed above, the main assumption of this model is the limitation on the possible values of CD_CTR and RQ_CTR. In the following the accuracy of this hypothesis is analyzed by evaluating, in a network of 10 nodes, the REQ loss probability, i.e., $R_{loss} = P\{CD\_CTR + RQ\_CTR = M\}$. This probability can be obtained by applying the methodology for the analysis of $M/G/1$-type Markov chains presented in Section 3.3.

Figure 8.19, for $M = 10$, plots R_{loss} for each network node and the various (symmetric) offered loads (i.e., 0.5, 0.6, 0.8). It clearly shows that the assumption M equal to the number of network nodes is a nice simplifying assumption. In fact, the assumption $CD\_CTR + RQ\_CTR \leq 10$ does not affect the accuracy of the performance indices computed with this model as the probability to reach such a boundary is very low even when the offered load is high.

Figure 8.20 plots the average response time $E[R]$ as a function of the node index for the various correlation levels among the slots arriving at the node (i.e., $n = 0, 1, 2$ in the L_NET process). The Bernoulli curve represents the case in which there is no correlation in the slots arriving at the node, i.e., $n = 0$.

The figure clearly shows that an accurate DQDB node model must take into consideration the correlation in the slot arrival process. As the correlation between busy slots is more significant for the nodes close to the end of the bus (see Section 8.3.1), the difference of the average response time between Bernoulli and the 2nd-order model is marked for nodes close to the end of the bus though it is not negligible for the other nodes.

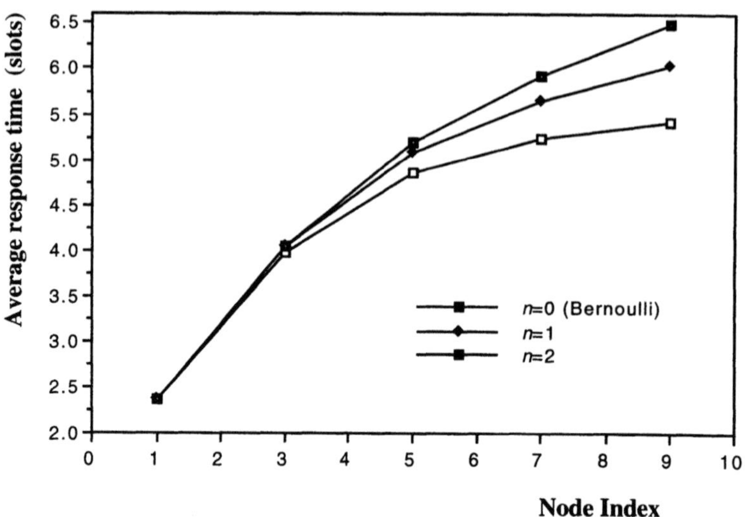

Figure 8.20   Average response time (OL=0.80)

The above model is the basis for the DQDB analyses performed in [142] and [115].

In [142] a simplified L_NET process is assumed by using a Bernoulli characterization. Under this additional assumption a Markovian model can be obtained by observing the system only at the packet departure instants. At the departure instants the CD_CTR is obviously zero and hence the size of the A_i matrices is significantly reduced.

In [115] the L_NET is modeled with a Markov chain with two states Y_t, $Y_t \in \{1, 2\}$. A slot is busy with probability γ_j, $j \in \{1, 2\}$ and is empty with probability $1 - \gamma_j$. Transitions in the underlying two state Markov chain occur every slot time. In the literature this process is referred to as a Switched Bernoulli Service Process (SBP) [83]. This means that the vector $(S_1, S_2, ..., S_n)$ previously introduced to characterize the L_NET process in this case is reduced to a single component (S_1).

Furthermore, the model includes a characterization of the BWB mechanism and the segment arrival process is a discrete time batch process with probability a_m, $(m = 0, 1, ...)$ to have m arrivals in a slot time.

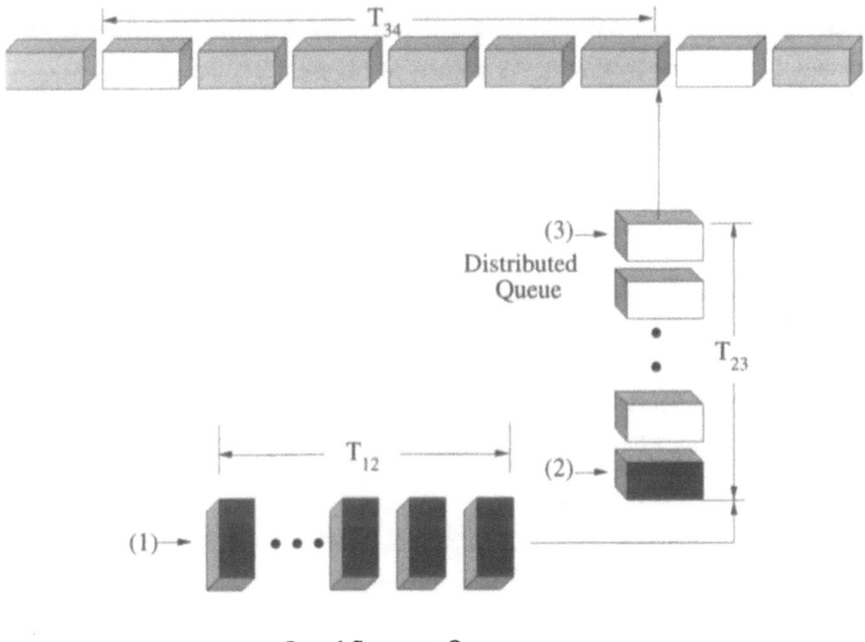

Local Segment Queue

Figure 8.21: Access delay decomposition

APPROXIMATE SOLUTIONS FOR INFINITE BUFFER MODELS. In [154], simple closed-formulas for approximating the performance figures of a tagged node in a DQDB network are derived. The L_NET process of a node{i} is modeled by a Bernoulli process with a probability q_i, $(q_i = \sum_{j=1}^{i-1} \lambda_j)$ to observe a busy slot, while the R_NET process of node$\{i\}$ is modeled by a Poisson process with rate Λ_i, $(\Lambda_i = \sum_{j=i+1}^{K} \lambda_j)$.

One of the key concepts in the analysis of this model is the decomposition of a tagged segment access delay (the time a segment spends within a node) into intervals identified by the following time instants (see Figure 8.21):

1. arrival epoch of the tagged segment;

2. time instant at which the tagged segment is inserted into the distributed queue;

3. time instant at which the tagged segment arrives at the top of the distributed queue (i.e., the node CD_CTR becomes zero);

4. end of the transmission of the tagged segment on the forward bus.

The segment access delay at node$\{i\}$ (T_{14}) is the interval between time instant 1. and time instant 4. As shown in Figure 8.21, T_{14} can be decomposed into the following random variables

- T_{12}: is the waiting time in node$\{i\}$ local queue;
- T_{23} : is the time a segment spends in the distributed queue from its insertion in the distributed queue to the time instant at which it arrives at the head of it;
- T_{34} : is the service time for node$\{i\}$ distributed queue (see Section 7.2.2), i.e., the time between successive empty slots observed by node$\{i\}$.

Since L_NET is modeled with a Bernoulli process, T_{34} has a geometric distribution and its LST is $\Phi_{34}(s) = [(1-q_i) \cdot e^{-s\Delta}] / [1 - q_i \cdot e^{-s\Delta}]$, where Δ is the slot duration.

As pointed out above, T_{34} is the time it takes to service a segment or a REQ queued in node $\{i\}$ distributed queue. Hence, the waiting time T_{23} is studied with a standard $M/G/1$ system with arrival rate Γ_i, $(\Gamma_i = \sum_{j=i}^{K} \lambda_j)$ and service time T_{34}. Thus the LST of T_{23} is (see (3.5))

$$\Phi_{23}(s) = \frac{s(1 - \Gamma_i \cdot E[T_{34}])}{s - \Gamma_i \cdot (1 - \Phi_{34}(s))} \, , \tag{8.55}$$

where $E[T_{34}] = \Delta / (1 - q_i)$ is the average of T_{34}.

Recalling that at most one segment can be in the distributed queue, $T_{24} = T_{23} + T_{34}$ can be regarded as the service time for segments queued in

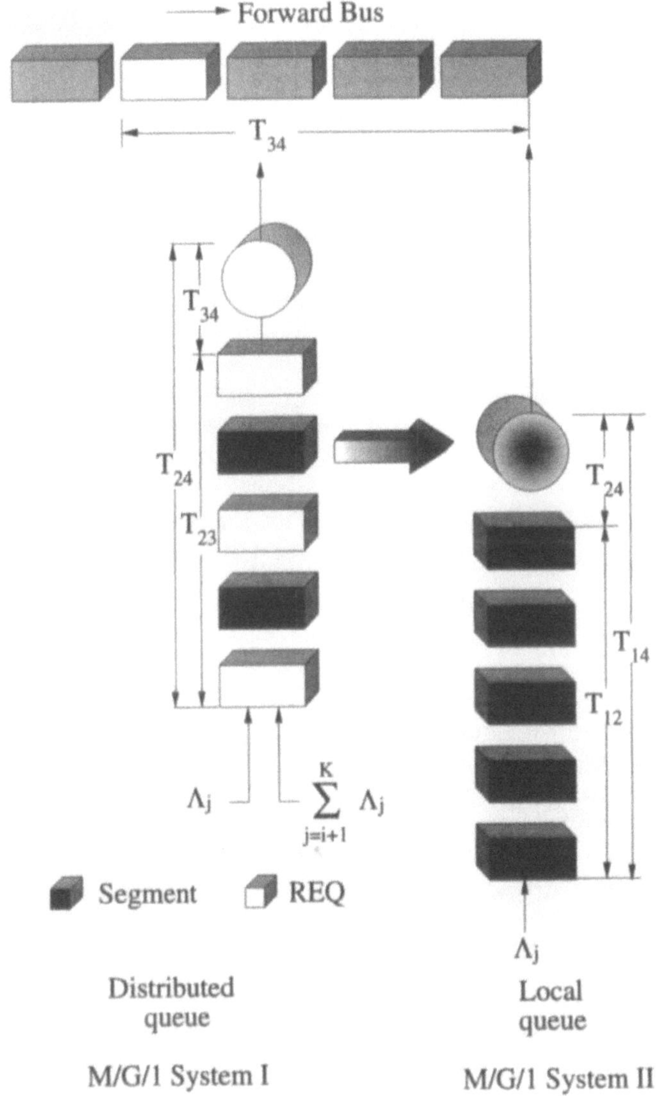

Figure 8.22: Nested systems

the local queue. Therefore T_{12} can be approximated with the waiting time experienced by the segments in an $M/G/1$ system with arrival rate λ_i and service time equal to T_{24}. Hence,

$$\Phi_{12}(s) = \frac{s\,(1 - \lambda_i \cdot E\,[T_{24\,(34)}])}{s - \lambda_i \cdot (1 - \Phi_{24}(s))}\quad,\tag{8.56}$$

where $\Phi_{24}(s)$ is the LST of T_{24}.

Finally, the LST of the access delay T_{14} is approximated by $\Phi_{12}(s) \cdot \Phi_{23}(s) \cdot \Phi_{34}(s)$.

In summary, as shown in Figure 8.22, two $M/G/1$ queueing systems are used to approximate the DQDB. The $M/G/1$ System I is used to estimate the segment waiting time in node$\{i\}$ distributed queue (T_{24}), while the $M/G/1$ System II uses the T_{24} estimation as a service time and provides the estimation of the delay experienced by a segment in the node$\{i\}$ local queue.

The main approximations introduced in this approach are related to the computation of the T_{24} distribution. In fact, in $M/G/1$ System I more than one segment can be queued. However, as was shown in Section 8.3.1 the probability of the event "more than one segment queued in node$\{i\}$" is almost negligible for a wide range of offered loads. Despite these approximations, the model is able to capture the dependence of the performance figures either from the node position or from the traffic pattern. The model proposed in [154] was extended in [42] to take into consideration a non memoryless distribution of the *length of the busy train* (T_{34}), i.e., the number of consecutive busy slots. In that paper the lengths of consecutive busy trains constitute a renewal process and hence those messages which arrive when the queue is empty are considered in a different manner (i.e., such messages experience an exceptional service). The waiting time of the messages (segments or requests) from the distributed queue (T_{23}) is therefore modeled by the waiting time in an $M/G/1$ system with an exceptional first service in a busy period [149].

9 Evolution Towards Gigabit Rates

FDDI and DQDB are technologies for LAN and MAN networks operating at speeds of 100-150 Mbps. For channel speeds increasing towards Gigabit/sec, the performance of both DQDB and FDDI becomes unacceptable. Specifically, the FDDI protocol capacity degrades (see Section 5.4), while DQDB unfairness increases (see Figure 7.29). Hence, new MAC protocols have been proposed for Gigabit/sec networks. In addition, switching-based networks are considered to be an alternative way to build Gigabit/sec networks [117]. In this chapter, these two lines of development are sketched.

9.1 SHARED MEDIUM GIGABIT NETWORKS

An extensive overview of MAC protocols designed to operate in the Gigabit/sec arena is presented in [49]. Two of them are sketched in this section: CRMA and MetaRing. These two MAC protocols have been selected because prototypes are available, and they represent two possible approaches to the problem: distributed (MetaRing) and centralized (CRMA).

9.1.1 Cyclic Reservation Multiple Access (CRMA)

CRMA is a MAC protocol for LANs and MANs operating at speeds of 1 Gigabit/sec and above ([124], [122]). It was originally designed for folded-bus and dual-bus configurations, but an extension of the protocol suitable for use in a ring configuration was later developed ([157]). In this section only the unidirectional, folded-bus topology (see Figure 9.1) is considered. The principal components of a folded-bus CRMA network are

- the Head-End node, responsible for generating empty slots and for the global access function;
- a folded bus; and
- a multiplicity of intermediate nodes, addressed by an integer number (*node index*) ranging from 1 to K, which access the outbound bus to transmit and the inbound bus to receive.

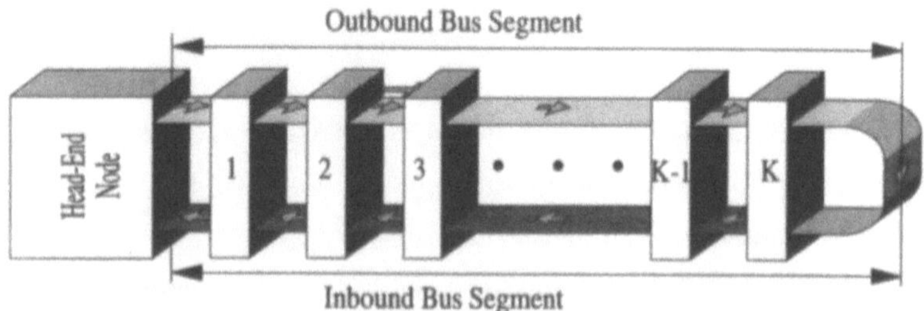

Figure 9.1: The CRMA unidirectional folded bus topology

The CRMA protocol is based on slots. Every slot is made up (see Figure 9.2) of an Access Control Field (*ACF*) and of a Segment Field (*SF*). The ACF includes a Busy/Free bit and an Access Command field. The Busy/Free bit indicates if the slot is busy or empty. The access command field contains an Access Command. Access Commands rule the access to the transmission medium.

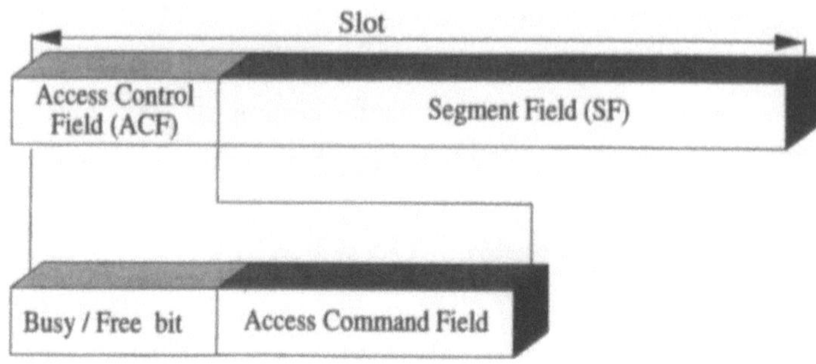

Figure 9.2: Slot format

In CRMA, access to the bus by the nodes is organized in cycles of slots. Figure 9.3 shows five cycles propagating on the folded bus. The cycles are explicitly numbered from zero to some maximum integer. When the cycle number reaches this maximum, it wraps back to zero. The nodes reserve slots in a future cycle and the Head-End node generates cycles sufficiently long to satisfy these reservations. Consequently, the cycle lengths are not fixed and are a function of node demands. Cycles for which there have been no reservations do not get generated. Reservations and the generation of transmission cycles are based on two Access Commands: *Reserve* (*RES*) and *Start* (*ST*), respectively. These commands are issued by the Head-End node. Every *Reserve* or *Start* command makes reference to a given cycle and carries the *cycle_number* as an argument. Each *Reserve* command also has the *cycle_length* as an argument, i.e., *RES(cycle_number,cycle_length)*.

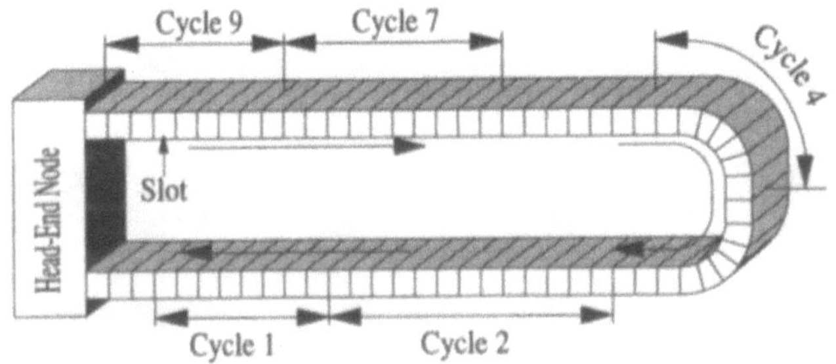

Figure 9.3: Slot cycles

The basic CRMA access mechanism is illustrated by the example given in Figure 9.4. The Head-End node periodically issues *Reserve* commands, with *cycle_length* set to zero. As the *Reserve* command of the cycle with number x passes a node (say node{i}) on the outbound bus segment, the node can reserve slots in that cycle by increasing the value of *cycle_length* for that cycle. Specifically, the number of slots reserved by node{i} in cycle x will be equal to the number by which node{i} increases *cycle_length*. Furthermore node{i} stores this number, as *reserve_length*, together with the *cycle_number* (x) in a local reservation queue. In Figure 9.4(a), nodes 1, 2, and 3 reserve 2, 1, and 3 slots in cycle x, respectively.

When a *Reserve* command passes the last node on the outbound bus segment, its *cycle_length* indicates the total number of slots requested by nodes in the corresponding cycle. The nodes do not modify the *cycle_length* on the inbound bus segment. Upon the return of the *Reserve* command to the Head-End node, a reservation, containing the *cycle_number* and *cycle_length*, is entered into a global reservation queue. This queue is served

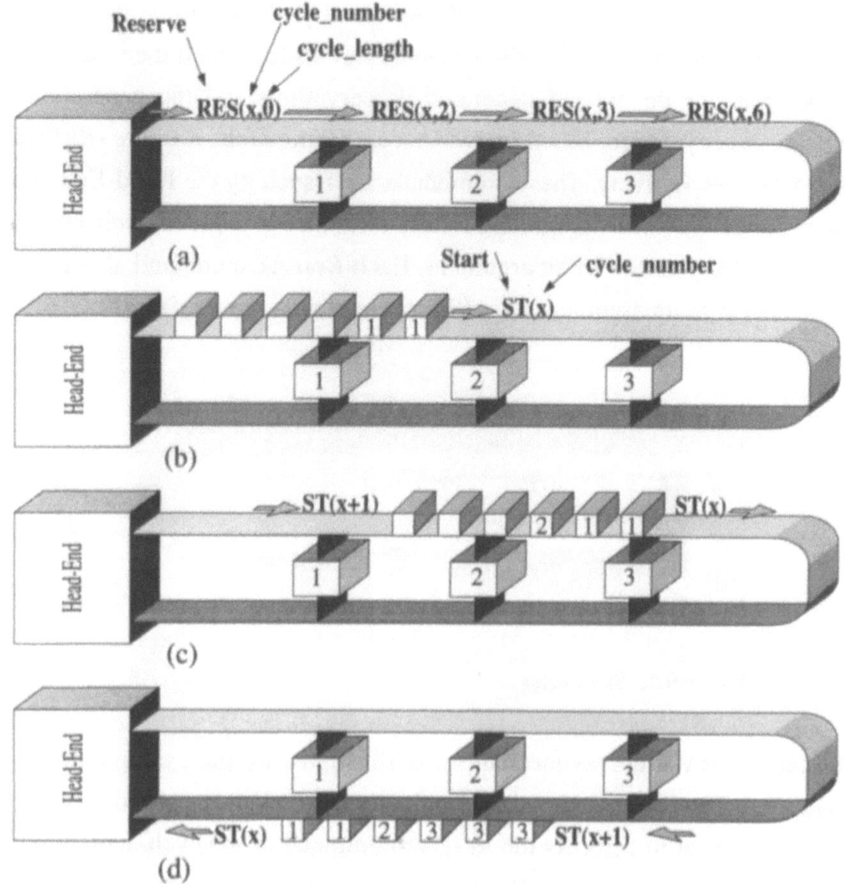

Figure 9.4: CRMA cyclic_reservation_access mechanism

according to a FIFO discipline. The Head-End node serves the reservation related to a cycle, say x, by issuing a *Start* command containing *cycle_number=x*, followed by as many empty slots as indicated by the *cycle_length* associated to cycle x.

After observing a *Start* command, e.g., $ST(x)$, a node checks its local reservation queue; if there is a reservation for *cycle_number=x*, the node waits for the first empty slot. It then uses all the slots it has reserved for that cycle (see Figure 9.4(b), (c), and (d)), this quantity is indicated by the *reserve_length* field related to cycle x in its local reservation queue. The slots used by a node in a given cycle are consecutive, due to the linear node-ordering on the outbound segment. Furthermore, every node reserves slots for all segments of a packet in the same cycle. This slot-contiguity property of CRMA substantially simplifies packet reassembly.

In order to prevent the global reservation queue from growing indefinitely, a backpressure mechanism is provided based on the *Confirm* and *Reject* commands. When a *Reserve* command comes back to the Head-End node, it is confirmed (by the Head-End node issuing a *Confirm* command) if the number of reservations in the global reservation queue does not exceed a given threshold. Otherwise, a *Reject* command is generated. This latter command notifies the nodes that all the reservations not yet confirmed must be rejected. Furthermore, when the threshold is exceeded, the periodic generation of *Reserve* commands is suspended until the number of reserved slots drops down below the threshold. The interval during which the generation of Reserve commands remains suspended is referred to as *backpressure period*. The threshold is generally set equal to the folded-bus latency, using the slot time as the time unit; with this choice it is guaranteed that the *Reserve* command issued after a backpressure period comes back to the Head-End node by the time that the number of slot reservations in the queue of reserved slots reaches zero. Therefore, the new reservations can be served immediately and no wastage of bandwidth occurs.

The reader can refer to [124] for more details on the CRMA MAC protocol, while its performance analysis can be found in [5], [6], [7], [155].

9.1.2 MetaRing MAC Protocol

The MetaRing is a MAC protocol for LANs and MANs operating at speed of 1 Gigabit/sec and above ([28]). It connects a set of nodes by means of two unidirectional counter-rotating rings made up of point-to-point serial links.

There are two versions of the MetaRing: *slotted*, for fixed-length cells, and *buffer insertion*, for variable-size packets. In this section only the slotted access mode is discussed.

In the slotted access mode, user data is segmented into cells, and cells are transmitted in one per slot. Slots are composed of a header and an information field. The header includes a busy bit which indicates whether the slot is empty or busy. A cell can only be transmitted in an empty slot. The MetaRing MAC protocol uses the shortest-path criterion to choose the direction in which a cell is transmitted. Cells are removed by the destination node, which frees the slot by resetting the busy bit to zero, and thus "the spatial reuse of bandwidth" is achieved. After cell removal, the slot is immediately available to transmit a new cell; hence, during one ring rotation, the same slot can transport several different cells. For example, in Figure 9.5 a

Figure 9.5   Full-duplex ring with "the spatial reuse of bandwidth"

slot made busy by station$\{i\}$ is freed by station$\{i+2\}$, hence the slot can be reused by station$\{i+4\}$ for its transmission to station$\{2\}$.

The MetaRing MAC protocol provides its users with two types of service: *asynchronous* and *synchronous*. Synchronous service requires a connection set-up phase, and synchronous connections are accepted only if peak-rate bandwidth can be guaranteed. Asynchronous traffic uses the remainder of the bandwidth. Hence the synchronous service is designed to support real-time applications, while asynchronous service is suitable for EDP applications.

Each node has two queues, one for synchronous traffic and the other for asynchronous traffic; the traffic in the synchronous queue has priority over the traffic in the asynchronous queue. Whenever a station observes an empty slot it can always transmit synchronous traffic, but an authorization is always required before transmitting asynchronous traffic. For example, in the SAT (from SATisfied) algorithm ([28]), asynchronous transmissions are authorized by a control signal, called SAT, which circulates in the opposite direction with respect to the flow of information it regulates (Figure 9.6). Each node has a counter, the value of which increases by one every time the node transmits an asynchronous cell. The node can transmit an asynchronous cell every time it observes an empty slot unless it has already transmitted k asynchronous cells. When the counter reaches the value of k, the node must refrain from sending new asynchronous cells until the SAT signal arrives at the node. There are three cases, depending on the counter value, which arriving SAT signal can encounter

(i)   the counter value is greater or equal than l, where l is a protocol parameter less or equal to k. In this case the node is *satisfied*.

(ii)   the counter value is less than l and the node asynchronous queue is empty. In this case the node is *satisfied*.

(iii)   the counter value is less than l and the node asynchronous queue is not empty. In this case the node is *not-satisfied*.

A not-satisfied node holds the SAT signal until the counter value reaches l, then it reaches the satisfied state. As soon as a node is satisfied it clears the

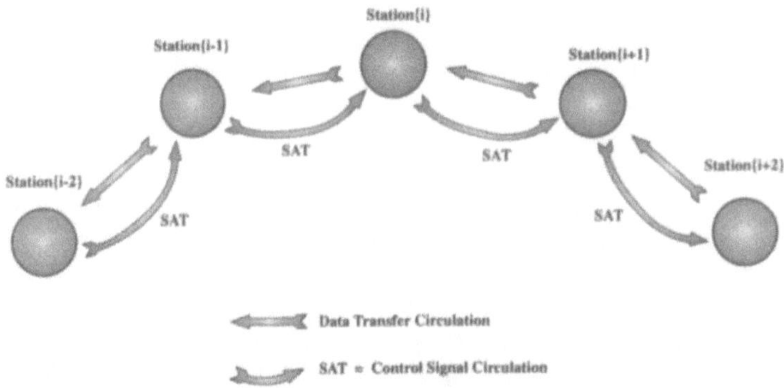

Figure 9.6: The relationship between data and SAT signal circulation

counter and immediately releases the SAT signal towards its upstream node. After releasing the SAT signal, a node is authorized to transmit up to k asynchronous cells.

The SAT mechanism guarantees that all stations, in each SAT signal rotation, can transmit at least l asynchronous cells. For this reason the SAT mechanism is called the "global" fairness mechanism [28]. Other fairness mechanisms have already been designed for MetaRing, e.g., the "local" fairness mechanism [34].

To prevent asynchronous traffic from degrading the QoS of the synchronous traffic, a mechanism which enables and disables asynchronous traffic transmission is included in the MAC protocol. This mechanism employs a control signal, named ASYNChronous-ENable (ASYNC-EN), which can carry one of the following three attributes: GREEN (ASYNC-EN(GR)), YELLOW (ASYNC-EN(YL)), and RED (ASYNC-EN(RD)). This signal circulates in the opposite direction with respect to the information it regulates. When the ASYNC-EN(GR) rotates around the ring, asynchronous traffic transmission can occur. Each node will forward the ASYNC-EN(GR) immediately after receiving it. After the ASYNC-EN(GR) has completed at least r rounds (r is a protocol parameter), a node with a backlog of synchronous traffic (i.e., a node for which the first packet in the synchronous queue has been waiting for more than a predefined threshold, which is an additional protocol parameter, hereafter called *Thres*) can change the attribute from GREEN to YELLOW. When nodes observe the ASYNC-EN(YL) signal, they must *a*) refrain from transmitting asynchronous cells and *b*) forward the signal to adjacent nodes. When after one round the ASYNC-EN(YL) returns to the node which set it (its origin node), that node switches the signal's attribute from YELLOW to RED. The ASYNC-EN(RD) is transferred once around the ring. A node forwards the ASYNC-EN(RD) to its upstream neighbour if it has no backlog of synchronous traffic; otherwise, it holds the ASYNC-EN(RD) until its backlog of synchronous traffic has been exhausted. When the ASYNC-EN(RD) signal returns to its origin node, that node will change its attribute back to GREEN. The ASYNC-EN(GR) signal should complete at least r rounds before an ASYNC-EN can be set to YELLOW again. The mechanism which integrates asynchronous and synchronous traffic therefore uses two protocol parameters: r and *Thres*.

The reader can refer to [28], [127], and [166] for more details on the MetaRing MAC protocol, while its performance analysis can be found in [8], [9], [28], [34], [52], and [166].

9.2 ATM-BASED GIGABIT NETWORKS

The limitation of shared-medium networks is mainly related to limited effective bandwidth per user. In these networks the peak bandwidth available to any single user is the same as the aggregate network bandwidth. Because the cost of a network rapidly increases with the medium speed, networks based on a shared medium will not be able to supply large aggregate bandwidth in a cost-effective way. Switch-based networks are generally considered to be the most economical way to build Gigabit networks [117].

The Asynchronous Transfer Mode (*ATM*) is the emerging technology for switch-based, Gigabit wide area networks. Nowadays, most vendors offer ATM-based systems as a cost-effective solution for satisfying the communication requirements of local area networks, as well. *ATM LAN* refers to the use of ATM techniques in a local environment. With this technology, each network station accesses the network via a dedicated link with a bandwidth of 622 Mbps or 155 Mbps (or, if desired, 45 Mbps or 1.544 Mbps).

Before describing the main features of an ATM LAN network, it is useful to outline the principal concepts of the ATM technology. ATM is a packet-oriented switching and multiplexing technique. ATM is asynchronous in the sense that cells containing information from an individual user do not have to appear at the receiving end at predictable times, as it would occur in traditional, time-division multiplexing (e.g., Synchronous Transfer Mode, *STM*).

ATM offers to its users an end-to-end connection oriented data service. As in traditional WANs, user information is packetized in fixed-size, 53-octet packets (*cells*). Information is transferred to a packet-switching node (*ATM switch*) by using a dedicated link. ATM switches using the well-known store-and-forward technique to route packets to their destination. Each cell includes a 5-byte header containing, among other information, the connec-

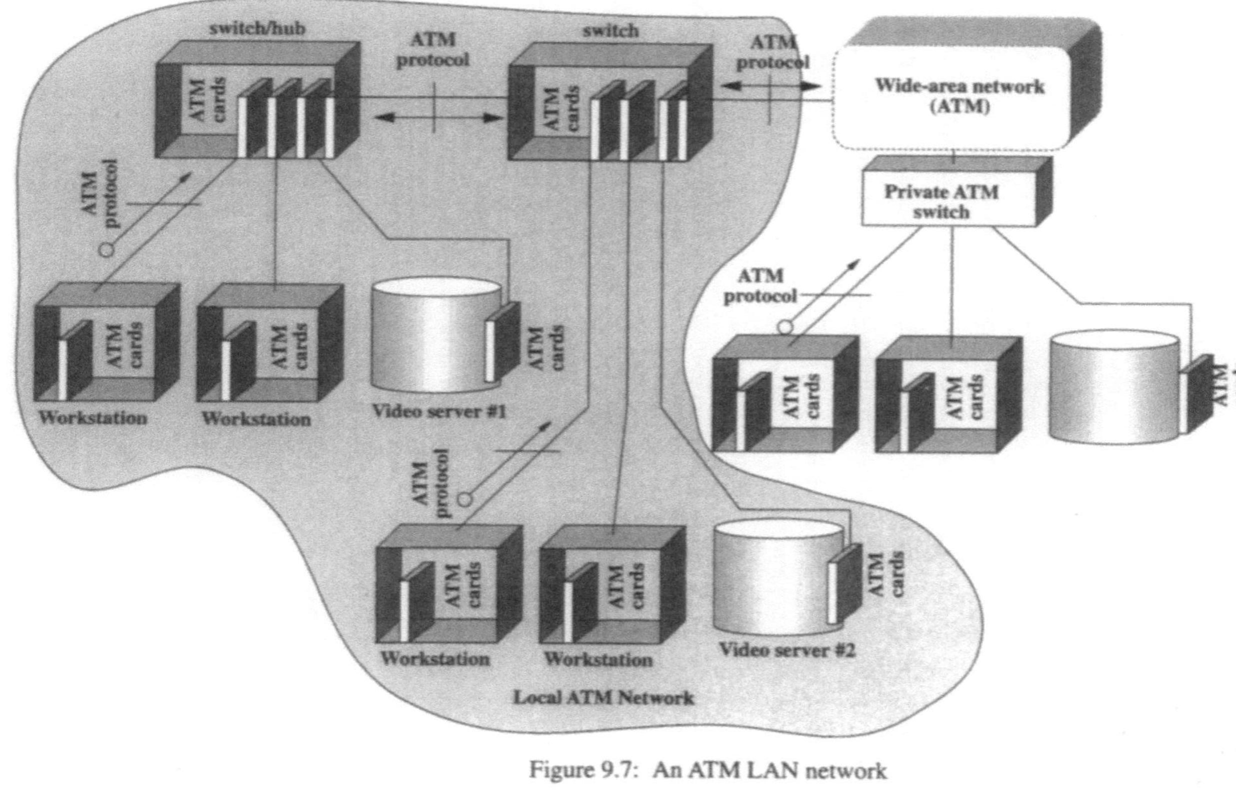

Figure 9.7: An ATM LAN network

tion identifier. Connection identifiers are assigned in the connection set-up phase, and they are used by the ATM switches to route the cells. Each user is assured that its data is reliably, rapidly, and securely transmitted over the network in a manner consistent with a subscribed QoS. Although statistical multiplexing is obtained by sharing a physical link among several virtual channels, cells coming from the user at a stipulated (subscription) rate are guaranteed delivery at the other end with very high probability and with low delay, almost as if the user had a dedicated high-speed link between the two points. Of course, the user does not have such a dedicated and expensive end-to-end facility.

ATM can support different speeds and provide different classes of service matched to the requirements of different traffic types (e.g., data, voice, and video). A precise description of ATM protocols can be found in several books (e.g., [10], [58], [116], [128], [79], and [135]).

ATM LAN standards were developed by the ATM Forum.[1] Figure 9.7 shows an example of an ATM LAN network. The network requires the following hardware components: host ATM interfaces (ATM cards), local ATM switches and ATM hubs. Physical point-to-point links exist between couples of network components: (switch, switch), (hub, switch), (host, switch) and (host, hub).

Depending on their bandwidth requirements, users access the ATM LAN via:

- a link to a local ATM switch; typically this is a 100-155 Mbps access link;
- a link to a local ATM hub or concentrator which is directly connected to a local ATM switch; typically this is a 10-30 Mbps access link; or
- a conventional LAN; in this case, a station of the LAN is directly connected to an ATM hub or switch.

Local ATM switches may also provide local users with a connection to an ATM WAN.

---

1. The ATM Forum is a worldwide organization, aimed at promoting ATM within the industry and the end-user community. Formed in October 1991, currently it includes more than 700 companies representing all sectors of the communication and computer industries, as well as a number of government agencies, research organizations and user.

9.2.1 ATM LAN

In this section, to better clarify the ATM LAN concepts, the case study presented in Figure 9.8 will be considered. Figure 9.8 presents a simple ATM

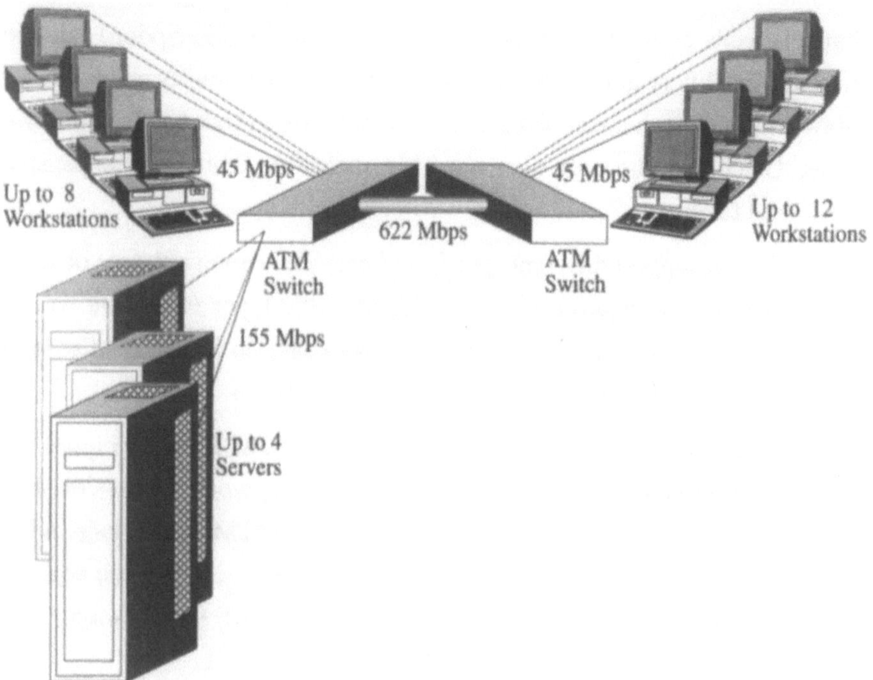

Figure 9.8: A simple ATM LAN

LAN network used in a distributed multimedia system. Several multimedia workstations use the network to access multimedia servers or to run distributed multimedia applications, e.g., teleconferencing, cooperative work, etc. As several workstations can require access to a server at the same time, a 155 Mbps access link has been chosen for the servers, while a 45 Mbps access link is assumed to be sufficient for each workstation. To avoid a potential bottleneck, the two ATM switches are connected by a 622 Mbps link.

When useful to enhance concreteness, reference is made to the FORE Systems technology. FORE was taken as a reference because details on its architecture are available in the scientific literature (see [11], [116]).

The two hardware components required by the network shown in Figure 9.8 are an ATM host interface in each network station, and two ATM switches. The ATM host interface is primarily in charge of ensuring the functionality of the ATM physical layer.

The ATM switch has two main functions: virtual connection management, performed mainly by software, and cell routing, implemented in hardware. Figure 9.9 shows the functional blocks of the FORE local ATM

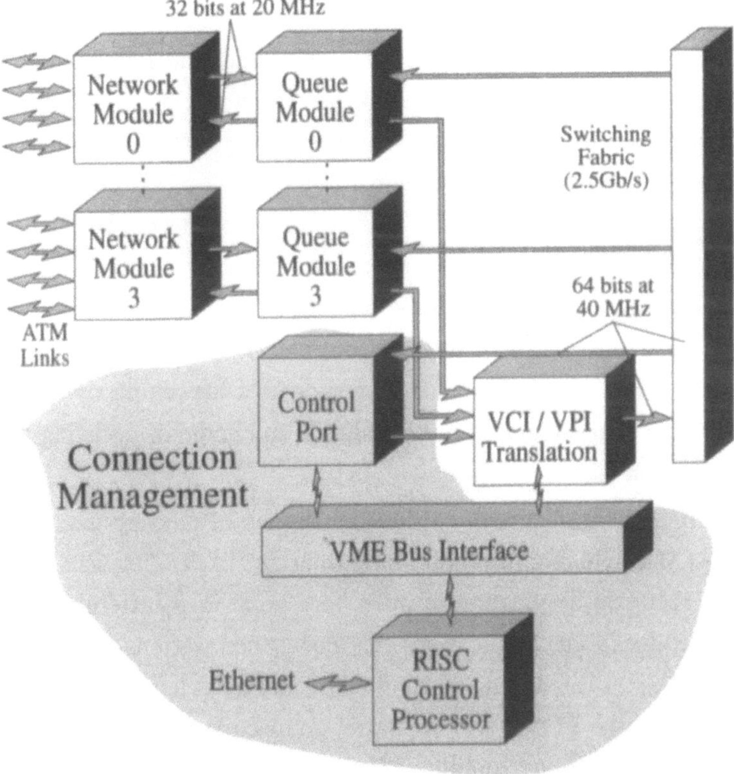

Figure 9.9: Functional architecture of the FORE ATM switch

switch. The switch provides a 2.5 Gbps aggregate throughput, and it is designed to support incoming and outgoing links with different speeds. As shown in the figure, there are up to four couples of Network and Queue modules. Each couple can support either 4 full-duplex links with a speed in the range 45-155 Mbps, or a single, full-duplex, 622 Mbps link. The ATM switching fabric transfers cells from the source network module to one or more destination network modules, hence it is in charge of providing the multicast service. The switching fabric is based on a non-blocking output-buffered architecture [58].

To better explain the routing of a cell, it is useful to remember that in ATM the connection identifier is made of two components, a Virtual Channel Identifier (*VCI*) and a Virtual Path Identifier (*VPI*). In addition, the connection identifier takes a different value on each link used by the connection. The mapping between the identifiers of a virtual channel in the incoming and outgoing links is performed in the switch using a translation table (see VCI/VPI Translation in Figure 9.9). Updating of the translation table occurs whenever a connection starts or terminates. A signalling protocol is implemented between the host and the switches involved in the connection to manage the connection set-up. The main function of this protocol is the setting up of the VCI/VPI translation table in each switch used by the new connection. Signalling messages are sent in the ATM LAN on predefined virtual circuits. Messages received on these predefined virtual circuits are delivered to the Control Processor using the Control Port of the ATM switch (see Figure 9.9). The Control Processor is responsible for setting up the connections crossing that switch; in particular, it is in charge of updating the VPI/VCI translation table.

Multimedia applications exploit the service offered by the ATM LAN via the ATM Application Programming Interface (*API*). Services offered by the API enable an application to open both point-to-point connections and multicast connections; applications can request connections with guaranteed bandwidth (e.g., peak and mean bandwidth, and burst length) and the appropriate QoS. In the FORE API software, connection set-up is requested by calling the *atm_connect* routine. The arguments of this routine allow the application to specify the destination ATM address and the QoS. In addition, an ATM Service Access Point is used at both sides of a connection in the same way as ports are used in the TCP/IP architecture [36].

Connections are managed according to the client/server model of distributed computing [36]. On the destination side, a server program uses the *atm_listen* routine to receive the request to open a connection. If an incoming connection is accepted the server invokes the *atm_accept* routine, which propagates back to the sender. When the connection is opened, data is sent by invoking the *atm_send* routine; data is received by invoking the *atm_receive* routine.

The API interface is designed for a new class of applications specifically tailored to the ATM technology. To smooth the migration to ATM, it is important that ATM LANs emulate many features of the existing LANs (*legacy* LANs). For this reason, the concept of *LAN emulation* was developed by the ATM Forum. Appropriate software, overlaid on an ATM network, emulates the services of legacy LANs, and this makes possible the use of existing applications without any modification also in the ATM environment. LAN emulation protocols for the Token Ring (IEEE 802.5) and Ethernet (IEEE 802.3) are already available.

... by inverting the horizontal routing, this is repeated by moving the receiver/driver.

The ATM interface is designed for a new class of applications with ... recently introduced to the ATM technology. To simplify the implementation it is important that ATM LANs simulate many features of the existing LANs for ...

Bibliography

[1]   Abeysundara, B.W., Kamal, A.E., (1991), High-Speed Local Area Networks and their Performance: a Survey, *ACM Computing Surveys*, **23**, 221-264.

[2]   Albert, B., Jayasumana, A.P., (1994), *FDDI and FDDI-II*, Artech House, Boston.

[3]   Altman, E, (1994), Analysing Timed-Token Ring Protocols Using the Power-Series Algorithm, *The Fundamental Role of Teletraffic in the Evolution of Telecommunications Networks*, J. Labetoulle and J.W. Roberts (Editors), Elsevier Science Publishers (North-Holland), Amsterdam, 961-971.

[4]   ANSI X3.139, (1987), *FDDI Media Access Control (MAC)*, ANSI.

[5]   Anastasi, G., Conti, M., Gregori, E., Lenzini, L., (1993), CRMA Media Access Control Protocol: A Simulative Analysis, *Computer Communications*, **16**, 39-47.

[6]   Anastasi, G., Conti, M., Gregori, E., Lenzini, L., (1993), Service Integration in CRMA: a Simulative Analysis, *Proceedings of the IEEE INFOCOM '93 Conference*, 715-721.

[7]   Anastasi, G., Conti, M., Gregori, E., Lenzini, L., (1995), Real-Time Applications in a CRMA Network: a Performance Analysis, *Computer Communications*, **18**, 871-879.

[8]   Anastasi, G., Lenzini, L., Motta, P., (1995), Performance Evaluations of a MetaRing MAC Protocol Integrating Video and Data Traffics in an Interconnected Environment, *Proceedings of the IEEE INFOCOM '95 Conference*, 1273-1281.

[9]   Anastasi, G., La Porta, M., Lenzini, L., (1996), A Performance Study of the Local Fairness Algorithm for the MetaRing MAC Protocol, *Proceedings of the International Zurich Seminar on Digital Communications*, 187-207.

[10]   ATM Forum, (1993), *ATM User-Network Interface Specification*, Version 3.0, Prentice-Hall, Englewood Cliffs, N.J.

[11]   Biagioni, E., Cooper, E., Sanson, R., (1993), Designing a Practical ATM LAN, *IEEE Network*, **7**, 32-39.

[12]   Bisdikian, C., (1990), Waiting Time Analysis in a Single Buffer DQDB (802.6) Network, *IEEE Journal on Selected Areas in Communications*, **8**, 1565-1573.

[13]   Bisdikian, C., (1992), A Performance Analysis of the IEEE 802.6 (DQDB) Subnetwork with the Bandwidth Balancing Mechanism, *Computer Networks and ISDN Systems*, **24**, 367-385.

[14]   Bisdikian, C., (1993), A Queueing Model with Applications to Bridges and the DQDB (IEEE802.6) MAN, *Computer Networks and ISDN Systems*, **25**, 1279-1289.

[15]   Blanc, J.P.C., (1987), On a Numerical Method for Calculating State Probabilities for Queueing Systems with More Than One Waiting Line, *Journal of Computational and Applied Mathematics*, **20**, 119-125.

[16]   Blanc, J.P.C., (1990), A Numerical Approach to Cyclic-Service Queueing Models, *Queueing Systems*, **6**, 173-188.

[17]   Blanc, J.P.C., (1991), The Power-Series Algorithm Applied to Cyclic Polling Systems, *Stochastic Models*, **7**, 527-545.

[18]   Blanc, J.P.C., (1992), An Algorithmic Solution of Polling Models with Limited Service Disciplines, *IEEE Transactions on Communications*, **40**, 1152-1155.

[19]   Blanc, J.P.C., (1992), Performance Evaluation of Polling Systems by Means of the Power-Series Algorithm, *Annals of Operation Research*, **35**, 155-186.

[20]   Blanc, J.P.C., (1993), Performance Analysis and Optimization with the Power-Series Algorithm, *Lecture Notes in Computer Science*, L. Donatiello and R. Nelson (Editors), **729**, 53-80.

[21]   Boudreau, P.E., Griffin Jr.,J.S., Kac, M., (1962) An Elementary Queueing Problem, *American Mathematical Monthly*, **69**, 713-724.

[22] Boxma, O.J., (1985), Two Symmetric Queues with Alternating Service and Switching Times, *Proceedings of the Performance '84 Conference*, E. Gelenbe (Editor), North-Holland, Amsterdam, 409-431.

[23] Boxma, O.J., Meister, B.W., (1987), Waiting-Time Approximations for Cyclic-Service Systems with Switchover Times, *Performance Evaluation*, **7**, 299-308.

[24] Boxma, O.J., Groenendijk, W.P., (1987), Pseudo-conservation Laws in Cyclic-Service Systems, *Journal of Applied Probability*, **24**, 949-964.

[25] Boxma, O.J., Groenendijk, W.P., (1988), Two Queues with Alternating Service and Switching Times, *Queueing Theory and its Applications* (*Liber Amicorum for J.W. Cohen*), O.J., Boxma, R. Syski, (Editors), North-Holland, Amsterdam, 261-282.

[26] Bruneel, H., Kim, B.G., (1993), *Discrete-Time Models for Communication Systems Including ATM*, Kluwer Academic Publisher.

[27] Burke, P.J., (1956), The Output of a Queueing System, *Operations Research*, **4**, 699-714.

[28] Cidon, I., Ofek, Y., (1993), MetaRing, a Full Duplex Ring with Fairness and Spatial Reuse, *IEEE Transaction on Communications*, **41**, 110-120.

[29] Cinlar, E., (1975), *Introduction to Stochastic Processes*, Prentice-Hall, Inc., Englewood Cliffs, N.J.

[30] Bondavalli, A., Conti, M., Gregori, E., Lenzini, L., Strigini, L., (1990), MAC Protocols for High-Speed MANs: Performance Comparisons for a Family of FASNET-based Protocols, *Computer Networks and ISDN Systems*, **18**, 97-113.

[31] Chang, K., Sandhu, D., (1990), Pseudo-Conservation Laws in Cyclic-Server, Multiqueue Systems with a Class of Limited Service Policies, *Proceedings of the IEEE INFOCOM '90 Conference*, 260-267.

[32] Chang, K., Sandhu, D., (1992), Mean Waiting Time Approximations in Cyclic-Service Systems with Exhaustive Limited Service Policy, *Performance Evaluation*, **15**, 21-40.

[33] Chang, K., Sandhu, D., (1994), Delay Analyses of Token-Passing Protocols with Limited Token Holding Times, *IEEE Transactions on Communications*, **42**, 2833-2842.

[34] Chen, J., Cidon, I., Ofek, Y., (1993), A Local Fairness Algorithm for Gigabit LAN's/MAN's with Spatial Reuse, *IEEE Journal on Selected Areas in Communications*, **11**, 1183-1192.

[35] Chiarawongse, J., Srinivasan, M. M., Teorey, T. J., (1991), The *M/G/*1 Queueing System with Vacations and Timer Controlled Service, *IEEE Transactions on Communications*, **42**, 1846-1855.

[36] Comer, D.E., (1991), *Internetworking with TCP/IP*, Prentice-Hall, Englewood Cliffs, N.J.

[37] Conti, M., Gregori, E., Lenzini, L., (1991), A Methodological Approach to an Extensive Analysis of DQDB Performance and Fairness, *IEEE Journal on Selected Areas in Communications*, **9**, 76-87.

[38] Conti, M., Gregori, E., Lenzini, L., (1991), DCP, A Novel Distributed-Control Polling MAC Protocol: Specifications and Comparison with DQDB, *IEEE Journal on Selected Areas in Communications*, **9**, 241-247.

[39] Conti, M., Gregori, E., Lenzini, L., (1991), A Comprehensive Analysis of DQDB, *European Transactions on Telecommunications*, **2**, 403-413.

[40] Conti, M., Grandoni, F., Gregori, E., Lenzini, L., Strigini, L., (1991), Interconnection of Dual Bus MANs: Architecture and Algorithms for Bandwidth Allocation, *International Journal of Internetworking*, **2**, 1-22.

[41] Conti, M., Gregori, E., Lenzini, L., (1992), A Vacation Model for Interconnected FDDI Networks, *Proceedings of the ACM 1992 Computer Science Conference*, 9-16.

[42] Conti, M., Gregori, E., Lenzini, L., (1992), DQDB Modeling: Reduction of the Complexity and a Solution via Markov Chains, *IFIP Transactions*, **C-5**, 47-62.

[43] Conti, M., Gregori, E., Lenzini, L., (1992), On the Approximation of the Slot-occupancy Pattern in a DQDB Network, *Performance Evaluation*, **16**, 159-176.

[44] Conti, M., Gregori, E., Lenzini, L., (1993), Metropolitan Area Networks (MANs): Protocols, Modeling and Performance Evaluation, *Lecture Notes in Computer Science*, L. Donatiello and R. Nelson (Editors), **729**, 81-120.

[45] Conti, M., Gregori, E., Lenzini, L., (1993), Analysis of a Medical Communication System Based on FDDI, *International Journal of Microcomputer Applications*, **12**, 118-127.

[46] Conti, M., (1993), Analysis of the Quality of Service in a MAN Environment, *IFIP Transaction*, **A-39**,137-148.

[47] Conti, M., (1993), The Influence of the Slot-Occupancy-Pattern-Process Model on the Performance Figures of a DQDB Subnetwork, *Proceedings of the Performance '93 Conference,* 395-399.

[48] Conti, M., Gregori, E., Lenzini, L., Neuts, M.F., (1994), An *M/G/*1 Type Approach to the Approximation of the Slot-occupancy Pattern in a DQDB Network, *Performance Evaluation*, **21**, 59-80.

[49] Conti, M., Gregori, E., Lenzini, L., (1994), E-DCP, an Extension of the Distributed-Control Polling MAC Protocol (DCP) for Integrated Services, *Computer Networks and ISDN Systems*, **26**, 711-719.

[50] Conti, M., Gregori, E., Lenzini, L., (1995), Estimating the Quality of Service of Token Passing MAC Protocols, *Computer Communications*, **18**, 15-23.

[51] Conti, M., Gregori, E., Lenzini, L., (1995), Influence of the BWB Mechanism on some performance figures of a DQDB subnetwork, *Computer Networks and ISDN Systems*, **27**, 1137-1161.

[52] Conti, M., Donatiello, L., Furini M., (1996), MetaRing^{+}: an Enhancement of the MetaRing Access Protocol for Supporting Real-time Applications, *Proceedings of the 8-th Euromicro Workshop on Real Time Systems*, 218-223.

[53] Cooper, R.B., (1970), Queues Served in Cyclic Order: Waiting Times, *Bell System Technical Journal*, **49**, 399-413.

[54] Cooper, R.B., (1990), *Introduction to Queueing Theory*, (3rd edition), CEE Press Books.

[55] Cox, D.R., (1955), The Analysis of Non-Markovian Stochastic Processes by the Inclusion of Supplementary Variables, *Proceedings of Cambridge Philosophical Society*, **51**, 433-441.

[56] Cox, D.R., P.A.W. Lewis, (1966), *The Statistical Analysis of Series of Events*, Methuen & Co. Ltd., London.

[57] Davids, P., Martini, P., (1990), Performance analysis of DQDB, *Proceedings of the IEEE 1990 Phoenix Conference*, 548-555.

[58] de Prycker, M., (1991), *Asynchronous Transfer Mode*, Ellis Horwood, New York.

[59] Doshi, B.T., (1986), Queueing System with Vacations - a Survey, *Queueing Systems*, **1**, 29-66.

[60] Doshi, B.T., (1990), Single Server Queues with Vacations, *Stochastic Analysis of Computer and Communication Systems*, Takagi, H., (Editor), Elsevier Science Publishers (North-Holland), Amsterdam 217-318.

[61] Dykeman, D., Bux, W., (1988), Analysis and Tuning of the FDDI Media Access Control Protocol, *IEEE Journal on Selected Areas in Communications*, **6**, 997-1010.

[62] Everitt, D., (1989), A Note on the Pseudo-Conservation Laws for Cyclic Service Systems with Limited Service Disciplines, *IEEE Transactions on Communications*, **37**, 781-783.

[63] Ferguson, M.J., Aminetzah, Y.J.,(1985), Exact Result for Asymmetric Token Ring System, *IEEE Transactions on Communications*, **33**, 223-231.

[64] Franken, P., Konig, D., Arndt, U., Schmidt, V., (1981), *Queues and Point Processes*, Akademie-Verlag, Berlin,

[65] Fuhrman, S.V., (1984), A Note on the $M/G/1$ Queue with Server Vacations, *Operations Research*, **32**, 1368-1373.

[66] Fuhrman, S.V., (1985), Symmetric Queues Served in a Cyclic Order, *Operations Research Letters*, **4**, 139-144.

[67] Fuhrman, S.V., Cooper, R.B., (1985), Stochastic Decomposition in an $M/G/1$ Queue with Generalized Vacations, *Operations Research*, **33**, 1117-1129.

[68] Fuhrman, S.V., Wang, Y., (1988), Analysis of Cyclic Service Systems with Limited Service: Bounds and Approximations, *Performance Evaluation*, **9**, 35-54.

[69] Garret, M.W, Li, S.Q., (1991), A Study of Slot Reuse in Dual Bus Multiple Access Networks, *IEEE Journal on Selected Areas in Communications*, **9**, 248-256.

[70] Genter, W.L., Vastola, K.S., (1990), Delay Analysis of the FDDI Synchronous Data Class, *Proceedings of the IEEE INFOCOM '90 Conference*, 766-773.

[71] Georgiadis, L., Szpankowski, W., (1992), Stability of Token Passing Rings, *Queueing Systems*, **11**, 7-33.

[72] Gerla, M., Chan, H.W., Boisson de Marca, J.R., (1985), Fairness in Computer Networks, *Proceedings of the ICC '85 Conference*, 1384-1389.

[73] Gnedenko, B. V., Kovalenko I. N., (1989), *Introduction to Queueing Theory*, (2nd edition), Birkhauser, Boston.

[74] Groenendijk, W.P., (1988), Waiting-Time Approximations for Cyclic-Service Systems with Mixed Service Strategies, *Teletraffic Science for New Cost-Effective Systems, Networks and Services*, M. Bonatti (Editor), Elsevier Science Publisher (North-Holland), Amsterdam, 1434-1441.

[75] Groenendijk, W.P., (1988), *A Conservation-law Based Approximation Algorithm for Waiting Times in Polling Systems*, Centre for Mathematics and Computer Science, Report OS-R8816.

[76] Groenendijk, W.P., (1990), *Conservation Laws in Polling Systems*, Ph. D. Dissertation, Centre for Mathematics and Computer Science, Amsterdam.

[77] Gross, D., Harris, C.M., (1985), *Fundamentals of Queueing Theory*, (2nd edition), John Wiley & Sons, New York.

[78] Grow, R.M., (1982), A Timed Token Protocol for Local-Area Networks, *Proceedings of the Electro '82 Conference*, paper 17/3.

[79] Haendel, R., Huber, M.N., (1991), *Integrated Broadband Network*, Addison-Wesley, Reading, MA.

[80] Hahne, E.L., Choudhury, A.K., Maxemchuck, N.F., (1992), DQDB Networks with and without Bandwidth Balancing, *IEEE Transactions on Communications*, **40**, 1192-1204.

[81] Halfin, S., (1983), Batch Delay Versus Customer Delays, *Bell System Technical Journal*, **62**(7), 2011-2015.

[82] Hammond, J.L.,O'Reilly, P.J.P., (1988), *Local Computer Networks*, Addison-Wesley, Reading, MA.

[83] Hashida, O., Takahashi, Y., Shimogawa, S., (1991), Switched Batch Bernoulli Process (SBPP) and Discrete-time *SBBP/G/*1 Queue with Application to Statistical Multiplexer Performance, *IEEE Journal on Selected Areas in Communications*, **9**, 394-401.

[84] Heymann, D.P., Sobel, M.J., (1982), *Stochastic Models in Operations Research*, Volume 1, McGraw-Hill, New York.

[85] Hoel, P.G., Port, S.C., Stone, C.J., (1971), *Introduction to Probability Theory*, Houghton Mifflin Company, Boston.

[86] Hooghiemstra, G., Keane, M., van de Ree, S., (1988), Power Series for Stationary Distributions of Coupled Processors Models, *SIAM Journal of Applied Mathematics*, **48**, 1159-1166.

[87] IEEE802.3 Standard, (1985), *Carrier Sense Multiple Access with Collision Detection*, IEEE.

[88] IEEE802.4 Standard, (1985), *Token Passing Bus Access Method*, IEEE.

[89] IEEE802.5 Standard, (1985), *Token Ring Access Method*, IEEE.

[90] IEEE802.6 Standard, (1990), *Distributed Queue Dual Bus (DQDB) Metropolitan Area Network*, IEEE.

[91] ISO 7498 *Basic Reference Model for OSI*.

[92] Jain, R., (1994), *FDDI Handbook - High-Speed Networking Using Fiber and Other Media*, Addison-Wesley, Reading, MA.

[93] Jing, W., Paterakis, M.,(1994) Message Delay Analysis of the DQDB Subnetwork Based on an Approximate Node Model, *IEEE Transactions on Communications*, **42**, 1120-1130.

[94] Jing, W., Paterakis, M.,(1995), Message Delay Analysis of the DQDB Subnetwork Based on an Approximate Node Model, *Computer Networks and ISDN Systems*, **27**, 653-676.

[95] Kabatepe, M, Vastola, K.S., (1992), Exact and Approximate Analysis of DQDB Under Heavy Load, *Proceedings of the INFOCOM'92 Conference*, 200-209.

[96] Kantorovich, L.V., Krylov, V.I., (1958), *Approximate Methods of Higher Analysis*, P. NoordHoff Ltd, Groningen, The Netherlands.

[97] Kim, C.K., Lee, S.H., Wu, L.T.,(1988), Circuit Emulation, *Journal of Digital and Analog Cabled Systems*, **1**, 245-256.

[98] Kleinrock, L., (1965), A Conservation Law for a Wide Class of Queueing Disciplines, *Naval Research Logistics Quarterly*, **12**, 181-192.

[99] Kleinrock, L., (1975), *Queueing Systems*, Vol. 1, John Wiley & Sons, New York.

[100] Kleinrock, L., (1976), *Queueing Systems*, Vol. 2, John Wiley & Sons, New York.

[101] Klessing, R.W., (1988), Overview of Metropolitan Area Networks, *IEEE Communication Magazine*, **24**, 9-15.

[102] Kuehn, P. J., (1979), Multiqueue Systems with Nonexhaustive Cyclic Service, *Bell System Technical Journal*, **58**, 671-699.

[103] Kuehn, P. J., (1979), Approximate Analysis of General Queueing Networks by Decomposition, *IEEE Transactions on Communications*, **27**, 113-126.

[104] Kurose, J.F., Schwartz, M., Yemini, Y., (1984), Multiple Access Protocols and Time-constrained Communication, *ACM Computing Surveys*, **16**, 43-70.

[105] Lam, S.S., (1980), A Carrier Sense Multiple Access Protocol for Local Networks, *Computer Networks*, **4**, 21-32.

[106] LaMaire, R. O., (1991), An M/G/1 Vacation Model of an FDDI Station, *IEEE Journal on Selected Areas in Communications*, **9**, 257-264.

[107] LaMaire, R. O., (1992), M/G/1/N Vacation Model with Varying E-limited Service Discipline, *Queueing Systems*, **11**, 357-375.

[108] Landry, R., Stavrakakis, I., (1993), Queueing Study of a 3-Priority Policy with Distinct Service Strategies, *IEEE/ACM Transactions on Networking*, **1**, 576-589.

[109] Lee, T.T., (1984), *M/G/1/N* Queue with Vacation Time and Exhaustive Service Discipline, *Operations Research*, **32**, 774-784.

[110] Lee, T.T., (1989) *M/G/1/N* Queue with Vacation Time and Limited Service Discipline, *Performance Evaluation*, **9**, 181-190.

[111] Leung, K.K., (1991), Cyclic-service Systems with Probabilistically-limited Service Discipline, *IEEE Journal on Selected Areas in Communications*, **9**, 185-193.

[112] Leung, K.K., Lucantoni, D., (1994), Two Vacation Models for Token-Ring Networks Where Service is Controlled by Timers, *Performance Evaluation*, **20**, 165-184.

[113] Limb, J.O., Flores, C., (1982), Description of Fasnet - A Unidirectional Local Area Communications Network, *Bell System Technical Journal*, **61**, 1413-1441.

[114] Martini, P., (1989), Fairness Issue of the DQDB Protocol, *Proceedings of the IEEE 4th Conference on Local Computer Networks*, 160-170.

[115] Matsumoto, Y., (1993), An Approximate Analysis Of DQDB Networks with the Bandwidth Balancing Mechanism, *Proceedings of the ICDDS 93 Conference*, 65-76.

[116] McDaysan, D.E., Spohn, D.E., (1994), *ATM*, McGraw-Hill, New York.

[117] Minoli, D., Keinath, R. (1993), *Distributed Multimedia Through Broadband Communications Services*, Artech House, London.

[118] Mirchandani, S., Khanna, R., (1993), *FDDI*, John Wiley & Sons, New York.

[119] Montuschi, P., Valenzano, A., Ciminiera, L., (1991), On the Equivalence of IEEE 802.4 and FDDI Timed Token Protocols, *Proceedings of the INFOCOM '91 Conference*, 435-440.

[120] Mukherjee, B., Bisdikian, C., (1992), A Journey Through the DQDB Network Literature, *Performance Evaluation*, **16**, 159-176.

[121] Mukherjee, B., Banerjee, S., (1993), Alternative Strategies for Improving the Fairness in and an Analytical Model of the DQDB Network, *IEEE Transactions on Computers*, **42**, 151-167.

[122] Muller, H.R., Nassehi, M.M., Wong, J.W., Zurfluh, E., Bux, W., Zafiropulo, P., (1990), DQMA and CRMA: New Access Schemes for Gbit/s LANs and MANs, *Proceedings of the INFOCOM '90 Conference*, 185-191.

[123] Nakumara, K., Takine, T., Takahashi, Y., and Hasegawa, T., (1991), Analysis of an Asymmetric Polling Model with Cycle-time Constraint, *High-Capacity Local and Metropolitan Area Networks*, G. Pujolle (Editor), Springer-Verlag, Berlin, 493-508.

[124] Nassehi, M.M., (1990), Cyclic-Reservation Multiple Access for Gbit/s LANs and MANs Based on Dual-Bus Configuration, *Proceedings of EFOC/LAN Conference*, Munich, 246-251.

[125] Neuts, M.F., (1981), *Matrix -Geometric Solution in Stochastic Models*, The Johns Hopkins University Press, Baltimora.

[126] Neuts, M.F., (1989), *Structured Stochastic Matrices of M/G/1 Type and Their Applications*, Marcel Dekker, Inc., New York.

[127] Ofek, Y., (1994), Overview of the MetaRing Architecture, *Computer Networks and ISDN Systems*, **26**, 817-830.

[128] Onvural, R.O., (1993), *Asynchronous Transfer Mode Networks*, Artech House, London.

[129] Potter, P.G., Zukerman, M., (1991), Analysis of a Discrete Multipriority Queueing System Involving a Central Shared Processor Serving Many Local Queues, *IEEE Journal on Selected Areas in Communications*, **9**, 194-202.

[130] Ramamurthy, G., Sengupta, B., (1992), Modeling and Analysis of a Variable Bit Rate Video Multiplexer, *Proceedings of the INFOCOM '92 Conference*, 817-827.

[131] Ramaswami, V., (1988), A Stable Recursion for the Steady-State Vector in Markov Chain of $M/G/1$-type, *Stochastic Models*, **4**, 183-188.

[132] Rodrigues, M.A., (1990), Erasure Node: Performance Improvements for the IEEE 802.6 MAN, *Proceedings of the IEEE INFOCOM '90 Conference*, 636-643.

[133] Rubin, I., Tsai, Z.H., (1989), Message Delay Analysis of Multiclass Priority TDMA, FDMA, and Discrete-time Queueing Systems, *IEEE Transactions Information Theory*, **25**, 1989.

[134] Rubin, I., and Wu, J.C.-H., (1992), Analysis of an $M/G/1/N$ Queue with Vacations and its Application to FDDI Asynchronous Timed-Token Service System, *Proceedings of IEEE GLOBECOM' 92 Conference*, 1630-1634.

[135] Saito, H., (1993), *Teletraffic Technologies in ATM Networks*, Artech House, London.

[136] Schrage, L., (1970), An Alternative Proof of a Conservation Law for the Queue $G/G/1$, *Operations Research*, **18**, 185-187.

[137] Schwartz, M., (1988), *Telecommunication Networks*, Addison-Wesley, Reading, MA.

[138] Sevcik, K., Johnson, M., (1987), Cycle Time Properties of the FDDI Token Ring Protocol, *IEEE Transactions on Software Engineering*, **13**, 376-385.

[139] Srinivasan, M.M., (1988), An Approximation for Mean Waiting Times in Cyclic Server Systems with Non-exhaustive Service, *Performance Evaluation*, **9**, 17-33.

[140] Stallings, W., (1984), *Local Networks*, Macmillian, New York.

[141] Stallings, W., (1993), *Networking Standards*, Addison-Wesley, Reading, MA.

[142] Stavrakakis, I., Tsakiridou, S., (1994), A Markov Service Policy with Application to the Queueing Study of a DQDB Station, *Computer Networks and ISDN Systems*, **26**, 1503-1522.

[143] Stidham, S.Jr., (1974), A Last Word on $L = \lambda W$, *Operations Research*, **22**, 417-421.

[144] Takagi, H., (1985), Mean Message Waiting Times in Symmetric Multi-Queue Systems with Cyclic Service, *Performance Evaluation*, **5**, 271-277.

[145] Takagi, H., (1986), *Analysis of Polling Systems*, The MIT Press, Cambridge Massachusetts.

[146] Takagi, H., (1988), Queueing Analysis of Polling Models, *ACM Computing Surveys*, **20**, 5-28.

[147] Takagi, H., (1990), Queueing Analysis of Polling Models: an Update, *Stochastic Analysis of Computer and Communication Systems*, H. Takagi (Editor), Elsevier Science Publishers (North-Holland), Amsterdam, 267-318.

[148] Takagi, H., (1990), Effects of the Target Token Rotation Time on the Performance of a Timed-Token Protocol, *Proceedings of the Performance '90 Conference*, 363-370.

[149] Takagi, H., (1991), *Queueing Analysis, Volume 1: Vacation and Priority Systems*, Elsevier Science Publishers (North-Holland), Amsterdam.

[150] Takagi, H., (1993) *Queueing Analysis, Volume 2: Finite Systems*, Elsevier Science Publishers (North-Holland), Amsterdam.

[151] Takagi, H., (1994), Queueing Analysis of Polling Models: Progress in 1990-1993, *Discussion Paper Series*, 584, Univeristy of Tsukuba, Institute of Socio-Economic Planning.

[152] Tangemann, M., Sauer, K., (1991), Performance Analysis of the Timed Token Protocol of FDDI and FDDI-II, *IEEE Journal on Selected Areas in Communications*, **9**, 271-278.

[153] Tran-Gia, P., Raith, T., (1988), Performance Analysis of Finite Capacity Polling Systems with Non-exhaustive Service, *Performance Evaluation*, **9**, 1-16.

[154] Tran-Gia, P., Stock, T., (1990), Approximate Performance of the DQDB Access Protocol, *Computer Networks and ISDN Systems*, **20**, 231-240.

[155] Tran-Gia, P., Dittmann, R., (1992), A Discrete-Time Analysis of the Cyclic Reservation Multiple Access Protocol, *Performance Evaluation*, **16**, 185-200.

[156] van As, H.R., Wong, J.W., Zafiropulo, P., (1990), Fairness, Priority and Predictability of the DQDB MAC Protocol Under Heavy Load, *Proceeding of International Zurich Seminar*, 410-417.

[157] van As, H.R., Lemppenau, W.W., Zafiropulo, P., Zurfluh, E.A., (1991), CRMA II: A Gbit/s MAC Protocol for Ring and Bus Networks with Immediate Access Capability, *Proceedings of EFOC/LAN Conference*, 262-277.

[158] Van den Hout, W.B., Blanc., J.P.C., (1993), The Power-Series Algorithm Extended to the *BMAP/PH/*1 Queue, *Center Discussion Paper 9360*, Tilburg University.

[159] Watson, K.S., (1984), Performance Evaluation of Cyclic Service Strategies: A Survey, *Proceedings of the Performance '84 Conference*, E. Gelenbe (Editor), 521-533, Amsterdam, North-Holland.

[160] Welch, P.D., (1964), On a Generalized *M/G/*1 Queueing Process in which the First Customer in a Busy Period Receives Exceptional Service, *Operation Research*, **12**, 736-752.

[161] Whitt, W., (1982), Approximating a Point Process by a Renewal Process, I: Two Basic Methods, *Operations Research*, **30**,125-147.

[162] Whitt, W., (1983), The Queueing Network Analyzer, *Bell System Technical Journal*, **62**, 2779-2847.

[163] Wolff, R.W., (1982), Poisson Arrivals See Time Averages, *Operations Research*, **30**, 223-231.

[164] Wong, J.W., Sauve', J.P., Field, J. A., (1982), A Study of Fairness in Packet-Switching Networks, *IEEE Transactions on Communications*, **30**, 346-353.

[165] Wong, J.W., (1989), Throughput of DQDB Networks Under Heavy Load, *Proceedings of EFOC/LAN'89 Conference*, 146-151.

[166] Wu, H.T., Ofek, Y., Sorhaby, K., (1992), Integration of Synchronous and Asynchronous Traffic on the MetaRing Architecture and its Analysis, *Proceedings of the ICC '92 Conference*, 147-153.

[167] Wynn, P., (1966), On the Convergence and Stability of the Epsilon Algorithm, *SIAM Journal of Numerical Analysis*, **3**, 91-122.

[168] Zukerman, M., (1988), QPSX - The Effect of Circuit Allocation on Segment Capacity Under Burst Switching, *Proceedings of the ICC '88 Conference*, 599-603.

[169] Zukerman, M., (1988), Queueing Performance of QPSX, *Proceedings of the 12th ITC Conference*, paper 2.2B.6.

[170] Zukerman, M., (1988), Overload Control of the Isochronous Traffic in QPSX, *Proceedings of the GLOBECOM '88 Conference*, 1241-1245.

Acronyms

ACF	Access Control Field
ANSI	American National Standard Institute
ATM	Asynchronous Transfer Mode
BWB	Bandwidth balancing mechanism
CBR	Constant Bit Rate
CDF	Cumulative Distribution Function
CSMA/CD	Carrier Sense Multiple Access with Collision Detection
DQDB	Distributed Queue Dual Bus
EDP	Electronic Data Processing
FDDI	Fiber Distributed Data Interface
FCFS	First-Come-First-Serve
FIFO	First-In-First-Out
Gbps	Gigabits per second
HOB	Head Of Bus
IEEE	Institute of Electrical and Electronics Engineers
i.i.d.	independent identically distributed
LAA	Lack of Anticipation Assumption
LAN	Local Area Network
LIFO	Last In First Out
LLC	Logical Link Control
LQ	Local Queue
LST	Laplace-Stieltjes transform
MAC	Medium Access Control
MAN	Metropolitan Area Network
Mbps	Megabits per second
OL	Offered Load

OSI	Open Systems Interconnection
OSI/RM	OSI Reference Model
PA	Pre-Arbitrated
PASTA	Poisson Arrivals See Time Averages
PDF	Probability Distribution Function
pdf	probability density function
PGF	Probability Generating Function
pmf	probability mass function
PS	Processor Sharing
PSA	Power Series Algorithm
QA	Queue Arbitrated
QBD	Quasi Birth Death
QoS	Quality of Service
RR	Round Robin
r.v.	random variable
SD	Scheduling Discipline
SONET	Synchronous Optical NETwork
THT	Token Holding Timer
TRT	Token Rotation Timer
TTRT	Target Token Rotation Time
VBR	Variable Bit Rate
WAN	Wide Area Network
w. p. 1	with probability 1

Glossary of Notation[*]

$\{A(t), t \geq 0\}$	Arrival process
A_n	Number of vacation packets that arrive during the n-th vacation
A	Generic random variable for $\{A_n\}$
a	End-to-end channel propagation delay normalized versus the average message length, i.e., $a = \tau/m$
B	Generic service time
B_k	Generic service time for k-type packets
$\{B_n\}$	Service process;
$B(x)$	Distribution function of B;
$B^*(s)$	LST of $B(x)$,
$b^{(i)}$	i-th moment of $B(x)$
b	Average service time, i.e., $b = b^{(1)}$
$b_k, b_k^{(2)}$	First and second moment of B_k, respectively
C_i	r.v. denoting the cycle length observed by Q_i
$C_{i,0+}$	r.v. denoting the residual cycle length observed by a tagged packet when arrives at Q_i
$C_{i\mid m}$	r.v. denoting the cycle length conditioned on the event "m Q_i transmissions"
$C_{i,n}$	Duration of the n-th cycle observed by Q_i, i.e., the time interval between the $(n-1)$th and n-th arrival of the server at Q_i
$\{C_{i,n}, n \geq 0\}$	The sequence of cycle length observed by Q_i

$\{C_{i,n|0}, n \geq 0\}$ The sequence of cycle length observed by Q_i conditioned on the event "zero Q_i transmissions"

$\{C_{i,n|1}, n \geq 0\}$ The sequence of cycle length observed by Q_i conditioned on the event "one Q_i transmission"

c Number of servers

$E[C]$ Average cycle length in a polling system

$\mathbf{e}$ Vector with all entries equal to one.

$\mathbf{e}_j$ Vector with zero entries except the j-th entry which is equal to one.

$I_{\{\bullet\}}$ Indicator function

$I_k:$ Intervisit lengths observed by Q_k

$I_{i,n}$ The length of intervisit period at Q_i during the n-th cycle

K Number of station in a MAN and in a polling system

L Work in the system

$L_{M/G/1}$ Work in an $M/G/1$ system

$L_{Polling}$ Work in a polling system

L_q Amount of work required by the packets waiting in the queues at a random point in time

$\{L_{SD}(t), t \geq 0\}$ Work in the system at time t

L_s Remaining service time for the packet in service at a random point in time

L_Y Work in a polling system at an arbitrary time during a switchover time

l_i Maximum number of packets served at Q_i during a service period

M Buffer size

$M/G/1_{GV}$ $M/G/1$ with generalized vacation service discipline

$M/G/1_{gated}$ $M/G/1_{GV}$ system with gated service discipline

$M/G/1/K_{VE}$ $M/G/1_{GV}$ with exhaustive service discipline and finite buffer

$M/G/1_{VE}$ $M/G/1_{GV}$ with exhaustive service discipline

$M/G/1_{1-limited}$ $M/G/1_{GV}$ system with 1-limited service discipline

$M/G/1_{E-limited}$ $M/G/1_{GV}$ system with exhaustive l-limited service discipline

$M_i^{(1)}$ Amount of work in Q_i when the server departs from this

m	Average message transmission time
$\mathbb{N}$	Set of natural numbers
N	r.v. describing the steady-state numbers of packets in the system
N_q	r.v. describing the steady-state numbers of packets in the queue
N_s	r.v. describing the steady-state numbers of packets in service
$\{N(t),\, t \geq 0\}$	Number of packets in the system at time t
$\{N_q(t),\, t \geq 0\}$	Number of packets in the queue at time t
$\{N_s(t),\, t \geq 0\}$	Number of packets in service at time t
OL	Offered Load, i.e., bit rate normalized with respect to the channel capacity
P_L	Packet loss, i.e., percentage of the bit rate which is not delivered to the receiver
p_i	Steady-state probability of i packets in the system, i.e., $p_i = \lim_{t \to \infty} P\{N(t) = i\} = P\{N = i\}$
Q_k	k-th stations in a polling system
$\mathbb{R}$	Set of real numbers
R	Response time, i.e., r.v. describing the time a packet spends in the system
R_x	Response time in the system x, where x specifies the system according to the Kendall notation
S	Total switchover time in a polling model
s and $s^{(2)}$	First and second moments of S, respectively
S_k	Switchover time from Q_k to Q_{k+1}
s_k and s_k^2	First and second moments of S_k, respectively
Sp_k	Service period of Q_k
$Sp_{i,n}$	Service period at Q_i during the n-th cycle
V_n	r.v. associated with the duration of the n-th vacation
V.	generic random variable for $\{V_n\}$
$V(t),\, v(t),\, V^*(s)$	Distribution function, density function and LST of V, respectively
$V_+,\, (V_-)$	Forward (backward) recurrence time of V
$V_+(t),\, V_+^*(s)$	Distribution function and LST of V_+, respectively

$V_-(t)$, $V_-^*(s)$	Distribution function and LST of V_-, respectively
Z_n	Number of packets in the system at the beginning of the n-th vacation
Z	Generic random variable for $\{Z_n\}$
W	r.v. describing the time a packet spends in the queue
W_k	r.v. describing the time a class-k packet spends in the queue
$W(x)$	Distribution function of W
$W^*(s)$	LST of $W(x)$
X_+, (X_-)	Residual (elapsed) time of a r.v. X
γ	Throughput, i.e., amount of the bit rate correctly delivered to the receiver
$\zeta(z)$	PGF of Z
$\Theta_c^{(n)}$	r.v. associated with the duration of the n-th service cycle
λ	Arrival rate
λ_k	Arrival rate for the class-k packets
μ	Service rate
μ_k	Service rate for the class-k packets
v	Speed of the signal in the transmission medium
π_i	Steady-state probability of i packets in the system at an arrival epoch
$\pi_{M/G/1}$	PGF of the number of packets in an $M/G/1$ system at an arrival epoch
$\pi_{M/G/1_{GV}}$	PGF of the number of packets in an $M/G/1_{GV}$ system at an arrival epoch
$\pi_{M/G/1_{VE}}$	PGF of the number of packets in an $M/G/1_{VE}$ system at an arrival epoch
ρ	Utilization factor, i.e., $\rho = \lambda \cdot b$
ρ_k	Utilization factor for the class-k packets, i.e., $\rho_k = \lambda_k \cdot b_k$
ρ_{max}	MAC protocol capacity
ρ_{single}	MAC protocol capacity with only one active node
τ	End-to-end channel propagation delay
$\tau_{i,j}$	Propagation delay between node$\{i\}$ and node$\{j\}$

Subject Index